DATE DUE

MAR 0 9 1993	
DEC 1 7 1993	
DEC 2 2 1997	

STUDIES IN INTERNATIONAL POLITICAL ECONOMY
Stephen D. Krasner, General Editor
Ernst B. Haas, Consulting Editor

Political Power
and the Arab Oil Weapon

POLITICAL POWER AND THE ARAB OIL WEAPON

The Experience of Five Industrial Nations

ROY LICKLIDER

UNIVERSITY OF CALIFORNIA PRESS
Berkeley · Los Angeles · London

University of California Press
Berkeley and Los Angeles, California

University of California Press, Ltd.
London, England

© 1988 by
The Regents of the University of California

Library of Congress Cataloging-in-Publication Data

Licklider, Roy E.
   Political power and the Arab oil weapon.

   (Studies in international political economy; 21)
   Bibliography: p.
   Includes index.
   1. Arab countries—Foreign relations.
2. Economic sanctions, Arab countries.   3. Petroleum
industry and trade—Political aspects—Arab countries.
4. Netherlands—Foreign relations.   5. Great Britain—
Foreign relations.   6. Canada—Foreign relations.
7. Japan—Foreign relations.   8. United States—
Foreign relations.   I. Title.   II. Series.
DS63.1.L53   1988   327'.0917'4927   87-43269
ISBN 0–520–06243–4 (alk. paper)

Printed in the United States of America
1   2   3   4   5   6   7   8   9

To my wife, Patricia, who knows why

CONTENTS

LIST OF TABLES

ACKNOWLEDGMENTS

Scholarly research is an odd combination of solitary and collective activities. This project would have been impossible without the assistance of many other people, and it is a personal pleasure as well as a professional obligation to acknowledge their assistance.

The Center for Computer and Information Services (CCIS) at Rutgers University provided computer time and supplied the needed data sets for analyzing the behavior of the industrial countries toward Israel and the Arab states, using events data. My colleagues Scott Keeter and Diana Owen were invaluable in helping me to use this material expeditiously. The analysis demonstrated that there were too few recorded events in the data sets to be useful and that extensive case studies would therefore be necessary. Gert Lewis at CCIS, through the center's connection with the Roper Center, also provided public-opinion data for different countries.

The lack of published materials meant this research was heavily dependent on personal interviews. I was able to visit all the countries except Japan, and rather to my surprise many academics, journalists, and past and present government officials were willing to talk to me. This very willingness, however, raised a delicate issue involving scholarly ethics. I guaranteed confidentiality to all those with whom I spoke. In fact, everyone I cited was sent a copy of the relevant material and asked if their views had been correctly represented and whether they would be willing to be quoted by name. In general the academics agreed to be quoted, while the others preferred to remain anonymous. Under the circumstances I felt obligated to accede to their wishes, even at the risk of frustrating interested readers. Because there were relatively few interviews for each country, I did not include the names of government officials, even in a bibliography, because confidentiality would be too easily breached.

Each of the case chapters essentially required a separate research project. I have done my best to acknowledge the important written works in my notes. For each country, however, I needed people who would point me in the right direction and who would talk to me to clarify various issues. One of the pleasures of this project was the people who generously furnished information and assistance to an unknown scholar; because I knew relatively little about each of the four foreign countries—Canada, Great Britain, Japan, and the Netherlands—this help was absolutely crucial to the project. I hope that those who helped think the results justified their confidence.

Leon Gordenker introduced me to Dutch politics and foreign policy, and J. L. Heldring was also very helpful. Gregory Flynn of the Atlantic Institute for International Affairs in Paris ransacked his organization's files to uncover copies of unpublished papers, one of which was by Heldring on the Dutch response. Although no full-length study of the Dutch reaction to the oil embargo is available in English, R. B. Soetendorp produced such a study in Dutch and was very generous with his time. I was also able to talk to the researchers of a major study on Dutch foreign policy headed by Philip Everts, Alfred van Staden, and Peter Baehr. One case in this study is the oil weapon, and Fred Grünfeld, who was responsible for this case in the larger study and had unusually good access to former officials, was very helpful indeed. This research is summarized in English (Grünfeld 1985), but the complete study is available only in Dutch (Grünfeld 1984). I therefore obtained much of the necessary information in interviews. In addition, a number of other Dutch academics, journalists, and government officials spoke candidly about the events. Grünfeld and Heldring read and critiqued the chapter. Like many other American academics, I found the Netherlands a very pleasant place to do research.

Britain posed an unanticipated research problem because almost nothing has been written on British policy toward the Arab-Israeli conflict since 1973. Memoirs of senior decision makers, while fairly common, tend to be a good deal less outspoken than their American counterparts, with the conspicuous exception of the diaries of Richard Crossman, a Labour cabinet minister from 1964 to 1970 and a fervent but not uncritical supporter of Israel. In particular, there was little ongoing academic research of the kind that was so helpful in other countries. I was thus forced to rely more heavily on personal interviews, and this in turn was somewhat hampered by the domination of British policymaking by civil servants and the draconian Official Secrets Act. One exception was another unpublished paper in the set provided by Gregory Flynn of the Atlantic Institute for International Affairs. Lynn Davis and Kenneth Hunt put me in touch with academics and former senior officials who generously shared their knowledge and insights with me. Among the former, I particularly appreciated the assistance of Avi Shlaim and Geoffrey Edwards, who read and critiqued the chapter; the latter preferred not to be named but are nonetheless warmly remembered. Rosemary Hollis sent me a copy of her unpublished work. Like so many other researchers, I found the facilities and personnel of the Royal Institute of International Affairs at Chatham House invaluable, especially that unique international scholarly resource, the Press Library; the current severe curtailment of its activities is particularly unfortunate.

There is not much literature on Canadian policy toward the Arab-Israeli conflict. The few, very impressive Canadian academics working in this area, however, were extraordinarily generous with their time and ideas. I am particularly grateful to Don Munton, who charted the field for me, and to Paul Noble, Howard Stanislawski, David Taras, John Benesh, Janice Stein, David Dewitt, and John Kirton, all of whom talked to me, or gave me unpublished materials, or both. Stanislawski read and critiqued the Canada chapter. A number of diplomats and political leaders, both active and retired, added a great deal but have requested anonymity.

Japan presents a major challenge to a researcher like myself who lacks extensive knowledge of its language and culture. Moreover, Japan was the only country I did not visit to conduct interviews, both because of the expense and because of the discovery of some unpublished research in English on the internal politics of the Japanese reaction to the oil weapon. I have relied particularly heavily on the work of Kenneth Juster (who spent several months in Japan researching a Harvard undergraduate honors thesis on the subject), Martha Caldwell (now M. C. Harris), and Michael Baker. I am indebted to Haruhiro Fukui, Nobutoshi Akao, and Yasumasa Kuroda for sending me unpublished manuscripts, to my colleague Ardath Burks for sharing his knowledge of Japan with me, and to a Japanese official for supplying me with publicly available government documents. Martha Caldwell Harris, Kenneth Juster, Henry Nau, Harold Solomon, and Harold C. Roser all read and critiqued the chapter on Japan.

A great deal of material is available on the foreign policy of the United States toward the Arab-Israeli conflict. I have relied primarily on the work of William Quandt and of Steven Spiegel. Allen Dowty, Steven Spiegel, Eytan Gilboa, and Robert Trice generously sent me copies of their work, much of it in prepublication form. Because of the amount of written material, there were fewer interviews than usual, but Henry Nau, Robert Lieber, and an anonymous former senior official were very helpful.

I found the theoretical analysis to be the hardest part of the project. My Rutgers colleagues Robert Kaufman, Charles Noble, and Harvey Waterman read drafts of the introduction and conclusion and greatly helped to add some clarity. A suggestion by Steven Krasner altered the entire thrust of the project in a way that I think improved it greatly.

Richard Mansbach read the entire manuscript at least twice, voluntarily; medals have been given for less. After reading the manuscript for the University of California Press, David Baldwin generously volunteered to abandon his anonymity, which enabled me to clarify his concerns and, I hope, better respond to them. Rutgers University reference librarians coped brilliantly with a blizzard of interlibrary loan requests. Virginia Licklider helped proofread the final manuscript and provided inspiration to think about the long-term implications of this research in a complex world. Naomi Schneider was gracious and supportive while shepherding the manuscript through the University of California Press approval process. Mary Renaud supervised the production of the book; and Kristen C. Stoever did an extraordinary job of editing the entire manuscript, greatly enhancing its readability. Unfortunately, all this assistance does not excuse me from responsibility for the errors and omissions that remain.

In the interest of full disclosure I should note that from January 1977 to June 1978, while on leave from Rutgers, I was program officer for the Exxon Education Foundation. This was a valuable and stimulating experience, but nothing I learned there was directly relevant to this research, and I am not conscious that working for the world's largest oil corporation affected my judgments on the issues of this study.

A long list of impressive organizations declined to provide financial support for this research. The Rutgers Faculty Assistance Study Program allowed me to spend a year working on it full-time, aided by a grant from the Rutgers Research Council. Funds for travel and other vital aspects of the work came from an inheritance from my grandmother, Fern Lawrence Eilers; I like to think she would have approved of my use of the money.

My major debt, however, is to my wife, Patricia. Her professional skills were helpful, but her personal support for a seemingly endless research project was critical for the study and particularly for its author's sanity. This book is therefore dedicated to her with all my love.

1

The Oil Weapon in Theoretical Context

The Arab use of the oil weapon during 1973–1974 is widely regarded as a turning point in modern international affairs. Economically it produced gas lines, changed the international distribution of wealth, and helped create a worldwide recession from which we have not yet recovered. Politically it fundamentally altered our perceptions of the international balance of power by showing that the Third World could directly influence the rest of the world by withholding important resources. Although we have learned a good deal about the economic consequences of the oil weapon, we remain largely ignorant of its political impact.

This situation is particularly ironic because the ostensible purpose of the Arab oil weapon was neither to raise oil prices nor to wreak havoc on Western economies, but rather to persuade the "international community [to] compel Israel to relinquish our occupied territories," to quote from the declaration of October 17, 1973, which initiated the event (*Oil and Security* 1974, 118). Moreover, much of the controversy over the oil weapon centered precisely on its utility as an instrument to coerce other states into altering their foreign policies. Most of the subsequent analysis,

1

however, has focused on the impact of the energy crisis on the national economies. This is not a trivial question, but the assumption that the most important results of the oil weapon were economic fundamentally misreads its significance.

Political analysis of the oil weapon has divided into two different approaches, which use different assumptions, draw on different sets of information, and reach different conclusions. One group of analysts starts from the accepted theory of international affairs, which concludes that economic sanctions do not by themselves force states to alter important policies. The other group starts with the obvious importance of Arab oil to the industrial economies and deduces that the Arabs have forced these governments to alter their policies toward Israel in the past and will be able to do so again if another international oil crisis occurs. The two schools thus derive very different lessons for both theory and policy from the experience.

This book is an attempt to resolve these differences or, failing that, at least to clarify the issues and furnish additional information to move the debate forward. We begin with a general discussion of power and influence in international affairs, followed by a brief historical overview of the Arab oil weapon. The reasons for adopting the case-study approach are set forth, along with an explanation of how the cases were chosen. This chapter concludes with an analysis of the very different lessons that people have drawn from the experience of the oil weapon, clarifying the point of dispute to which the book is addressed.

Unlike most studies of economic sanctions, this work focuses on the behavior of the target rather than on the actor governments. Most of the book is therefore composed of case studies of the Arab-Israeli policies of five countries, along with systematic discussions of these policies. These chapters represent the most detailed analysis now available of the actual impact of the oil weapon on the target countries' policies toward the Arab-Israeli conflict. Their general conclusion is that there was relatively little policy change during the time of the oil weapon itself but that in the decade or so afterward all five governments have showed somewhat more sympathy for the Arab position.

The conclusion attempts to use the historical materials of the separate case chapters to address more general theoretical and policy issues. Perhaps the most striking finding is the ultimate lack of difference among the target governments. The governments originally had different attitudes toward the issue, different levels of dependence on Middle Eastern oil, and domestic pressure groups with different degrees of influence. Some countries were embargoed by the Arabs, some had their Arab oil imports reduced, and some were allocated as much Arab oil as they wanted. Nonetheless, they all followed the same pattern. None made any significant policy change during the period of the oil weapon itself. Moreover, the long-term change of each country can be plausibly explained by factors other than fear of cutting off oil supplies, in particular the major increase in Arab wealth that resulted from the rising price of oil. Historically this major increase in wealth coincident with the levying of a major economic sanction is unprecedented. This in turn implies that the more general weaknesses of international coercion were more important in explaining the outcomes of this particular event than its idiosyncrasies and that this is likely to be true of other events in the future as well.

Concepts of Political Influence

Perhaps the most fundamental theoretical issue of international politics is how one state can exert power over another. Ideas about influence underlie most foreign policies and the notion of power has been central to the study of international relations ever since the rise of realism during World War II. Nonetheless, we still know remarkably little about the circumstances under which one state can alter the behavior of another.

Problems of Theory

There are at least two reasons for this difficulty. The first is theoretical. The concept of power is one of the most difficult in social science, and scholars of international relations have often implied that power is an attribute of particular actors that can in principle be measured but is at any rate relatively constant. The classic definition of power is the ability to get people to do what you want them to do. This is an attractively simple notion, which also seems intuitively plausible; we think routinely about great powers and superpowers, and everyone seems to know what we mean.

The notion of power as a single quality is in fact not very useful, however. There is no agreement on how to measure power, and, as often happens in social science, this apparently abstract problem presents at root a more serious conceptual difficulty. Qualities such as military strength, level of technological development, size of population, and consumption of electrical power are all qualities that may or may not be used to influence other people's behavior, but the utility of each may depend on the particular situation; for example, American military power is useful in persuading the Soviet Union not to invade Berlin but not in persuading the Canadians to stop asking us to do something about acid rain. In addition, these various qualities cannot readily be substituted for one another; different countries have different mixes. Also, we know that these qualities must be *used* by governments to influence the behavior of others, and we know that having more capacity does not guarantee success. Vietnam remains the classic example for the United States: a country that possessed much less military and economic capacity was able to defeat the Americans. How useful is it, then, to talk about power as a single quality?

Second, power is relational, not absolute. For example, whether India is a powerful state depends on whom you ask. The U.S. government might not think so, but the government of Sri Lanka would disagree. Power, then, is relative, depending on the qualities of the actors involved. Moreover, even among the same pair of actors, power varies with the issue. Vietnam could successfully persuade the U.S. government to withdraw from its territory, but it could never hope to convince Americans to accept Vietnamese control of the continental United States. Power depends on the issue as well as the actors.

The single most important conclusion of modern theory about power is that it is a relationship between two or more entities at a particular time with particular qualities rather than a static quality of one state (Lasswell and Kaplan 1950; Dahl 1963; Deutsch 1968; Baldwin 1985). Four major dimensions of any such relationship seem particularly important: its *domain* (the actors and the targets); its *scope* (the kind of behavior involved); the *range* of positive or negative sanctions; and the *weight* of the power relationship, i.e., the change of the probability that the target will do what the actor wants.

This approach is clearly more complex than the first. One hopes it may allow systematic comparison of different power relationships to determine the circumstances under which state behavior is affected by others. So far, however, we have not seen much of this sort of analysis. This approach allows us to view the oil weapon as an instrument used to change some dimensions of the existing power relationship between the actors (the Arab oil-producing states) and the targets (the "international community" named in the Arab states' 1973 declaration, which we have taken to mean the industrial states) in order to increase the likelihood that the targets would apply sufficient pressure on Israel to force the policy changes the actors desire.

Lack of Empirical Research

The second reason for the difficulty surrounding the question of power is a lack of empirical research. An enormous amount of effort has gone into studying how various kinds of behavior useful in establishing power relationships are best carried out. The massive literature on military policy is one example, along with smaller but still quite substantial literatures on intelligence and covert operations, economic sanctions, and diplomacy. Clearly, the capability to organize and carry out these kinds of activities shapes the power relationships of a state. But there has been remarkably little effort to study the *actual effect* of these activities on the foreign policies of target states.

Almost all of the foreign policy literature implicitly assumes a target state that acts like a unified, rational actor with known preferences. Again, the clearest example is the strategic literature, which so often predicts with enormous confidence the behavior of target states that are not even identified, much less analyzed. Similarly, the literature on economic sanctions is concerned with timing; the impact of such sanctions is usually assessed by assuming the target is a "black box" whose behavior can be predicted without discussion.

As a result of all this, we have relatively little theoretical or empirical literature on the question of how the acts of one state actually influence the policies of others. Baldwin (1985, 13–38) at least lays out a useful framework for analysis. He develops the notion of statecraft as "governmental influence attempts directed at other actors in the international system" and classifies policy instruments to exert such influence as propaganda, diplomacy, economic statecraft, and military statecraft. Another set of materials comes from comparative foreign policy, which attempts to order the discussion of influences on foreign policy. Relatively little research has been done, however, on the "external" variable cluster (Harf, Hoovler, and James 1974, 235–250; Rosenau 1980, 317–339). (The recent work of the CREON [Comparative Research on the Events of Nations] project on the "external predisposition component" is a much more sophisticated intellectual construct than the "external" variable cluster of Rosenau's pretheory [V. Hudson 1984].)

One major effort to close this gap has been the recent literature on the limited use of force to persuade a government to change its policy, variously called coercive diplomacy, armed suasion, or (the term I prefer) coercive bargaining (Schelling 1966; George, Hall, and Simons 1971; Young 1968; May 1973, 125–142; Luttwak 1974, 1976; Lauren 1979; Mack 1975; Thies 1980; Blechman and Kaplan 1978; Kaplan 1981; Craig and George 1983; Karsten, Howell, and Allen 1984; Mandel 1985; and Scolnick 1985). This literature has marked a major step forward in our ability to analyze the utility of foreign policy instruments. "Coercive bargaining" has been limited to military force, however, depriving policymakers of information about the instruments they are most likely to use (propaganda, diplomacy, and economics). Such limitations also make it difficult for academics to develop theories of how different instruments affect target states.

Economic sanctions constitute another instrument that has been widely used and studied. Scholars have employed two different definitions of the term. The more limited meaning is economic penalties levied on a country by an international organization. But economic sanctions have increasingly become synonymous with economic pressures exerted by one country or group of countries on another. This study uses this latter, broader definition.

The impressive literature on economic sanctions can readily be traced back at least fifty years and includes contributions by individuals from different nations with different political ideologies using different research methods (see, among others, Guichard 1930; Clark 1932; Royal Institute of International Affairs 1938; Galtung 1968; Adler-Karlsson 1968; Wallensteen 1968, 1969; Freedman 1970; Doxey 1971, 1980; Knorr 1975; Barber 1976; Losman 1979; Olson 1979; Renwick 1981; Weintraub 1982; Ayubi et al. 1982; Nincic and Wallensteen 1983; Hufbauer and Schott 1983, 1985; Hill and Mayall 1983; Daoudi and Dajani 1983,

1985; McCormick and Bissell 1984; Ellings 1985; Baldwin 1985; Paarlberg 1985; Wagner 1986; Goldsmith 1986; Leyton-Brown 1987).

For our purposes, however, this literature shares two failings. First, it focuses almost entirely on the *actor's* decisions. Thus, as in David A. Baldwin's major new study (1985), it typically asks the actor's major question: Are sanctions the most appropriate policy instrument available? Almost none of this work focuses on the *target's* chief question: How do sanctions affect my alternatives and my policy decisions? Second, as a result, even when conflicting goals within the actor government are noted, the literature still assumes that target governments behave like unified, rational actors whose primary values are economic. This assumption often turns the analysis of the impact of sanctions into a quasi-economic exercise, ignoring the fact that the response to such sanctions is a supremely *political* decision by the target state's leadership. This study will attempt to address these shortcomings by focusing on the responses of several target governments to a particular set of economic sanctions, the Arab oil weapon of 1973–1974.

Why Do Governments Change Their Foreign Policies?

If we are interested in how governments try to persuade one another to change their policies, we must look at the process of policy change. Interestingly, the literature does not seem to provide a full-blown model of such a process, although recently scholars have shown a burst of interest in the broader question of change in the international system (Gilpin 1981; O. Holsti, Siverson, and George 1980; Buzan and Jones 1981; and Zinnes 1984) and how such change in the international environment affects foreign policy as an input. But relatively little work focuses on what brings about foreign policy change as an output and how that change actually occurs. We will therefore start by sketching out such a model.

This somewhat oversimplified model makes several assumptions. First, the government as a whole has a policy on this issue—that is, a general rule implemented over time by the state's decision makers. (This excludes several types of issues: new issues for which no general policy has yet been formulated; trivial issues in which the leading decision makers have not been involved; and issues where the disagreement within the government is so strong that no coherent policy has been followed.) At the same time, this general policy need not be supported by all decision makers, be followed rigidly at all times, or remain unaltered. The critical point is that the controlling decision makers *believe* such a policy exists.

Second, because of this policy history decision makers are not likely to change policy without some reason. Like objects in Newton's Third Law, governments and individuals in motion tend to stay in motion; inertia is usually the best predictor of future activity. Our task is to suggest factors that seem likely to produce such change. Third, the change must be made by the government's controlling decision makers, and they must recognize that the change is occurring. This approach does not apply, for example, to a decision that decision makers think follows their previous policy when in fact it breaks with it. Fourth, the political system is not totally dominated by a single person, and coalitions among the elites are therefore important. Finally, the policy changes, once made by top decision makers, will in fact be implemented. We know this is no small assumption when implementation requires changed behaviors by complex organizations. Nonetheless, this study focuses on why decision makers change their minds; problems of implementation require a different sort of analysis.

Policy Change and New Decision Makers

Foreign policy change seems more likely when those holding decision-making positions are replaced than when they remain in office. One excellent, if somewhat overlooked, analysis of this phenomenon is Philip M. Burgess's study (1968) of Norwegian and Swedish alliance policies. Before World War II both countries pursued policies of neutrality. During the war Norway was invaded and occupied by Germany, while Sweden successfully maintained its neutrality. After World War II Norway joined the Western alliance, while Sweden remained neutral.

Inasmuch as Norway and Sweden are so similar in both their domestic composition and geographic location, they make a particularly good case for comparative analysis. Burgess argued persuasively that different policies occurred because Norway's previous neutrality policy had failed, whereas Sweden's had succeeded. He contended, however, that the crucial *process* for this alteration of policy was leadership change. Prewar Norwegian leaders were either captured by the Germans or discredited by the failure of their policy; after the war, the new leaders were skeptical rather than sympathetic about arguments in favor of neutrality. In Sweden, in contrast, the leadership had remained in power and its original arguments were strengthened. Thus, even though each group of leaders could refer to the experiences of the other country, they drew directly opposite conclusions from the same set of "facts" and adopted divergent policies as a result.

Ernest May (1973) makes a similar argument. In a fascinating effort to show how Americans should have decided whether or not to bomb North Vietnam in 1965, he asks why the use of aerial bombing to coerce a nation into negotiating has only sometimes succeeded. Looking at seven cases (Mussolini in Ethiopia, the nationalists in the Spanish civil war, Japan in China, Germany against Britain, the World War II Allies against Italy, the United States against Japan, and the United States in Korea), he concludes that the bombing itself did not change the attitudes of individual decision makers: "In the instances in which bombing probably produced a political effect, there occurred changes of government" (May 1973, 130).

Why is policy more likely to change with new leaders? There seem to be several overlapping explanations. First, when decision makers change policies they themselves have implemented, they are implicitly admitting their previous actions had been in error. Individuals are usually reluctant to make such admissions, and organizations seem even more so. Second, when new people take office, they often find it politically useful to distinguish themselves from their predecessors by altering policy, or at least saying they will do so. In addition, to the extent that the personal qualities of the decision maker influence the substance of policy, new people are likely to possess personal qualities that will produce different policies. Such qualities may translate into different policy preferences or different priorities for various policy areas.

Having said this, we also know that the replacement of decision makers does not guarantee foreign policy change. Policy change is more likely to accompany such a personnel change under any of the following circumstances:

a. The leadership change is the result of systemic (revolutionary) rather than routine change of government.

b. The government has few alternate power centers. Thus, countries with permanent bureaucracies, independent legislatures, and effective political oppositions are less likely to have major policy changes linked to personnel changes than states where the executive rules largely unchallenged. A related condition is:

c. The issue in question is under the direct control of the new decision makers. This, for example, goes a long way toward explaining why Ronald Reagan had more success in changing the rhetoric of U.S. foreign policy (which he directly controlled) than its substance, which required changes in behavior of other groups and individuals over whom his control was sometimes tenuous.

d. The issue in question is believed to have contributed to the previous personnel change, either by undermining the previous incumbents or by bolstering the political strength of the new leaders.

e. The new officers see such a change as important and are prepared to expend significant political resources to bring it about.

Policy Change with the Same Decision Makers

Rumors to the contrary notwithstanding, people do change their minds. The sources of such change are likely to be found in the following situations:

One such situation is when the leaders in office decide to change a policy. This decision usually results from new information, a changing international situation, or a domestic political shift. (For an interesting, although rather abstract, discussion of foreign policy change in unitary actors, see Mansbach and Vasquez 1981, chaps. 6–8.) If the leaders decide not to change, political elites may either replace them with new leaders committed to a policy change or use this threat to force the incumbents to change their positions. Political elites are defined as groups that can influence the leading decision makers by face-to-face contact and by some perceived control over their tenure in office. In democratic regimes the elites usually include members of the dominant political parties, some members of the national legislature, and representatives of influential interest groups. In other societies military officers and regional politicians may be more important.

If the leaders and the elites resist change, the politically relevant public may either replace the current elites and leaders with new ones committed to a policy change or use this threat to force the old leaders and elites to change their position. This influence may be exerted by elections, demonstrations, riots, or full-scale revolution; the politically relevant public includes everyone who has the potential to participate in such activities.

This description of possible processes of foreign policy change identifies a number of possible targets within the target government for an actor government. One can concentrate on replacing the current decision makers, affecting their calculus of decision, encouraging elites to pressure or replace them, or encouraging the public to pressure or replace them. We know relatively little about this process; it is likely that there are different strategies for each target, for example. The five cases will be studied in order to suggest hypotheses about strategies that can be most readily used and are most effective.

The purpose of this study is to apply the concept of political power to the Arab oil weapon to start to correct these weaknesses in the study of political power and foreign policy change. But we must start by defining the oil weapon itself.

The Arab Oil Weapon

The fourth Arab-Israeli war broke out in October 1973. This triggered the oil crisis, which consisted of a number of separate but related events. First, the Arab oil producers (loosely organized as the Organization of Arab Petroleum Exporting Countries, or *OAPEC* as opposed to OPEC, the Organization of Petroleum Exporting Countries) embargoed oil shipments to the United States and the Netherlands. Second, most OAPEC members reduced their total oil exports so that embargoed countries would not be able to purchase oil from other importers. These events produced an apparent oil shortage worldwide that allowed OPEC (composed of the Arab oil producers and others, such as Nigeria, Iran, and Venezuela) to escalate earlier price hikes, increasing the price of oil fourfold in a short period of time.

The embargo and the production cutbacks differed from the price increases in two important ways: They were carried out only by Arab states, and they had explicit political motives—to force Israel to return territories captured in the 1967 war, obtain "legitimate rights" for the Palestinians (interpreted by one Arab analyst as an independent Palestinian state [Ali 1986, 129]), and alter the status of Jerusalem. These two separate acts (the embargo and the production cutbacks) are called the "oil weapon," although the distinction remains unclear in common discussion. There was no "OPEC embargo," for example, because only Arab states embargoed oil shipments; indeed, the diversion of non-Arab OPEC oil fundamentally undercut the oil weapon. (Strictly speaking, there was no OAPEC embargo either, because it had been imposed by separate states meeting as the Arab oil ministers rather than as the OAPEC Council of Ministers in order to avoid the risk of splitting the organization [Maachou 1983, 81–82]. Nonetheless, even Arab accounts refer to it as an OAPEC embargo [al-Sowayegh 1984, 101, 129]. The term "Arab oil weapon" thus seems most appropriate, even though Iraq, Oman, and Tunisia did not participate.)

Two types of sanctions were applied—embargoes and production cutbacks. The embargo totally cut off exports of oil to certain countries. In mid-October all ten actor governments embargoed the United States. The Netherlands was embargoed at the end of the month, and Portugal, Rhodesia, and South Africa were eventually added to the list as well. Indirect embargoes also were instituted against countries that processed or transshipped oil for the Netherlands or the United States, particularly the U.S. Navy. Saudi Arabia listed countries in this category as Trinidad, the Bahamas, the Dutch Antilles, Canada, Puerto Rico, Guam, Singapore, and Bahrain (which apparently became both an actor and a target government at the same time); this list was later reduced, however. A few European refineries in Italy, France, and Greece were embargoed for the same reason (Lenczowski 1976b, 65). On March 18, 1974, the Arab oil ministers decided, over the protests of Libya and Syria, to lift the embargo on the United States "provisionally"; on June 3 it was removed altogether. The embargo against the Netherlands ended on July 11.

The second sanction was a series of general oil-export reductions. On October 17, 1973, the OAPEC oil ministers agreed to cut October exports by a minimum of 5 percent from the September level, with another 5 percent cut to follow in November. Consumer countries could be exempted entirely or in part from the cutbacks depending on their positions vis-à-vis the Arab-Israeli conflict. On November 4 it was decided to reduce November exports by 25 percent of the September level, with a further 5 percent cut scheduled for December. This figure included oil not being sent to embargoed countries, so, presumably, exports to other countries would be cut less than 25 percent.

By December Saudi Arabia had developed a very complex classification system for consumer nations, with at least five different categories. Most-favored countries, for instance, were to get as much oil as they required; this category included Great Britain, France, Spain, the Arab importing countries, African countries that had broken diplomatic relations with Israel, and Islamic countries. Friendly countries, a second category, were neutrals who had modified their policy in favor of the Arab positions (Belgium and Japan); they were entitled to their pre-embargo level of imports. Yet another group—the other European Community (EC) countries, except for the Netherlands and perhaps Denmark—were subjected to cutbacks but exempted from the December 5 percent reduction, presumably because of their November 6 declaration that asked Israel to end its "territorial occupation" dating from the 1967 war and supported the "legitimate rights of the Palestinians." Denmark and the Netherlands had both signed this statement, but this did not save them from the embargo. A fourth category of consumer-nations, neutrals, was subject to all cutbacks. Embargoed countries (the United States, the Netherlands, South Africa, Rhodesia, and Portugal) comprised a fifth category; these nations were to get no Arab oil at all (Lenczowski 1976b, 66). Denmark apparently was subjected to a partial embargo. On December 25 OAPEC decided to reduce the December cut to 15 percent of the September total and not to impose the January 5 percent cut. All export restrictions were ended in June 1974.

We discussed earlier some of the limitations of theory and research on the process of state influence. The Arab oil weapon of 1973–1974 presents an unusual opportunity to help remedy this situation. As an example of economic coercion, the oil weapon can be analyzed in a way that moves beyond the literature on military statecraft. The oil weapon involved similar economic sanctions aimed at different groups of states, allowing scholars to move away from the limitations of the single-case approach and to look for generalized patterns of behavior. The unit of analysis here is not the single case of the oil weapon but the many states that imported oil from the Arab countries. For the purposes of this study, the oil weapon presents a very "clean" case of economic statecraft; the Arab oil producers could not threaten the industrial countries militarily. In addition, many sophisticated observers viewed the oil weapon as an example of successful economic coercion, even though this was unlikely in the face of the theory and research on economic sanctions. Indeed, there was some suggestion that the use of the oil weapon was a watershed in world affairs concerning the utility of military versus economic force. Because the targets were democracies, it was possible to research a number of them fairly easily.

Design of the Study

The oil crisis has generated a literature impressive both in quantity and quality. But this work has concentrated almost exclusively on economic issues such as energy policy, petrodollars, the international debt crisis, and Middle Eastern development (a useful summary of much of this work is found in Kemezis and Wilson 1984), with hardly any discussion of the significance of the oil crisis for foreign policy and international politics. This is particularly ironic given the Arab governments' stated goal to influence the policies of other countries toward Israel. This, in turn, leads us to the first major question of the study: *Did the target countries alter their policies toward the Arab-Israeli dispute as the Arab governments apparently wished?*

As Baldwin (1985, 28) notes, this requires a counterfactual analysis, that is, how Arab-Israeli policies of the industrial countries would have developed had the oil weapon *not* been employed. By definition we can never know the answer with certainty, but the format of each case involves a fairly detailed history of the country's policy toward the Arab-Israeli conflict before 1973, during the oil crisis, and afterward in order to give some perspective on whatever changes occurred around 1973–1974.

As usual in foreign policy studies, the dependent variable proved more difficult to conceptualize than it seemed. Although most of the analysis focuses on governments' foreign policy acts, I have also included their policies toward the Arab economic boycott as part of the issue, even though these might be regarded as economic policy. To facilitate the distinction between these two economic weapons, I have used the term "embargo" to refer to a decision by an actor not to *sell* to a target and the term "boycott" for a decision not to *buy* from a target. (For a more sophisticated distinction between these terms, see Daoudi and Dajani 1983, 2–9.) Accordingly, the oil *embargo* is the policy of the OAPEC states not to sell oil to the United States and the Netherlands for a few months in 1973–1974; the Arab economic *boycott* is the long-standing policy (albeit irregularly practiced) of many Arab countries not to buy goods and services from

Israel, from firms or individuals that do business with Israel (secondary boycott), or from firms or individuals that do business with firms or individuals that do business with Israel (tertiary boycott).

Although this factual question clearly is central to this analysis, it is important primarily as it contributes to an understanding of the second major question: *Why did the target states change (or not change) their policies toward the Arab-Israeli dispute?*

This second question is central to our analysis of the weight of the power relationship established by the oil weapon. I have used two techniques to try to clarify it. First, the historical analysis of each case is followed by a systematic discussion of alternate explanations for this policy development in an attempt to determine how significant the oil weapon was. The discussion is organized along the lines of James Rosenau's comparative foreign policy variable clusters: external, societal, governmental, and individual (1980, 115–169).

The effect of the international system is not discussed in each chapter because the analysis focuses on different governments acting at the same time; by definition the systemic variables have the same value for all countries at the same time (Harf, Hoovler, and James 1974). Nonetheless, it would be a mistake to overlook this group of variables because much of the oil exporters' influence potential stemmed from a number of qualities then extant in the international system. Technologically the industrial states depended heavily on oil. Economically many of these states imported a large percentage of this oil, and recent increases in international demand had created an increasingly tight world market for petroleum. Geologically OAPEC controlled a majority of the world's proven reserves. Politically, although the two superpowers remained militarily dominant, they had increasing difficulty in using their military power to politically control other states; in particular, a military takeover of the oil fields was not an attractive strategy. This perceived inability of the industrial powers to apply military force was crucial in allowing the militarily weak oil exporters to reduce oil supplies and raise oil prices with impunity.

The U.S. government went to considerable difficulty to alter one of these systemic variables, the relative power between oil producers and consumers. For this purpose it supported the International Energy Agency (IEA), whose task was to reduce the dependence of the Organization of Economic Cooperation and Development (OECD) members on imported oil by supporting research, energy conservation, exploration of new energy sources, emergency stockpiles, and especially a program of sharing among states in the event of future embargoes or shortages. Potentially this was an important proposal, and its establishment in 1974 was a minor triumph for U.S. diplomacy. It seems to have had very little impact on the perceptions of policy elites, however; no one to whom I talked in any of the five countries thought it had materially affected the vulnerability of the oil-consuming states. This perception, of course, may be in error, but until it is contradicted by reality, it is likely to shape governmental action.

The other variable clusters are discussed separately for each target state. External variables include all attempts by actors outside the state (usually other governments) to influence policy; the oil weapon itself was one such external variable. Societal variables are those factors within a society but outside the formal governmental structure that influence policy: public opinion, the press, and interest groups, among others. Governmental variables include party structures, national legislatures, governmental bureaucracies, and the general process by which these interact to produce government policy. Finally, the impact of individual personalities on the decisions is assessed.

This rather complex analysis seeks (a) to systematically evaluate possible explanations, other than the oil weapon, for a particular government's behavior and (b) to allow for comparisons among the countries in an attempt to identify patterns of influence.

One strategy to determine the political impact of the oil weapon is, thus, to focus on different explanations for the policy of individual countries. A second involves the selection of cases. The book focuses on the industrial countries because their many similarities (level of development, democratic governments, etc.) make it easier to trace the reasons for any differences in their reactions to Arab pressure. Moreover, these countries were the primary targets of the Arabs themselves, as only the United States could reasonably be expected to persuade Israel to alter its foreign policy, and only the industrial countries could expect to persuade the United States to do so. Finally, the lack of relevant information about how countries reacted to the oil weapon, discussed below, is probably more severe for Third World countries.

Ideally this study would consider all of the industrial countries to see how and why they altered their policies toward the Arab-Israeli conflict as a result of the oil weapon, but international events data sets assembled by WEIS (World Events Interaction Survey) and COPDAB (Conflict and Peace Data Bank) reveal that too few acts of the industrial countries toward either Israel or the Arab states were recorded to allow reliable analysis. My early work (1982a, 1982b, 1987) confirms that critical information was simply not available publicly. In particular, it was impossible to readily determine what specific demands and threats were made by the Arab governments to what particular targets, who decided how to respond within the target governments, and what policies were changed as a result. Case studies were therefore necessary.

It is important to note that we are not dealing with one case, the oil weapon. Instead, the unit of analysis is countries that import oil, potentially more than a hundred, although I focus on a particular subset here. This large number of cases is useful for building and testing generalizations. It also means, however, that case selection is crucial because the cases comprise most of the book in addition to serving as the basis for broader analysis. The twenty-four members of the OECD were the "population" I wished to study.

The United States was central to the political rationale for the oil weapon because it was the only state capable of persuading Israel to alter its policies; it was therefore included. I chose other cases using two criteria: the degree of pre-1973 energy dependence on Middle Eastern oil (a measure of potential vulnerability), and the Arabs' own classification of states ranging from friendly to embargoed. Thus I included the most dependent country in the OECD (Japan) and the least dependent (Canada), along with the Netherlands (embargoed) and Great Britain (classified as friendly by the Arabs at the beginning of the embargo). The ability to compare the responses of different states to the same phenomenon is crucial in any attempt to answer the second question.

Economic vulnerability was used as a criterion because, according to the supply theory of sanctions, the more vulnerable targets would be more likely to make concessions than the less vulnerable states. Economic-sanctions theory is ambivalent on the question of country vulnerability. On the one hand, much of the literature is concerned with how such economic vulnerabilities occur and how they can be reduced. On the other hand, the implicit conclusion of the realpolitik school is that it doesn't really matter, that, ultimately, sanctions don't work anyway.

The issue of vulnerability is more complex than it seems. Keohane and Nye (1977, 11–19), among others, distinguish between "sensitivity" (the economic importance of imports from a particular country) and "vulnerability" (the costs involved in replacing such imports), noting that a target country may import a great deal from an actor but that if it can be readily replaced, the target's vulnerability is low. In the case of the oil weapon, however, the difficulties of replacing a great deal of imported oil were very great, and there was enormous uncertainty about the economic impact of ending such shipments. Thus, one could argue that sensitivity and vulnerability were essentially identical in this case. In order to help determine the impact of vulnerability on behavior, the OECD countries that drew the largest and smallest percentages of their total energy supplies from Middle Eastern oil (Japan and Canada respectively) are included in this study as cases.

But why should anyone care? The 1973–1974 oil weapon is now history, and the precise circumstances that shaped it are unlikely to be repeated. As I write, we are in the midst of the second worldwide oil glut since the embargo, Israel and Egypt remain in a "cold peace," Henry Kissinger and Pierre Trudeau are out of power while Ruud Lubbers and Yasuhiro Nakasone head their respective governments, the Middle East is convulsed by the Iran-Iraq war, and the Palestinian Liberation Organization is a shadow of its former self. Aside from the dubious pleasure of reliving our fairly recent past, why undertake an in-depth study of the impact of the oil weapon?

As a social scientist, I believe knowledge of the oil weapon's impact is useful if it helps to clarify our reasoned expectations of how actors in the international system behave. These expectations in turn come from theory, a set of general principles derived from some idea of what causes things to happen. We do not usually feel the need to explicate our underlying theories when we explain the past and predict the future (although I continue to believe such explication is useful and necessary to meaningfully evaluate the explanations and predictions). Nonetheless, our theories guide our thinking. The common assertion that my reaction will depend on the particular situation is untrue; my reaction will depend on my evaluation of the situation (which will often not be shared by others), and this evaluation in turn is shaped by my underlying ideas and beliefs. Thus, the effort to develop and test theory—the central activity of modern political science—is not simply an abstract intellectual exercise but, rather, is an attempt to clarify for us how and why we expect certain things to happen. Or, as Graham T. Allison puts it, explanation is the common ground among foreign policy analysts of various methodological persuasions—why things happen as they do and, by inference, how they are likely to happen (1971, vi).

The oil weapon attracted an unusual amount of public attention precisely because it signaled the need to change some expectations about the way the world worked. Many people worried about the oil weapon as a precedent. Could the Arabs repeat their apparent successes with impunity? Could other countries duplicate the feat with other commodities? Was military power obsolescent in the face of new forms of economic coercion? What did this mean for the future?

We cannot answer these questions by considering the oil weapon in isolation. We must compare it to other events, looking for behaviors that seem to repeat themselves over time. If there are no patterns, then international behavior is random, which no one really believes. But if we are to look for patterns, then we need theory to guide our selection of events for comparison from the infinite number available. To put it differently, we must state the general category in which we believe the oil weapon belongs. This will allow us to see whether the experiences of the oil weapon suggest possible changes in our theories about how events in this category work.

Obviously any particular set of events can be placed into as many categories as the analyst desires. As noted above, the oil weapon has generally been seen as a part of a set of economic issues. This book will focus on it as an example of the use of political power by one set of states to influence the behavior of another. Therefore, we pose the third major question: *What do these experiences tell us about the conditions under which one group of states is more likely to be able to influence others?*

As noted above, we care about the theoretical utility of the oil weapon in part because it allows us to refine our expectations about the behavior of states. Thus, the importance of our fourth question: *What are the policy implications of the Arabs' use of the oil weapon?*

Several subsidiary questions suggest themselves. How likely is another Arab oil embargo? Would such an embargo cause the industrial countries to significantly alter their policies on the Arab-Israeli dispute or on other foreign policy issues? Can other resources be used successfully by Third World states to influence the industrial countries on foreign policy issues? What steps can target countries take to make themselves less vulnerable to pressures such as the oil embargo? The policy utility of an intensive study of the oil weapon lies in its ability to provide answers for these kinds of questions that, if not definitive, can at least advance our knowledge.

This study rests on the assumption that the best way to study the actual and likely future impact of any foreign policy instrument is to look at the *targets* rather than the *actors*. For reasons that are not entirely clear to me, this strategy is unusual among foreign policy analysts. More important, it explains why there is no in-depth treatment of the motives and decisions of the Arab oil-producing states in this study of the Arab oil weapon. Such a major study is certainly needed (the best full-scale discussion is found in Daoudi and Dajani 1985, but it appears to be based mostly on published materials; for an excellent analysis of the decision to impose the oil weapon by Saudi Arabia based on the personal documents of one of the key Egyptian players, see Korany 1986, 87–112). But someone else will have to write it. Lacking such a study, I cannot say with confidence that the oil weapon was a success for the Arab states. I can talk with more confidence about its effect, however, which is the point of this analysis.

Should the Oil Weapon Have Altered the Policies of the Targets?

Was it reasonable to expect the Arab oil weapon to have influenced the target governments? Two major sources of predictions about the likely outcome of the oil weapon come from (a) the theories and cases from the literature on economic sanctions and coercive bargaining—based on general patterns of other cases—and (b) judgments by well-informed, sophisticated observers that stressed the unique aspects of the Arab oil weapon. As expected, neither group was unanimous, but both had a fairly clear central tendency: The theories tended to predict little or no success, although most observers at the time thought the oil weapon would have considerable impact on the foreign policies especially of Western Europe and Japan.

Previous Cases and Theories

In a major review of the economic-sanctions literature, Daoudi and Dajani contend the field has been dominated by two opposing sets of assumptions (1983, 18–55, 174–188; see also Daoudi 1981, 37–61). (Their analysis has persuaded me to retract my earlier assertion [1982b, 366] of unanimity within this literature.) The first assumption, known as the Geneva school of thought, holds that the threat of economic sanctions imposed by an international organization would deter aggression and thus prevent war. Such thinking grew out of the "lessons" of World War I; the label "Geneva" refers to the League of Nations. Indeed, the League of Nations had some success with sanctions, particularly in the case of the Yugoslavian invasion of Albania in 1921. Adherents of the Geneva school attributed the failure of sanctions against Italy to a failure of nerve by the Western democracies rather than to the inherent limitations of sanctions.

More recent literature has been dominated by a realpolitik approach that involves a second set of assumptions. Based on theory and historical case studies, this school of thought asserts that one government is almost never able to persuade another to alter its foreign policy significantly by the use of economic sanctions alone. For most scholars, this is counterintuitive: Economic sanctions make sense as a foreign policy instrument. But the general acceptance of the contrary view, that they are essentially empty gestures, has grown from careful analysis of a number of case studies. Sanctions by international organizations include the League of Nations against Italy and the United Nations against Portugal, Rhodesia, and South Africa. Many sanctions have been instituted by individual states and alliances; Hufbauer and Schott list 103 cases from 1914 to 1984 (1985, 13–20). Prominent examples would include Soviet sanctions against Yugoslavia, the Arab economic boycott of Israel, and U.S. sanctions against practically everyone, including prewar Japan, the Netherlands, the Soviet Union and its allies, China, Britain, Cuba, Iran (twice), Nicaragua, and Poland.

Recently scholars have agreed on this latter conclusion with a unanimity that is unprecedented in the social sciences. Indeed, the evidence is so strong that it has become one of the few conclusions of applied social science that appear to have become part of the belief system of Western foreign policy–makers; even when the Reagan administration applied sanctions to the Soviet Union over the issue of Poland, for example, few officials expected Soviet policy to change as a result (personal interviews in the U.S. State Department at the time). Renwick (1981, 1) and Knorr (1984, 205–206) both maintain the history of economic sanctions has been ignored by policymakers, but the realpolitik interpretation of this history has apparently been learned by government officials and the attentive public. (The fact that sanctions are still applied suggests they are used for reasons other than altering the behavior of the targets.)

This consensus in turn has been questioned and refined by a number of very recent studies that have featured theory (McCormick and Bissell 1984; Olson 1979; Lindsay 1986; R. Wagner 1986), cases (Weintraub 1982; Paarlberg 1985; Goldsmith 1986), and aggregate-data analysis (Hufbauer and Schott 1983, 1985). The trend has culminated in Baldwin's study. These authors all work within the realpolitik school but argue that sanctions are more useful than previous realpolitik authors have contended. These authors are all revisionist realists, however, rather than full-blown Geneva school theorists; none of them sees sanctions as a useful general alternative to war, for example.

These two schools share a (largely implicit) set of assumptions about sanctions, which may be called the "supply theory" of sanctions. According to this theory, the actor state deprives or threatens to deprive the target of economic benefits. (Most existing theory assumes the deprivation consists of reducing the supply of goods to the target, but more recently a variety of financial sanctions have also been used [Hufbauer and Schott 1985, 28–29, 89; Shubik and Bracken 1984]. The underlying logic remains the same.) The target state in turn makes a set of rational utility calculations about the costs and benefits of proceeding on its original course or altering its policy. Like much theory in international politics and economics, this approach assumes the target states are unified, rational actors.

In reaction to this rather simplistic notion, Baldwin (1985, 130–138) develops a much more sophisticated set of ideas about how economic sanctions influence targets. He argues that any one of the Kaplan and Lasswell eight base values (power, respect, rectitude, affection, well-being, wealth, skill, and enlightenment) can be manipulated by economic state-craft as an instrument of power. We need not assume, therefore, that economic values are primary for the target. He also notes success should be defined not simply in terms of forcing a target to change a policy but also in terms of increasing the target's cost of maintaining the policy; success thus becomes a continuous rather than a dichotomous variable (albeit considerably more difficult to measure).

As often happens in social science, the two schools of thought never really meet. The Geneva school argues that, *if sanctions are universally applied,* the target state will have no choice but to alter its behavior. The heart of the realpolitik critique is that economic sanctions cannot be successfully applied in practice, regardless of their theoretical utility. History demonstrates, realist scholars contend, that sanctions cannot and will not be applied universally. The Geneva approach is thus dismissed as utopian and irrelevant, not as necessarily wrong.

Indeed, the central theoretical question of realist scholarship on economic sanctions is why sanctions do not work. Hakan Wiberg (1969, 14) argues that a successful embargo involves three quite separate things in the face of presumed resistance by the target: effectiveness, efficiency, and success. First, an "effective" embargo denies significant amounts of the selected *product(s)* to the target; the target can counter by stockpiling, developing alternate sources, paying a higher price, and so forth. Second, an "efficient" embargo causes the desired *economic impact* on the target state, presumably by denying it the product(s). The target may respond by finding substitutes, altering its production processes, or making more fundamental changes in its economy. Third, a "successful" embargo, however, is defined in political, not economic, terms by the target *altering its policy* as desired by the actor. The target can presumably resist even efficient sanctions by appealing to nationalism in order to increase the willingness of its populace to pay high economic costs, shifting the costs within its society to groups with little political power, or changing the terms of the dispute by resorting to military force.

The realpolitik analysis notes that each of these three steps is difficult by itself; therefore sanctions that accomplish all three are unlikely to occur, particularly when the targets are wealthy states or superpowers. Most work has centered on the first two aspects of economic sanctions; the political questions have generally received less attention (interesting exceptions include Losman 1979, 128; Sylvan 1983, 216; Hufbauer and Schott 1985, 81–82; and Milward 1984, 109), perhaps because the economic preconditions for political influence have been so difficult to put in place.

The realpolitik approach is generally reinforced by the literature on coercive bargaining that concludes that it is very difficult for one government to
influence another, particularly on a policy the target believes is important.

> The attempt to modify behavior across national boundaries is perhaps the
> purest of all political acts. Unlike their domestic counterparts, foreign policy
> officials cannot appeal to the common ties of culture and history to secure the
> compliance of those whose behavior they are attempting to modify. Unlike do-
> mestic officials, they cannot merely rely on the structure from which their own
> authority is derived to induce compliance. The foreign policy official is the only
> politician whose actions are directed toward persons and situations that are nor-
> mally responsive to cultural standards, historical aspirations, and sources of au-
> thority that are different from his own. Hence the foreign policy undertaking is
> the most delicate of political actions and the most fragile of political relation-
> ships. It involves a degree of manipulation of symbols that is unmatched in any
> other political situation. It requires a balance between the use of persuasion on
> the one hand and the use or threat of force on the other that is more precarious
> than it is in any other kind of politics. (Rosenau 1980, 103)

These scholars generally find that external pressures often change the
issue from a substantive question (here the Arab-Israeli dispute) to a question of the target state's independence, producing internal unity and increasing the political risks of capitulation. William Coplin and Michael
O'Leary (cited in Mansbach and Vasquez 1981, 189–190) make a related
argument that policy change is the result of costs, benefits, and affect
(friendship or hostility) toward the other member of the power relationship, and that external pressures increase the level of hostility, making
change more difficult and less likely to be gradual or incremental.

We should also keep in mind that when coercion is undertaken by a
major power (as in Vietnam), the actor is normally pursuing a number of
other policies in different countries at the same time. The target state,
however, may regard this question as its single most important issue and
expend its resources accordingly. To wit, the United States fought a limited war in Vietnam, while for the North Vietnamese it was a total war;
North Vietnam was therefore prepared to pay a much higher price than
the United States in order to attain victory (Mack 1975).

Finally, coercion is a bargaining process, so the target retains the ability
to gain some concessions, forcing the actor to clarify its own goals and
communicate a willingness to bargain for a solution acceptable to the target. Luttwak observes that "if either the political leaders or the relevant

public of the target country do not accept the compulsion of necessity and choose to defy the threat, [coercion] will fail, *regardless of the balance of forces between the two sides*" (1974, 57, emphasis in original). To put it differently, a state cannot be coerced unless it agrees to be, because it always retains the ability to escalate the conflict. The actor's task is to produce a compromise solution that is more attractive to the target than further escalation—simply applying punishment is unlikely to succeed.

The Arab oil weapon, therefore, seemed likely to be another case of the failure of economic sanctions to coerce. The Arabs were making public demands on a group of wealthy, democratic states, including a superpower. Under these circumstances, it seemed unlikely they could cause and maintain a basic policy change.

The Oil Weapon as a Unique Case

Thus, in general, analysts studying the ability of states to influence other states' policies using economic sanctions and limited military force had independently concluded that such actions were unlikely to succeed. This conclusion was based on previous cases, however, and by definition no single case is exactly similar to previous ones.

There was indeed a reasonable argument that because the oil weapon was so different from other economic sanctions it *should* have succeeded. In terms of the power relationship between the Arab oil producers and the Western target governments, the range of sanctions was quite extensive and the scope of the demands was not; it therefore seemed logical that the weight of the relationship should be fairly great.

The potential impact of the oil weapon was considerable for several reasons. In the first place, oil is a uniquely potent commodity for sanctions; crucial for industrial economies, it cannot easily be replaced by substitutes and is consumed in such quantities that it is difficult to stockpile. Also, by 1973 the industrial states had become increasingly dependent on oil. As the price of Middle Eastern oil dropped during the 1950s and 1960s, Western Europe and Japan shifted from coal to oil as their major energy source. In 1950 oil constituted only 14 percent of Europe's energy and 5 percent of Japan's; by 1972 the figures were 60 percent and 73 percent respectively. Because these countries had no significant domestic supplies (North Sea oil was just coming on stream in the 1970s), they became increasingly dependent on oil from the Middle East. By 1972 47 percent of Europe's total energy consumption came from Middle Eastern oil; the comparable figure for Japan was 57 percent, although for the United States it was only 2 percent (Darmstadter and Landsberg 1976, 20–22).

This potential vulnerability was increased because Middle Eastern oil production was concentrated in relatively few states, many of which had sufficient financial resources to be able to endure the economic costs of temporarily shutting down oil exports. Moreover, the trigger for the embargo was a war between Israel and the Arabs, the one political issue capable of producing political unity among very disparate Arab governments. By 1973 the excess supply of world petroleum, which had foiled the 1967 Arab embargo, had ended. In addition, there was enormous uncertainty in the West as to the real economic impact of the oil weapon. Reasonable people predicted that the industrial economies simply could not survive unless supply was increased and prices were reduced. A considerable literature on crisis argues that policy change is much more likely under such a threat than under normal circumstances (Hermann 1972; Snyder and Diesing 1977; Williams 1976; Lebow 1981).

In terms of the scope of the relationship, the Arab governments had made only limited demands on the targets, asking for rhetorical concessions on a relatively unimportant foreign policy point for Western Europe and Japan. This combination of relatively insignificant demands with a very large threat (magnified by uncertainty) made it reasonable to expect the Arabs to be able to extract major political concessions—at least from the Western Europeans and the Japanese—despite the experience of earlier cases. Indeed, the impression lingers among a number of experienced observers that the Arabs were able to do so. Klaus Knorr, one of the most sophisticated analysts of economic sanctions, lists this set of events as one of four examples of successful economic coercion, although he also is skeptical of the particular impact of the embargo (1984, 187, 198). He is hardly alone in his position.

> Western Europe was shown to be at the mercy of international forces over which it had little or no control. Arab oil-producing states exercised unprecedented political influence over Western Europe and Japan. The "oil weapon" was used by producer states to insure that consumer countries adopted policies supportive of the Arab position on Middle East issues. The specter of "Finlandization," to which reference is often made in discussions of Soviet–West European relations,

confronted Western Europe. But this was a new form of Finlandization brought about by Arab oil-producing states—against whose pressures West European governments appeared powerless. (Pfaltzgraff 1978, 159)

When in 1973 the Arab oil-producing states decided to cut production and stop supplies to certain countries altogether, their decision had an enormous impact on West European policy and it affected America's course of action to a much greater extent than ever before. . . . Whoever ruled the oil fields potentially ruled Europe and Japan. (Laqueur 1974, 223, 248)

The result was change, drastic not only by recent standards but even in some respects by those of the two centuries since the Industrial Revolution. . . . This was the first time that major industrial states had to bow to pressure from pre-industrial ones. . . . Their victory upset the hierarchies of power long enjoyed, or resented, according to one's station, and opened up prospects of quite new political balances. By the same token, it was by far the biggest extension of the world's effective political arena since the Chinese Revolution. . . . The Arab oil policy registered the first success ever obtained at the highest level of politics by economic sanctions. (International Institute for Strategic Studies 1974, 1–2)

Indeed the oil embargo may have been the most decisive part of the [1973 Arab-Israeli] war—the part that led to an unfavorable diplomatic outcome for Israel, and that continues to have the most far-reaching consequences. (Friedman, Seabury, and Wildavsky 1975, 155; cf. Seabury 1974; al-Sowayegh 1980, 222–223; Manoharan 1974, 95, 103; Itayim 1974, 4)

The oil weapon's power to influence target governments to alter their policies remains open for debate, both in individual countries and as a general proposition. Case studies are therefore necessary to resolve the issue.

2

The Netherlands: Moral Fervor and the Oil Weapon

The Netherlands was the only OECD country other than the United States to be fully embargoed by the Arab oil-producers. As such, it is a crucial case in any effort to evaluate the utility of the oil weapon.

Historical Background

Interestingly, Holland was one of the first European nations to establish relations with Saudi Arabia. As colonial rulers of Indonesia, the Dutch found themselves managing the Moslem pilgrimages from there to Mecca and established a consulate in Jedda more than a hundred years ago. Despite this isolated incident, however, Dutch involvement in the Middle East is essentially a recent phenomenon.

The Netherlands has usually been regarded as a firm ally of Israel in the Middle Eastern conflict; indeed, this reputation probably was one reason it was embargoed in 1973 (Lenczowski 1976b, 15). In fact, as R. B. Soetendorp has demonstrated, this reputation is not entirely deserved. The Dutch were somewhat reluctant to support the partition of Palestine in the United Nations (they abstained in the committee vote, supporting it only on the floor of the General Assembly), and they did not recognize Israel de jure until January 1950, well after other Western European countries had done so.

Soetendorp persuasively argues that these actions resulted from the Dutch government's overriding concerns with other areas of foreign policy, particularly with the nationalist revolution in Indonesia. Thus, because Indonesia was Moslem, the Dutch government was reluctant to antagonize Moslems by supporting Israel, and in fact recognized Israel de facto only after transferring sovereignty to Indonesia (C. B. Arriens, personal letter to the author, May 10, 1984). The government was also susceptible to pressure from its allies, in part because of its need for assistance, or at least understanding, in Southeast Asia; thus, it supported partition because the United States asked it to do so and delayed recognition because of a request by Great Britain. The Netherlands and Great Britain gave de facto recognition on the same day. Similarly, Soetendorp views Dutch support of the British, French, and Israelis in Suez in 1956 as based on supporting the United Nations and the principle of freedom of the seas, a strong consideration in a traditionally trading nation such as the Netherlands. The central thrust of his argument is that Dutch foreign policy toward the Middle East has been based on Dutch national interests rather than on the moral concerns that are its normal rationale. (The best summary of this argument in English is Soetendorp 1984; for more detailed discussions in Dutch, see Soetendorp 1982, 1983.)

Although Soetendorp is probably too harsh in rejecting public sympathy for Israel as a factor in Dutch policymaking, his important analysis certainly alters the traditional image of the Netherlands as Israel's faithful ally. As long as the Middle East was regarded as a secondary area of foreign policy, it could be subordinated without much thought to more important policy questions. Thus, Soetendorp's analysis is more useful in explaining Dutch policy before 1967 than afterward when the importance of the area began to increase, a process that culminated in the 1973 embargo. At any rate, it is interesting that during the first few years of Israel's existence, when national interest clashed with support for Israel, Dutch governments seemed quite willing to stress their own interests. During the 1960s national interest and support for Israel reinforced one another. It was not until 1973 that these interests again came into direct conflict.

The high point of Dutch public support for Israel was the 1967 war. Dutch volunteers were recruited to join the Israeli army, and employers' and employees' organizations united with the government to send the wages of three working hours free of taxation to Israel (Baehr et al. 1982, 21). This attitude was reflected in government policy. During the 1967 war Holland supplied Israel with "large amounts of weaponry" (Grünfeld 1985, 158). In the United Nations it followed the English version of Security Council Resolution 242 calling for the return of "occupied territories," as opposed to the French and Spanish version of "the occupied territories"; the English interpretation presumably allowed the Israelis to retain some of the land involved (Voorhoeve 1979, 236–237; for a more detailed discussion of this issue, see Caradon et al. 1981).

Dutch support for Israel declined after 1967 for a variety of reasons, one of which was developments within the European Community. The Netherlands has traditionally supported the uniting of Europe (Baehr 1980, 247–251). One important development in this regard was the attempt to produce a common European foreign policy through the process known as European Political Cooperation (EPC), or the "Davignon machinery." Established in 1969, this process involved regular meetings of the prime ministers or foreign ministers and more frequent meetings of the political directors of the foreign offices of the European Community governments in an attempt to formulate policy on particular topics (see Allen, Rummel, and Wessels 1982).

The Middle East was one of the first two issues in this process; the other issue was the Conference on Security and Cooperation in Europe (Garratt 1982, 90). Over time the EPC process was fairly successful in this difficult area. Although agreement was not reached on all aspects of the issue,

> the Six did reach agreement on how to approach the most delicate issue of the Middle East conflict: the question of Israel's withdrawal from areas occupied in 1967 (with "minor modifications"); the modalities of "recognized and guaranteed borders" (international guarantees, demilitarized and internationally controlled security zones); the demand for an "administrative internationalization"

of Jerusalem; the settlement of the question of the Palestinian "refugees"; as well as shipping rights through the Suez Canal and the Gulf of Aqaba. (Steinbach 1979, 41–42)

The level of agreement remained limited, however. Several observers assert that in May 1971 the foreign ministers considered a joint position paper on the Middle East that was never published but whose contents were leaked to the press. Reportedly, it set forth a comprehensive proposal for ending the Arab-Israeli conflict, including an endorsement of the Arab interpretation of Resolution 242 and recognition of the "legitimate rights of the Palestinians" (Maull 1976, 118; Sicherman 1980, 847; Reich and Coquillon 1987, 162–163; Miller 1974, 10). It has been suggested that the Dutch were instrumental in opposing the draft because of Israeli pressure and that this was one reason the Arabs later embargoed Holland. In any event, a former senior Dutch official noted that EPC was never a decision-making process, so it could not have produced a position paper of such an outspoken nature (personal interview, June 1983). In 1972 the European Commission forwarded a memorandum to the EPC Council of Ministers suggesting a coordination of energy policies and the creation of agreements with oil-producing countries to trade development for guaranteed prices. Nothing was done, however (Allen 1978, 323). In July 1973 EPC also refused a request from the Irish and Italian representatives to "consider its long-term position on the Middle East," because agreement seemed so unlikely (Wallace and Allen 1977, 238).

The Dutch were torn between their sympathy for the Israelis and their desire to support European integration. It is not clear that the choice was seen in these terms, but, quite reasonably, they seem to have felt that European integration was more important to them than was Israel. The result was a significant shift in Dutch Middle East policy starting about 1971, supported internally by Foreign Minister Joseph Luns and more especially by his successor, Norbert Schmelzer (personal interviews, June 1983 and July 1984). In 1972 the Netherlands therefore voted in favor of General Assembly Resolution 2949, which not only asserted that recognition of Palestinian rights was indispensable in any peace settlement but also "reaffirmed" support for Resolution 2628, which included similar language, although the Netherlands had voted against it in 1970 (Djonovich 1978, 74, 242–243). The Netherlands supported Resolution 2949 despite the abstention of EPC member Denmark. In Parliament this vote was explicitly justified as the result of EPC negotiations (Soetendorp 1984, 40).

Heldring argues that European integration is more popular in the Dutch Parliament than in either the Foreign Ministry or the public at large (1978, 312–322), although a study of Dutch elites found government officials slightly more sympathetic than parliamentarians toward European unification (Baehr 1980, 248). At any rate this policy shift on the Middle East was apparently not politically popular. Indeed, while the Dutch government adopted a Middle East policy linked to that of its EC partners, its rhetoric for domestic consumption stressed some small differences of interpretation, trying to be "European in Brussels and Dutch in Amsterdam." Not surprisingly, the result was a good deal of confusion about what Dutch Middle East policy actually was—an issue the oil embargo did not entirely resolve.

The Oil Crisis

Despite this policy change, however, the initial Dutch response to the outbreak of the 1973 war was strongly supportive of Israel. The war began on October 6. The Dutch government thought the Egyptian-Syrian attack, which began the war, violated international law (Soetendorp 1982, 269) and issued a public statement on October 9 that condemned the attack and called for a return to the prewar boundaries and negotiations based on Resolution 242 (Voorhoeve 1979, 238). No other EC country made such a formal statement. The Arabs were unhappy about being told to withdraw from their own territories that they had just managed to partially reconquer, and they regarded Resolution 242, particularly the English version supported by the Netherlands, as an inappropriate basis for negotiations. The Dutch govenment, like most outside observers, expected a quick Israeli victory; thus, a call to return to the prewar boundaries was not necessarily seen as pro-Israel (personal interview with a senior Dutch government official, July 1984).

During an EPC meeting in Copenhagen on October 11–13, the Dutch were willing to support a cease-fire in place. In conjunction with the Danes, however, they vetoed an unusual proposal that would have allowed the British and French to speak for the Community at the United Nations on the subject of the Middle East. The veto was presumably because Great

Britain and France, long among the most pro-Arab members of the Community, would have moved Europe's rhetorical position somewhat more toward the Arabs. Although the meetings were supposed to be confidential, the Dutch and Danish positions were leaked to the Arabs, reportedly by the French, the Italians, or both (Pearson 1979, 122, who cites Boon 1976, ch. 10; Baehr et al. 1982, 23; personal interview with a senior Dutch government official, July 1984; Grünfeld 1985, 164).

The Dutch government seems to have had little warning that these actions would lead to an Arab oil embargo (Voorhoeve 1979, 239). Dr. H. N. Boon, then Dutch ambassador to Italy, has asserted that the Dutch embargo had been planned by the Arabs as early as June and that he had warned The Hague of this shortly after the war broke out, but that his warnings were ignored (Pearson 1979, 121, 137; see also Boon 1976). A senior official has stated that the warnings were sent only after the embargo had begun (personal letter to the author, November 1984). Even if Boon is correct, there would have been very little warning.

On October 17 four Arab ambassadors to The Hague visited the foreign minister, Max van der Stoel, to explain the situation. There are conflicting reports of what happened at this meeting. One version quotes one of the ambassadors as saying he had been "thrown out the door" by the foreign minister, which would certainly have been a diplomatic innovation. In fact, one Arab ambassador was so concerned about this report that he called on van der Stoel to state that they had been well treated (personal interview, July 1984). The Dutch contend that van der Stoel simply restated his government's position in a routine manner. Grünfeld states that his manner was seen as "dry and undiplomatic" (1985, 164). This may not have improved the Dutch image among the Arabs, but it seems unlikely to have caused the embargo.

Domestic support for the government's position was strong. There was basically no dissent in the parliament, and a few members urged going further by providing Israel material support. A newspaper advertisement in favor of Israel was signed by all postwar Dutch prime ministers, and a number of Dutch political parties held pro-Israel rallies attended by high-level political leaders, including cabinet members. By coincidence, on October 7, the day after the 1973 war started, a rally had been scheduled to support Soviet Jews. It was used to protest Austria's decision to close the Jewish transit camp to Israel in response to Arab terrorism; the prime minister, among others, attended. On October 13 the defense minister (apparently without informing the foreign minister [personal interview, July 1984]) attended a mass meeting in support of Israel that was addressed by leaders of nearly all political parties and churches; the Arabs took particular note of it (Balta 1974, 16). Trade unions and the press were also sympathetic (Baehr et al. 1982, 23).

The Arab governments announced their decisions to embargo the United States on October 18–21. On October 21 Algeria announced it would also embargo the Netherlands, and the next day the Arab League advised its member governments to follow suit. On the same day Iraq announced the nationalization of Royal Dutch Shell properties. On October 22 a senior Shell official visited the ministries of economic affairs and foreign affairs to ask for a clarification of Dutch Middle East policy. On October 23 the Dutch government reaffirmed its prewar position of support for the English version of Resolution 242—the version already opposed by the Arabs for a number of reasons, including the fact that it did not directly address the Palestinian issue. Within the next three days Kuwait, Abu Dhabi, and Oman embargoed the Netherlands.

On October 26 the Saudi government issued an ultimatum to the Dutch government asking it to condemn Israeli aggression, recognize the Palestinians' right of self-determination, and demand that Israel abandon all occupied territories. (These were essentially identical to the demands in the October 17 OAPEC embargo resolution [Szyliowicz 1975, 185].) The next day the Dutch foreign minister told the Saudi ambassador that the Dutch government had not received a formal response to its October 23 statement and that it would not do anything until it had. On November 2, after giving the Dutch government twenty-four hours' notice, Saudi Arabia embargoed the Netherlands.

There is still active debate over the real reason for the Arab embargo of the Netherlands. One position is that it was done because of pro-Israeli actions and statements by the Dutch government and leading politicians (Lenczowski 1976a, 15). An anonymous Algerian official provides the closest approximation of an authoritative Arab statement in print:

En réalité, la Hollande est un des rares pays européens dont la politique s'est toujours affirmée sans réserve, ni nuance, en faveur des thèses israéliennes et contre, toute prise position susceptible de gêner d'une façon ou d'une autre la politique de Tel Aviv. (Balta 1974, 14) (Actually, Holland is an unusual European country because it has consistently supported, without reserve or distinction, the positions of Israel and has opposed any position which would hinder Israeli policy in any way.)

Others argue that because Holland is a small country, it would be unable to retaliate, and because Rotterdam is a major port for Europe-bound oil, an embargo on Holland could exert additional leverage on Europe as a whole (Kelly 1980a, 400; Laqueur 1974, 243).

Regardless of the reason for the embargo, the Dutch government's initial reaction was surprise and shock. In previous years even stronger support of Israel had not produced an oil embargo. Moreover, unlike the United States, which was formally warned of the likelihood of embargo if it persisted in arming Israel, the Netherlands had received no formal warnings from the Arabs until the Saudi statement of October 26, which was issued only after the other Arab governments had already declared an embargo.

The available evidence indicates that the Dutch government regarded the oil weapon as a serious potential threat to its economic security, at least for a time. The embargo produced great uncertainty—initially it was not even clear that the Netherlands had been embargoed (Pearson 1979, 120). No one knew how much oil would be available, and no one could predict with confidence the impact of the loss of a certain amount of oil on the Dutch economy.

The sense of crisis was heightened by the refusal of Holland's European allies to share existing stocks of oil; because the Dutch had adopted a divergent Middle East policy, other countries saw no reason to help. One Belgian diplomat is quoted as saying: "Eight countries were strictly neutral, the ninth has followed public opinion and these are the consequences" (Baehr et al. 1982, 24). In addition, one could argue that the Dutch could afford to be embargoed because only 40 percent of their total energy came from oil, while the comparable figure for France was 80 percent (Pearson 1979, 131). A plan for oil sharing among the Europeans had been established within the OECD after the Suez crisis (Turner 1978b, 152–153), and on October 30 the Dutch requested that it be implemented. Britain and France rejected the request (Garratt 1982, 85). The oil-sharing plan might not have been too useful, however, because it excluded the Japanese and U.S. markets (Turner 1978b, 177). Nevertheless, the rejection of their request had a significant psychological impact on the Dutch.

An EPC meeting was called on November 6 in Brussels to consider the Dutch request for oil sharing. Instead, it produced a new policy statement on the Middle East. The declaration was elegantly crafted, calling for Israel to end "the territorial occupation" resulting from the 1967 war. This formulation skirted the problem of Israel's withdrawal from all the territory or only some of it, the central ambiguity of Resolution 242. (I disagree with Garfinkle's [1983, 3] contention that this was equivalent to requiring Israel to withdraw from all occupied territories.) It also asserted that the "legitimate rights" of the Palestinians should be taken into account in any peace settlement in the Middle East but did not specify those rights. The Dutch government signed the agreement (after "heated phone calls" to The Hague, Pearson [1979, 123] says; a former senior official does not recall them [personal interview, July 1984]), and the European Community for the first time had a unanimous position on the Middle East, although the Dutch, and reportedly other governments as well, promptly expressed contradictory private interpretations of the declaration (Grünfeld 1985, 165).

What did this decision mean? Had Dutch policy changed or not? There was a good deal of controversy over the subject at the time, and it is still difficult to tell. Because this is the only policy statement capable of revealing actual coercion of the Dutch during the oil embargo, the issue is important. Part of the problem is that Dutch policy on the Palestinians was ambiguous before the embargo, as noted in the previous section. The central thrust of that policy had been to treat the Palestinian problem as one of refugees rather than of self-determination. As earlier noted, however, there had been some movement on this issue even under the conservative governments, particularly with regard to the 1972 Palestinian resolution in the U.N. General Assembly.

In 1971, while still out of power, van der Stoel had been involved in preparing the Labour party policy document called "Flashpoint Middle East," which advocated such a policy change. Evidently he did not see the Brussels statement as a major shift because he did not consult the prime minister before agreeing to it (Fred Grünfeld, personal interview, February 1983). Van der Stoel later stated publicly that Dutch policy had not been changed by the November 6 statement except in referring to the legitimate rights of the Palestinians. A French analyst notes that the statement simply repeated U.N. declarations (de la Serre 1974, 81), and two

British observers suggest that the oil weapon had simply allowed European leaders to say what they had thought all along (Adams and Mayhew 1975, 133). Perhaps the best single analysis of the oil weapon concludes that "the use of the oil weapon speeded up and crystallized the EC position, rather than fundamentally changing it" (Maull 1975, 8).

At the same time, as Robert Lieber points out, regardless of the content of the November 6 declaration, "the context in which it appeared was substantially one of appeasement of the Arabs. It angered both Secretary Kissinger and Egyptian President Sadat by calling for greater Israeli concessions than Sadat was seeking at the time" (Lieber 1976, 13; see also Rustow 1977, 513). Much of the Dutch press and public opinion also interpreted the declaration as appeasement, a significant change in Dutch policy that had been brought about by the oil embargo (Grünfeld 1985, 165–166). It may well be that this change would have come about in any event, but at a minimum the oil embargo furnished the occasion to set it forth.

If the November 6 statement was designed to appease, it was a clear failure. The Arab reaction to the statement only made the Dutch situation worse: They lifted the 5 percent production cutback scheduled for December, helping the other European countries, but retained the Dutch embargo, leaving the Dutch in a worse comparative position. To make things worse, even though the Dutch had agreed to the November 6 statement, their European allies still refused to share oil. This raises an interesting point: Although it appears that the Dutch government was worried about oil supplies at least into December, there is no evidence of any new initiative toward the Arabs when the November 6 statement failed to end the embargo. This suggests that either the Dutch did not expect the November 6 statement to end it or that there were severe limits on the concessions the Dutch government was prepared to grant. One senior official reports that in October two Dutch envoys were sent to the Arab capitals to explain the government position and that this had served to strengthen rather than mitigate Arab demands. The government concluded that further concessions would simply lead to renewed pressures (personal interview, July 1984).

Toward the end of November the oil-supply situation began to ease. A senior official said the Dutch were relieved by Britain's decision not to implement an Order in Council to require oil companies to deliver contracted supplies (Voorhoeve 1979, 241); this allowed the informal oil sharing worked out by the companies to continue (personal interview, July 1984). On November 21 Henry Kissinger said at a press conference that the United States should "alleviate difficulties that have been caused . . . by policies which we consider responsible during the war [sic], which could be the case with the Netherlands." Dutch officials were relieved by this remark, although they chose not to say so in public (Heldring 1975, 4).

Earlier, in October, it had been noted in the parliament that the Netherlands exported as well as imported energy, in particular that it supplied about 40 percent of France's natural gas and lesser amounts to West Germany, Belgium, and Italy. If its European partners refused to help, some had intimated that the Netherlands would have to reconsider its policy of exporting this gas. This threat was also indirectly mentioned in speeches by the prime minister in the second half of November, thus putting Holland in the unusual situation of simultaneously being embargoed and tacitly threatening an embargo of its own. At about the same time, the West German chancellor called for solidarity with the Dutch; Heldring also refers to surreptitious assistance from West Germany and Belgium (1975, 4). It appears that, for these or other reasons, some quiet agreements were reached; on November 20 the Dutch government "express[ed] satisfaction with the 'common position' of the Nine," even though the OECD and EC both refused to adopt formal oil-sharing (Pearson 1979, 132, 134; Lieber 1976, 15).

On December 1 the oil ministers of Saudi Arabia and Algeria were in Brussels; they were not prepared to come to Holland. Van der Stoel refused to go to Brussels to see them, saying it would be a "journey to Canossa." The minister of economic affairs, R. F. M. Lubbers, did go, accompanied by D. W. van Lynden (director-general of political affairs) from the foreign office. The Arabs did not seem very interested in ending the embargo and reportedly proposed that the Dutch end diplomatic relations with Israel, demand that Israel leave all occupied territories, and formally link the Dutch position to that of the rest of the EC (Fred Grünfeld, personal interview, February 1983). (Presumably the last provision was a response to van der Stoel's assertion that Dutch policy had not changed despite the November 6 statement.) It was also suggested that the Dutch make some special gesture such as the provision of military assistance to the Arabs. Lubbers rejected these demands out of hand, although he reported publicly only the Arabs' request that the Dutch formally link themselves to the European Community position. Arab-Dutch negotiations on policy toward Israel ended with this meeting.

At about the same time, the Arab demands began to change, although not always consistently. Lenczowski (1976a, 19–20) reports, citing the Lebanese newspaper *Arab World* of December 4, that at the Algiers summit conference on November 26–28, the Arabs reached a secret agreement to ask Western Europe to end military and economic assistance to Israel and to lift their embargoes on military assistance to the Arabs; Asian countries (presumably Japan) would be asked to sever all political, economic, and cultural relations with Israel. In retrospect, this was perhaps the high point of Arab militancy. On December 9 in Vienna the oil ministers decided to accept, instead of full withdrawal of Israel from the occupied territories, a timetable for such withdrawal guaranteed by the United States (Shwadran 1977, 73; Lenczowski 1976a, 20–21).

By the end of December it was fairly clear that the Dutch would not suffer much from the Arab oil embargo. Rationing was prepared for January but was never implemented. Relations between the Dutch and Arab governments improved as the new year began, and in January the Arabs agreed to let oil go through Rotterdam en route to the rest of Europe, particularly Belgium (Lieber 1976, 15–16). The Dutch airline, KLM, was once again allowed to fly to Damascus. As usual, Libya went its own way, purchasing six Fokker aircraft from Holland in March while withholding oil supplies and not allowing KLM back until October (Sahr 1974, 1336). Although the embargo was not formally ended until July, it became a purely symbolic issue.

Dutch Middle East Policy Since the Crisis

Since 1974 the Dutch government has pursued its Middle East policy strictly within the framework of European Political Cooperation, marking a major change from its earlier policy. Within the confidential discussions of the EPC process, Holland reportedly remains more likely to support Israel than most other members. But in public statements, U.N. votes, and similar activities, the Dutch government stands by the decisions reached by the Community. It will differ publicly from its allies only if at least one other country does so as well. The Netherlands is not prepared to be isolated on a Middle East issue again (Soetendorp 1984, 41).

The Dutch were therefore involved in a decision, taken within EPC, to initiate the Euro-Arab dialogue between the EC and the Arab League. Its genesis was a decision at the foreign ministers' meeting in Copenhagen in December 1973, which several Arab ministers attended, "to propose a basic program of cooperation and mutual assistance between the members of the Market and . . . the Arab League countries." At the end of January 1974, while Holland was still under Arab embargo, the foreign ministers at a meeting in Brussels decided to initiate the dialogue. The Dutch refused to agree while still under Arab embargo, and the British were unhappy about proceeding in the face of U.S. opposition. By June the embargo was lifted, however, and Kissinger had announced that the United States no longer objected to such a dialogue (Steinbach 1979, 42; Shwadran 1977, 92–93). A permanent secretariat, employing 350 people, was established in Paris (al-Mani 1983, 50–51).

Despite some fears, particularly in the United States, the European willingness to talk with the Arabs did not indicate capitulation to all Arab demands. Beate Lindemann notes that "the cooperation of the Nine in the Euro-Arab Dialogue has little influence on the shaping of Community attitudes to the central issues of the Middle East" (1982, 122). After a good deal of internal controversy, the Europeans decided not to include political issues; the agenda was thus "science, technology, meteorology, ecology problems, and education." They also refused to accept a PLO delegation, resulting in a lengthy stalemate that was finally resolved in February 1975 by the "Dublin formula" of abandoning national representation altogether. Instead, the Europeans and Arabs each would send one large delegation, with the Arab delegation including individual members of the PLO (Shwadran 1977, 93–94; A. Taylor 1978, 432; Ismael 1986, 106; Reich and Coquillon 1987, 167).

Just before these meetings began in 1975 the EC announced a new trade agreement with Israel that cut tariffs on Israel's industrial products by 60 percent immediately and pledged to eliminate them entirely by 1977 (Hurewitz 1976, 295); it also included provisions to establish joint ventures in industry and science. Although it is true that this agreement was the result of the Community's "Mediterranean policy" (R. Bailey 1983, 187–191) and became the pattern for later ones with Egypt, Lebanon, and Syria, it was in fact the first of its kind. The EC thus was able to maintain an "evenhanded approach in commercial relations with all Middle East countries bordering the Mediterranean" despite its shifting political position on the Arab-Israeli issue (Feld 1978, 80–81; see also Minerbi

1974); the EC refused to extend similar treatment to non-Mediterranean Arab countries (al-Mani 1983, 6–7). Indeed, Garfinkle (1983, 6) argues that this display of independence caused the Arabs to accept European terms on the opening of the Euro-Arab dialogue. There is apparently no extensive study of the politics of this agreement within the European Community, but it certainly demonstrates the limits of Arab control of the situation. The Arabs did nothing to interfere, and the Euro-Arab dialogue was only briefly interrupted.

In 1976 at a meeting in Luxembourg, the PLO issue came up again in an almost comic-opera fashion.

> While the European side had assumed that the members of the Arab League were represented jointly in the delegations at Luxembourg—as had been the case in the groups of experts—the Arabs explained in their introduction that the member-states were represented individually; since the Community accepted this statement silently, a quasi-recognition of the PLO was thereby achieved. (Steinbach 1979, 44–45)

This is an example of the lack of substance of most of the meetings. In fact, the dialogue was inconclusive; it led to the International Conference for Economic Cooperation (CIEC), which in turn ended in stalemate in 1977. The Arabs escalated the pressure again in 1977 (Ramazani 1978, 53) without much more success. The dialogue was resumed, but just as it seemed to be making progress it was stopped by Sadat's dramatic trip to Israel; the Arab League requested that the dialogue be "suspended" in April 1980 (Steinbach 1979, 45; al-Mani 1983, 63–67; Ismael 1986, 107). Several other efforts to restart it failed, although by 1984 some talks were again under way (Reich and Coquillon 1987, 167).

In general, both sides seemed to have been more interested in "atmosphere" than projects (Allen 1978, 327–338); the main result was improved contacts and interpersonal relations and not much else (Feld 1978, 79; al-Mani [1983] gives a detailed analysis). A European analyst attributes the lack of substantive agreement on nonpolitical topics to "different approaches" to issues such as Arab industrialization, technology transfer, trade policy, financial cooperation, and Arab migrant workers in Europe (A. Taylor 1978, 436–439); an Arab scholar attributes it to European unwillingness (al-Mani 1983, 5–7). The Arabs didn't get much except for the Europeans' willingness to talk.

In the broader arenas of public declarations and U.N. voting, the EPC countries have produced a measured but clear policy shift in favor of the Palestinians, reportedly slowed by concern for "American sensitivity" (Kohler 1982, 89). In 1974 the Palestinian right to a national identity was affirmed. In 1976 it was agreed that the Palestinian people should have a territorial base (Sicherman 1980, 847–848). In June 1977 the right of the Palestinians to a homeland was added (Ramazani 1978, 52). In 1979 the foreign ministers asserted that Palestinians must be involved in the negotiations (al-Mani 1983, 113–114). Finally, in June 1980 the Venice Declaration advocated that the Palestinian people be allowed to exercise self-determination and that the PLO be "associated" with the negotiations. The next logical step would seem to be recognition of the PLO, but this has not yet happened. Holland has had to compromise on this last issue as well: Its membership in the United Nations Interim Force in Lebanon (UNIFIL) and its foreign minister's Middle East visit as president of the EC Council of Ministers both required the Dutch government to deal directly with the PLO. In practice it has proved willing to do so while continuing to maintain that these contacts did not constitute diplomatic recognition (Soetendorp 1984, 43–44).

One potential problem with the EPC declarations on the Middle East is that they conflict with U.S. foreign policy—a particular dilemma for the Dutch, who traditionally have close ties with the United States. The Dutch government has responded to this problem by supporting the EC rhetoric while opposing any European actions that would interfere with the Camp David process (Soetendorp 1984, 44–45).

The Venice Declaration seemed to make this problem acute, although the Dutch situation was greatly simplified because the EPC did not follow Venice with a statement recognizing the PLO; in fact, the final declaration was watered down in response to American concerns (Allen and Smith 1984, 45). Indeed, the Europeans generally seem to be backing away from the logical implications of their statements, mostly because of a lack of concrete results and hopes that the U.S. government can do something useful in the area. The EC has, however, condemned both Israel's annexation of the Golan Heights and the invasion of Lebanon.

Several elections also made further initiatives unattractive in the short term: François Mitterrand made France less aggressively pro-Arab and reduced internal pressures within the EC to push further in that direction; Margaret Thatcher developed close ties with Ronald Reagan and has a particular concern with terrorism; and Menachem Begin's reelection made

Israel unlikely to respond favorably. In addition, Reagan's foreign policy team proved to be significantly less sympathetic toward the Palestinians than Carter's, although in the longer run the addition of Greece and Spain to the EC moved the Community in a more pro-Arab direction (Garfinkle 1983, 71–74; Ismael 1986, 125; Reich and Coquillon 1987, 171–175). Increased terrorism in Western Europe and a common sense of the difficulty involved in influencing the Middle East have also contributed to what a French observer has called "the end of a specific European approach, presenting itself as an alternative to U.S. diplomacy" (Moïsi 1984, 219).

Aside from EPC declarations and U.N. votes, the Dutch were forced to act publicly on the issue of their embassy's location in Israel. Alone among the European powers, the Dutch had kept their embassy in Jerusalem rather than moving it to Tel Aviv—a problem because Israel claims Jerusalem as its capital, while the Arabs contend Israel is there only because of armed aggression. A former Dutch ambassador to Israel attributes the location of the embassy more to the personality of the Dutch envoy of the time than to a considered government position (C. B. Arriens, letter to the author, May 1984). In 1980 the Knesset passed a law asserting Israeli sovereignty over the entire city of Jerusalem, affirming it as the capital of Israel. Arab countries demanded that the few countries with embassies in Jerusalem move them to Tel Aviv. The Dutch government at first refused. The U.N. Security Council called on all members to comply, and the Netherlands finally agreed. In parliament the foreign minister "acknowledged the fact that Arab pressure weighed very heavily in this matter" (Soetendorp 1984, 45; see also Grünfeld 1985, 170–172); Arriens refers to the Security Council resolution as a "fig leaf" to cover the coercion of the Dutch government (letter to the author, May 1984).

During the same period, the Arab economic boycott of Israel became a public issue in the Netherlands for the first time. In 1978 a doctoral dissertation by R. M. Naftaniel, published by a Dutch Zionist group, showed that many Dutch corporations had acceded to Arab boycott demands concerning the origins of products as well as the ethnic backgrounds of employees and that the Dutch government had done nothing about these activities. There was considerable public and press uproar, followed by a parliamentary inquiry. Despite all the furor, however, in the end no laws were passed on the subject (Phillips 1979, 23–28, 64–73).

Dutch policy on the Arab-Israeli issue since the oil weapon can be divided into two different time periods. The Dutch government made little or no significant policy change on the Arab-Israeli issue while the oil weapon was being applied (the only instance of such change is the November 6, 1973, statement, which may have accelerated policy development but did not fundamentally alter it). In the longer term, however, the Netherlands abandoned its semi-independent policy on the Middle East in favor of closer links with the European Community, did not act on the Arab boycott, and increasingly supported the Arab position vis-à-vis the Palestinian issue.

External Influences on Dutch Policy

In a sense this entire book is about one important example of an external factor, the Arab oil weapon. Leaving that aside, what other external influences shaped Dutch policy toward the Middle East? Pressure from the other states of the European Community, as embodied in the EPC process, was clearly important. The goal of the process is to produce a "European" foreign policy to which all member governments can adhere. This is particularly difficult on an issue such as the Arab-Israeli conflict, where different governments have traditionally taken very different positions. Although EPC members have been remarkably successful in keeping their discussions secret, with only occasional conspicuous exceptions (Wessels 1982, 3), it is generally believed that France, Britain, Italy, and recently Greece tend to advocate moving the EPC positions closer to those of the Arab governments; Holland and Denmark are more inclined to resist, and West Germany and Belgium seem to wind up in the middle. Holland thus finds itself in a minority position within EPC.

The Dutch also displayed some ambivalence at the time about the whole process of European Political Cooperation (Wallace 1982, 47–48). One strand of Dutch policy stresses the advantages of a united Europe. Another important, historic aspect of Dutch foreign policy, however, has been its desire to stand as a moral exemplar for the rest of the world, what is called in Holland the "leading-country" role. Traditionally this idea was thought to come from the strong role of Calvinist parties in Dutch politics. In recent years, however, Calvinist influence has waned, but similar ideas continue to be observed both in Dutch foreign policy elites (Baehr 1980, 239–241) and among young Dutch leftists (Deboutte and van

Staden 1978, 75), suggesting that this idea is a national characteristic. If the Netherlands adopted EPC fully, it would be unable to fulfill this moral leadership role except within confidential government circles. There was some opposition to EPC in Holland at the time of the embargo because it might have forced changes in Dutch policies, and this remains an issue (Wallace 1982, 13).

Ironically, the EPC process also seemed to threaten established European institutions. Some of these organizations had attempted to move away from direct control by political heads of state toward accountability to the populace and reliance on majority rule rather than unanimity; EPC had been deliberately established outside the European Community institutions precisely to ensure tight control by the political heads of states (Deboutte and van Staden 1978, 68). Thus, people who supported independent Dutch policies and those who favored closer integration of Europe both had reason to be wary of EPC. One opponent of the EPC Middle East position was Jan Meijer, a senior official in the Foreign Ministry, whose opposition stemmed in part from a sense that the proper forum in which to resolve the Middle Eastern problem was the United Nations (Grünfeld 1985, 163–164). This distaste for EPC as a process—differentiated from dislike of certain policies it had produced—was a significant factor in Dutch reaction at the time and helps explain the willingness of very different Dutch governments to maintain a semi-independent policy toward the Middle East before 1973.

The Arab oil weapon clearly strengthened the hand of the European governments that were pressuring the Dutch to change their policies. On the surface this occurred because the Dutch needed help from their allies in order to get oil and had to pay for it by policy concessions, notably the November 6 statement. In fact, however, this is probably too simple an explanation, for if it were true, the Dutch would have made further concessions when the November 6 statement failed to get cooperation from either the Europeans or the Arabs. The psychological impact of being singled out and isolated by the embargo may have been the more potent device, particularly for a government that has traditionally supported European unity.

These pressures were not successful in the short run; no significant policy changes were made while the embargo was in force. The changes occurred later for reasons we will explore. It is clear, however, that the Arab position was greatly strengthened by the fact that the Dutch were in the minority among their own allies. Indeed, the implicit alliance between the Europeans and the Arabs was a potent force, but perhaps not until after the embargo had been lifted.

> The Arab countries realized and recognized that the Dutch position had been *strengthened* and that it would be up to the Europeans to persuade the Nether-lands to change its policy. Two years later, in 1976, the Netherlands govern-ment began to change its policy towards the Middle East in accordance with the views of the other members of the European Community. (Baehr et al. 1982, 28, emphasis added; see also Pijpers 1983, 176–177)

The United States might have taken some action to offset the Euro-peans and Arabs but apparently did not do so. At any rate, there is no evi-dence that the Americans made a systematic effort to force the Dutch to maintain their policies or that, if they did so, it significantly influenced Dutch foreign policy. The closest approach to this was probably Kissinger's offer mentioned earlier to help the Netherlands if necessary. This offer was certainly welcomed by the Dutch government and may well have strength-ened its hand in bargaining with its EPC partners (Grünfeld 1985, 166), but it came after the worst of the uncertainty had subsided.

Conversely, David Allen and Michael Smith (1984, 45) maintain that the United States is the single most important influence on Europe's Mid-dle East policy as a whole. They argue that (1) only the United States has the resources to exert a significant impact on the Arab-Israeli dispute, and (2) the Europeans react to U.S. initiatives rather than devising a truly in-dependent diplomacy. This influence is thought to limit the pro-Arab ten-dencies of some of the European governments, strengthening the hand of those arguing for restraint, such as the Dutch.

Societal Influences on Dutch Policy

Public Opinion

The external pressures on the Dutch government clearly favored a shift to a more pro-Arab position, as indeed they had all along. Opposition to such change came predominantly from domestic sources. Indeed, external factors usually can influence foreign policy only if they are reinforced by domestic actors. The links between external pressure and internal policy

Table 1.   Dutch Public Opinion: Sympathies in the Arab-Israeli Dispute

Do you consider yourself neutral in the conflict between Israel and the Arab states, or do you take sides—are you on the side of Israel or on the side of the Arab states?

	Neutral (%)	Favor Israel (%)	Favor Arabs (%)	No Answer (%)
June 12, 1967	26	67	0[a]	7
June 19, 1967	25	64	0[a]	11
February 1968	31	52	0[a]	17
December 1969	27	51	1	21
January 1973	59	36	1	4
November 1973	51	40	1	8
November 1974	56	37	3	4

SOURCES: Soetendorp 1982, 217; Erskine 1969–1970, 631.

[a] Less than 1 percent.

advocacy are difficult to trace: Domestic actors are reluctant to admit they are acting as a proxy for an outside power and may not be conscious of the degree to which they are influenced by foreign pressure. When the actors exhibit such confusion, the analyst must tread warily.

The single most important domestic factor explaining Dutch Middle East policy appears to be the widespread public support for Israel (see table 1; Everts 1985, 157–158). Several different reasons can be given for this tradition of support. First, Dutch colonies were located in Southeast Asia and the Caribbean; unlike some other important European countries, the Netherlands had no colonial involvement in the Middle East. Hence, Dutch foreign policy officials possessed less expertise about the Middle East than the British and French. At the same time, the Dutch officials were less pro-Arab, lacking a history of intensive contact with the Arabs.

Second, the almost total destruction of the Dutch Jewish community during World War II is also an important factor. Dutch Jews had been politically and socially prominent before the war and were almost entirely exterminated—a particularly traumatic fact for Holland, which has prided itself on the moral underpinnings of its policymaking. The consequent sense of guilt and obligation to Jews, combined with the lack of a colonial tradition in the Middle East, produced a Middle East policy dominated by concern for the survival of Israel, at least rhetorically. Third, the Netherlands is unusual among European countries in the significant role of evangelical protestantism in public affairs. The Calvinist emphasis on the Old Testament translated into support for the return of the Jews to the Middle East and the establishment of the state of Israel.

Fourth, the Dutch have traditionally used foreign policy to further their foreign trade. Trade to the Middle East was not a significant part of Dutch imports and exports before the embargo, but within this relatively unimportant sector, Israel was the single most important trading partner of the Netherlands before 1973. Finally, the long-standing Dutch commitment to the United Nations also played a role. During the 1960s this tradition promoted support of Israel. The Arabs were seen as aggressive powers violating international norms by supporting terrorism and favoring the annihilation of Israel, which the United Nations had created.

For several reasons, however, the almost automatic Dutch public support for Israel began to decline between 1967 and 1973. The 1967 war established Israel as the dominant Middle Eastern power. No longer the valiant underdog, Israel was now an occupying power. Support for Israel thus became less obviously necessary to guarantee its survival. Also, many in the Netherlands who were interested in foreign affairs became committed to Third World national liberation movements as the best way to reduce inequality and injustice. As the Palestinian issue became more prominent, these people increasingly saw the Arab-Israeli problem in terms of local nationalism competing with European colonialism. In addition, as those whose formative experience had been World War II grew older and fewer, the sense of guilt for the extermination of Jews during World War II diminished.

Dutch public support thus has changed over time, but in some ways the shift is more apparent than real. Even at its height, public support was never absolute. It could be overridden by the government when an important Dutch issue was at stake, such as the Indonesian revolution or European Political Cooperation. Nonetheless, there was a tendency to support Israel, at least as long as the Middle East remained a relatively unimportant area for Dutch foreign policy.

If it is easy to overestimate the degree of the Netherlands' public support for Israel, the decline in such support should not be exaggerated either. Anti-Israeli feeling never seems to have made much headway among the general public, although support for the Palestinians has risen over time, particularly among members of the elite who directly influence policymaking. The result has been a growing tendency to favor a neutral rather than a pro-Israeli position. Soetendorp cites a group of public opinion polls that, despite some methodology problems, show a fairly consistent increase in neutrality (see table 1).

Public support for Israel actually *increased* slightly during and after the embargo. Indeed, a 1977 poll asked: "To which of the parties does your sympathy go?" Nearly half (49 percent) of the respondents answered "Israel," 25 percent said they didn't know, 12 percent said "no party," and 9 percent responded "all parties." Four percent answered "the Palestinians," with 2 percent saying their sympathies lay with the Arab states (Soetendorp 1982, 218). It is hardly surprising, therefore, that there was considerable public uproar over Dutch government decisions to sign the November 6, 1973, statement and to move its embassy to Tel Aviv. A number of informed observers have suggested that the policies of the Begin government, particularly the gradual annexation of the West Bank, have significantly reduced Dutch public support for Israel, even among Dutch Jews (Grünfeld 1985, 160; personal interviews, February 1983), but I have seen no data on this.

But if this public support can be overridden by the government, does it really matter? Public opinion traditionally plays little direct role in Dutch foreign policy—making, which is dominated by the executive. At the same time, public interest in foreign affairs has increased in the past few years, particularly in areas such as South Africa or Chile; in general, however, there is little evidence that its actual impact on foreign policy has increased perceptibly (Baehr 1980, 235–236). We do not completely understand the influence of public opinion in foreign policy in a democracy, but it does seem to restrict the governments' freedom to make policy. The Dutch government got itself into trouble in the Middle East in the first place because it followed public opinion (although, as we shall see, these views were also held by many in the government, particularly the politicians). Public support largely explains the government's refusal to make significant concessions during the embargo. Thus, when I asked a number of informed observers if the Dutch government had considered ending diplomatic recognition of Israel (one demand issued by the Arab oil ministers in Brussels), they were unanimous that such an act was unthinkable. This response appears to reflect a societal consensus that although the Dutch re-

tain the right to disagree with Israel, Israel's existence is not subject to discussion, even if high costs result for the Netherlands.

Public support for Israel has also influenced Dutch foreign policy because of the tradition of concern for morality and the resulting political
style (Baehr 1978, 7). We must remember Soetendorp's caution that this
concern may disguise national interests, but it remains a significant curb
on government postures. As one sophisticated observer and practicing
politician put it:

> You don't create respect, certainly in the Dutch political system, by being busi-
> nesslike and accommodating and giving in. Much of the political fight in this
> country is a struggle in public relations policy in which you depict yourself as
> the more reliable, the more moral, the more ethical, the more principled party
> than all the others by taking more into account human rights and all that. In
> such a political culture, you can't be cynical like the French sometimes can. It
> simply doesn't pay. It's not good politics. (personal interview, February 1983)

Interest Groups

Another societal factor may push the Dutch government toward a more
pro-Arab position, however, and indeed may explain much of the shift in
that direction since the embargo. Holland has traditionally been quite
sympathetic to its business interests. Historically a trading nation, the
Netherlands has followed the British model of allowing private corporations considerable freedom to operate rather than the continental tradition
of state control. Indeed, one of the major oil corporations, Shell, is closely
linked to Holland.

The oil companies are usually seen as a force pressing for foreign policies more congenial to the Arabs, and although Shell (reflecting significant
differences in corporate style) has not been as outspoken as, say, Mobil, its
management almost certainly would have preferred that the Western countries, including Holland, had been less pro-Israeli over the years. Indeed, at
one point during the crisis, Gerritt Wagner, president of Shell, had a
"lively" discussion with van der Stoel about the "risky course" of Dutch
Middle East policy (personal interview, July 1984). But neither this particular discussion nor the company's general preference seems to have had
much impact on Dutch policy until after 1973, although we need not to-

tally credit Wagner's assertion that the oil companies did not apply any pressure but simply supplied information (Grünfeld 1985, 164).

Why the difference before and after the embargo? One result of the embargo, probably unanticipated by the Arabs, was a tremendous increase in their disposable wealth. Before the embargo Israel was the Netherlands' single largest trading partner in the Middle East. This changed radically as the Arabs began to spend money for development during the 1970s: Dutch construction firms obtained big contracts in the Middle East, and exports of all kinds from Holland increased drastically. The Arabs became major customers.

This political disengagement for economic motives implied compromise on both sides. The Arabs, here as elsewhere, generally allowed economic rather than political considerations to dominate international trade. Using total OECD exports to OPEC (not Arab) countries, Robert Lieber notes that in 1972–1973, 10.7 percent of these exports came from France, and 3.4 percent came from Holland. By 1980 the French share had dropped slightly to 9.8 percent, whereas the Dutch share had increased marginally to 4.0 percent (*OECD Economic Outlook*, December 1981, table 51, p. 115, cited by Lieber 1982, 338). Concomitantly, the Dutch government became significantly less willing to antagonize the Arabs on foreign policy issues (Grünfeld 1985, 161). The Dutch populace seems to have accepted this rationale. Certainly the opinion polls previously mentioned suggest that the public has remained fundamentally in sympathy with Israel but prefers to take a neutralist position in the Arab-Israeli conflict.

The theoretical literature on economic sanctions assumes that sanctions influence the policy of the target state by persuasively threatening to withhold the supply of some commodity, thereby damaging the target's economy. But the Dutch do not seem particularly worried about oil supply; the lesson they apparently learned from the embargo is that the Arabs cannot cut off their oil supply, given cooperative oil companies and the close geographic and economic links to other European countries. Anxiety about oil supply is therefore not a persuasive explanation for the shift in Dutch policy toward the Middle East. The more plausible and potent factor in the Netherlands seems to be a general sense that one does not make unnecessary trouble for good customers. The embargo indirectly transformed the Arabs into good customers, and in this way the oil weapon did indeed influence Dutch policy.

This influence was graphically demonstrated during the 1980 debate over the Dutch embassy in Israel, one of the few times that Dutch foreign policy was forced to move out from under the protective cover of the EPC process. The business community did not formally ask the government to move the embassy; this was not because business lacked interest, but because the request was unnecessary. At the behest of a member of parliament, civil servants in the Department of Economic Affairs drafted a report on Dutch economic interests in the Middle East, which also included a set of political arguments in favor of moving the embassy. The report was sent directly to parliament without being cleared by the economic affairs minister or the foreign minister, much to their discomfiture (Grünfeld 1985, 172–173, 177)—a remarkable example of the impact that increased Dutch exports to the Middle East had on its foreign policy.

Two other potential domestic interest groups should also be examined briefly. One is the "action group" supporting Palestine. Action groups are a relatively recent phenomenon in Dutch foreign policy–making, reflecting both the democratization of foreign policy and the increasing lack of fit between the regular political parties and particular interests and concerns. The American equivalent would be the single-interest group (Cohen 1977–1978, 206–208; Grünfeld 1985, 159). One pro-Arab demonstration was staged with about two thousand participants (Sus 1974, 71). Informed observers generally agree, however, that the Dutch committee on Palestine had virtually no impact on Dutch foreign policy at this time. It has been unable to exploit Dutch concern for the Palestinians much in the intervening years because it has been unable to break from the PLO's opposition to the existence of the state of Israel (Grünfeld 1985, 159).

The Dutch Jewish community has been a potential counter to this group. It is relatively small, however, and has not been particularly influential in recent Dutch elections. Moreover, because of the general Dutch support for Israel there had been no need to organize separate pressure groups (Grünfeld 1985, 159). At any rate, little organized response came from the Jewish community, which was generally satisfied with government policy, even after the November 6 statement. In fact, Grünfeld (personal interview, February 1983) reported that the Jewish community considered publicly thanking the Dutch government for its support of Israel but finally decided the government would be needlessly embarrassed. There are exceptions, however, to the Jewish community's general satisfaction with Dutch policy toward the Middle East: In 1980 when the Dutch embassy was moved from Jerusalem to Tel Aviv, several Jewish interest groups tried unsuccessfully to prevent this policy change (Grünfeld 1985, 174–175).

Governmental Influences on Dutch Policy

Parliament and Parties

We can divide our discussion of governmental factors into two sections, one dealing with political parties and parliament, the other with the bureaucratic organizations within the government itself. The Dutch political system is characterized by a great many parties. From 1918 to the present members of between seven and fourteen parties have held seats in parliament at any one time (Cohen 1977–1978, 200). Coalition governments and the problem of central control are thus inevitable. The Dutch tradition of a weak prime minister evolved as a result. In fact, strictly speaking, Holland has no prime minister—the office is more properly called "minister-president," reflecting the constitutional fact that its only power is to chair the cabinet. These two traditions make for very independent departments because the prime minister has no power to coordinate them and because, in a coalition government, any attempt to discipline a minister may well bring down the government.

The legislature (States-General) does not exercise much control either. In theory, ministers are individually responsible to it, but in practice any controversial decision is likely to be brought before the cabinet and turned into an issue of confidence for the entire government. Given the difficulty of forming coalition governments, it is very unlikely that a majority of parliament would vote out a minister, particularly on a foreign affairs issue, which is usually seen as relatively unimportant. Cohen quotes a Dutch M.P. as saying: "You don't bring a government down on a foreign affairs issue" (1977–1978, 201; see also Grünfeld 1985, 101–105). Perhaps as a result, parliament's credibility in the area of foreign policy is not high. A survey of the Dutch foreign policy elite found only 23 percent of members

of parliament felt the organization carried out its foreign affairs role
"well"; 62 percent said "rather poorly," and 15 percent said "badly." Respondents from other parts of the elite gave similar or more critical responses (Baehr 1980, 235). Thus, there are relatively few checks on the
authority of the foreign minister. As Baehr says, "The foreign minister is
unquestionably *the* dominating figure in the field of Dutch external relations" (1974, 172–173; see also 1982, 6; emphasis in originals).

This situation allowed Dutch foreign policy to be dominated for twenty
years by Joseph Luns, foreign minister in a number of different centrist
coalition governments. His position, according to one authority, was so
strong he could have defied a majority of the cabinet and still remained in
office (Baehr 1978, 4, 15). Luns resigned in 1971 to become secretary-
general of NATO.

In May 1973 a leftist coalition government led by Socialists was
formed. It held a fairly narrow majority (82 of the 150 Second Chamber
seats), and, in fact, its constituent parties had garnered only 49.8 percent
of the votes (Lijphart 1975, 207). The new foreign minister was Max van
der Stoel of the Labour party. Although van der Stoel was widely regarded
as a leftist (certainly as compared to his predecessors), Heldring points out
that he was under considerable pressure from the left wing of his own
party, which felt he was too "Atlanticist":

> Suggesting that he was abandoning Israel was for the left wing, despite its own
> sympathies for the Arab cause, a fair means to get rid of a minister whom it
> thought too pliant to the U.S., Israel's protector. Van der Stoel's position
> within the government is only secure because its survival is dependent on the
> support of two Christian Democratic parties, who sustain him precisely [be-
> cause] of his Atlanticist convictions. These paradoxes of Dutch domestic poli-
> tics are partly responsible for the opacity of Van der Stoel's Middle East policy.
> (1975, 4)

Grünfeld notes, however, that van der Stoel's position was strengthened
because much of his opposition came from outside the parliamentary frac-
tion of the Socialist party, which tended to be somewhat pro-Israel. Moreover, as criticism from the conservative opposition increased, the Socialists
became less likely to replace van der Stoel (Grünfeld, letter to the author,
December 1983).

In fact, parliament was not involved in the decisions during the oil crisis. It was not consulted about the November 6 declaration (nor, appar-

ently, was the prime minister). The only relevant legislation concerned domestic oil distribution. No formal hearings took place on the foreign policy implications of the crisis. In general, parliamentary support for van der Stoel's policies remained strong; he was even criticized by the Foreign Affairs Committee for not supporting Israel more strongly before the embargo was declared, which one authority sees as an indication of the difficulties a more pro-Arab policy might have encountered (Grünfeld 1985, 167).

The 1980 decision to move the Dutch embassy to Tel Aviv seems to have been an extraordinarily sloppy process, with open disagreement among ministers and between political officials and civil servants. As part of this, parliament was consulted so late that the actual decision had apparently already been made and implemented first. Although this caused a certain amount of uproar, the decision was not rescinded, and it is not clear that such a situation could not recur (Grünfeld 1985, 170–176).

In a multiparty system such as that of the Netherlands, which also has a tradition of parliamentary debate on foreign affairs, one might have expected sharp political opposition to a Middle East policy that resulted in an oil embargo. Actually, little or no organized opposition emerged from either side. No party argued either that the government had been too pro-Israel, and had thus brought the Arab oil embargo on the Netherlands, or that the Netherlands could not afford to allow itself to be coerced by oil pressure. These points of view were certainly articulated in public debate, but neither became an important part of the ongoing Dutch political contest.

In fact, the parties found themselves caught in a paradox. In the first place, historically, the parties have not differed much on the issue of Israel (Grünfeld 1985, 161), although the conservative parties had been somewhat more enthusiastic in their support of Israel. Thus, even though they were presumably closest to business interests that would worry about the effects of the Dutch posture, the conservatives could not very well criticize the government for being too pro-Israel. Groups on the left were inhibited from criticism because the policies had been carried out by their own government, and there was a real reluctance to threaten to bring down the first leftist government in many years just after it had taken office. Moreover, many influential members of the Labour party were strong supporters of Israel. These factors seem to have combined with the strong public support of Israel already discussed to keep the issue out of serious political discussion. Debate in parliament centered around whether the government, and particularly van der Stoel, had competently executed Dutch policy, not whether that policy was correct.

As noted above, van der Stoel came to office committed by a Labour party position paper (which he had written) to adopt a more "even-handed" posture in the Middle East. Important members of his party also wanted to reduce the dependence of Dutch foreign policy on the United States. As Heldring notes, however, the oil embargo forced a reversal of this policy: "This had the ironic side-effect of making a left-of-center government more dependent on the goodwill of the United States, from which at least the Labour Party had wanted to detach itself somewhat, and on the multinational oil companies which had also formed a target of socialist attacks" (1978, 320).

Bureaucratic Organizations

The oil embargo is an issue where politics and economics interact. Thus, it follows that it would fall somewhere between the areas of responsibility of government agencies concerned with foreign affairs and domestic economic welfare. Such areas have long been bones of contention in the Dutch government. For reasons already discussed, each ministry tends to operate independently. As a result, it is difficult to coordinate their activity in areas for which responsibility is not clearly defined. This definitely includes foreign economic affairs, and the result has been some unusual governmental institutions, two of which are relevant here.

After World War I responsibility for international economic policy was taken by the Foreign Ministry. In the 1930s, however, it passed to the Ministry of Economic Affairs as a result of the depression and the high prestige of economic affairs ministers. In 1946 the Directorate-General of Foreign Economic Relations was created as an interdepartmental coordinating unit. However, it was eventually placed under the Ministry of Economic Affairs and not surprisingly became linked to its parent institution. This unit is traditionally a powerful force within the government both because of its history of close ties with multinational corporations and because a good deal of Dutch foreign policy addresses issues such as the European Common Market. In 1972 the cabinet decided that Foreign Affairs would be responsible for coordinating and supervising Common Market policy by means of an interministerial committee chaired by the minister of state in the Foreign Ministry; nonetheless, the underlying problem remained (Wels 1983, 376–377; Baehr 1978, 11–12; Deboutte and van Staden 1978, 67; Grünfeld 1985, 88–89, 91–94).

Foreign aid also combines international politics and economics and is a particularly important part of Dutch foreign policy. The Foreign Ministry is divided into three sections: political affairs, European cooperation, and international cooperation (for greater detail and a diagram, see Deboutte and van Staden 1978, 59). Political affairs and European cooperation report to the foreign minister. The director-general of international cooperation, however, handles international organizations (including the United Nations) and foreign aid; he reports both to the foreign minister and to a separate minister-without-portfolio for international development. (There is also a secretary-general, the highest civil servant in the department, but his role is administrative and he seems to have played no particular role in this crisis.) This dual command gives the international cooperation section a good deal of independence, and it now seems to be developing as a separate organization (Baehr et al. 1982, 12; Grünfeld 1985, 69–76; Wels 1983, 383–384).

The scene appears to have been set for a classic bureaucratic struggle over how to respond to the oil weapon, pitting the ministries of Foreign Affairs and Economic Affairs against one another, with the additional complication of a semi-independent group within Foreign Affairs. In fact, nothing like this developed. The debate took place almost entirely within the Foreign Ministry itself. Why did this happen? The minister of economic affairs, R. F. M. Lubbers, was a new appointee, relatively young, who had come from the business sector rather than government. His ministry was deeply involved in efforts to cope with the domestic economic impact of the oil embargo, and he does not seem to have felt competent to give advice on Middle East policy. As a result, the Ministry of Economic Affairs played no significant role in the debate over the Dutch position on the Arab-Israeli conflict (personal interviews, Fred Grünfeld, February 1983, and a senior Dutch official, July 1984).

The situation in the international cooperation section was somewhat different. The minister for international development was Jan Pronk; he and van der Stoel were strong competitors (Baehr et al. 1982, 12–13;

Grünfeld 1985, 73), and he seems not to have taken part in the decisions. The director-general of European cooperation had only recently taken office and also was not deeply involved.

The four key individuals seem to have been the foreign minister (van der Stoel), the directors-general of political affairs (van Lynden) and international cooperation (Jan Meijer), and the ambassador-at-large (Scheltema). Of these, only van Lynden and Scheltema were apparently speaking for their organizations. Meijer's preference for the United Nations over EPC may have been partly the result of institutional factors, but pro-Israeli sympathies seem to have been dominant. Van der Stoel was clearly responsive to forces other than his own department. Van Lynden, in contrast, was primarily concerned with the negative reactions of other European governments to the Dutch Middle East position. The European division of his department was in charge of day-to-day diplomatic relations with these governments, so it is hardly surprising that he encouraged the government to be more sympathetic to the Arab position. Also, as political director, he was deeply involved in the EPC process, which was one major vehicle for this pressure. (Dutch diplomats stationed in the Middle East, who would have made similar arguments, seem to have carried little weight. The Division of Africa and the Middle East, which falls under the political division, was relatively small and was not regarded as an important institution [Grünfeld, personal interview, February 1983].) The bureaucratic struggle over Dutch policy therefore pitted the Foreign Ministry's political division, which advocated a more pro-Arab position, against individual and societal factors that encouraged support of Israel.

The Influence of Individuals on Dutch Policy

Jan Meijer was at this time director-general of the international cooperation section. Widely regarded as a brilliant and somewhat Machiavellian figure, he was also a Socialist civil servant, an unusual combination. Thus, when van der Stoel became foreign minister in 1973, Meijer's influence increased considerably (although Meijer himself denies that his party membership had anything to do with it and regards such an implication as an attack on his integrity as a civil servant [Grünfeld, letter to the author, December 1983]). A strong supporter of Israel, Meijer disliked the EPC process. He is thought by many to have written the government's October

9 statement accusing Egypt and Syria of violating the armistice and asking all parties to return to their prewar borders (Pearson [1979, 123] does not name him, but the description fits; Grünfeld [1985, 164] says it was prepared in Meijer's directorate). Meijer also opposed, in connection with the Copenhagen EPC meeting in October, any declaration on the Middle East ostensibly because he thought the decision should be made at the United Nations rather than in the European Community (Grünfeld 1985, 164). Although he was not able to persuade van der Stoel to adopt this position, Meijer presumably influenced the Dutch decision to deny the British and the French authority to speak for the European Community in the United Nations. These positions, and in particular the importance he attached to them, seem to have stemmed from his personal background rather than his position; it is hard to imagine another person in the same job taking these stands so vigorously. Thus, a number of Dutch acts that may have triggered the embargo can be traced in part to his influence.

At the same time, van der Stoel was hardly a helpless puppet in Meijer's hands, as some have portrayed him. After all, he had considerable knowledge of foreign affairs (although not much experience with the Middle East) and had been state secretary for foreign affairs a few years earlier. Van der Stoel had advocated some movement in Dutch Middle East policy before he came to power, but he was also a strong supporter of Israel and an anti-Gaullist. It is also worth remembering that, as a Socialist, van der Stoel was something of an outsider in the Foreign Ministry, despite his experience. Indeed, there was serious discussion about the ability of diplomats and civil servants in the ministry to work effectively with a Socialist foreign minister, although, in fact, there seem to have been no serious problems; at least van der Stoel never found it necessary to create an institution staffed by people personally loyal to him within the Foreign Ministry (Deboutte and van Staden 1978, 63–64). We must recall that the Middle East was not a particularly important foreign policy area before 1973, and within the Foreign Ministry there were few knowledgeable people who shared van der Stoel's attitudes on the Arab-Israeli question. Van der Stoel had good reason, therefore, to rely on Meijer's advice at the time.

Over time, Meijer's influence over Middle East policy waned. Policy moved in a somewhat pro-Arab direction, and both Meijer and van Lynden left their posts within a year of the crisis. The ultimate question, however, is whether a different foreign minister would have produced a different policy. The strongest case is undoubtedly that another foreign minister might have made the embargo less likely by, for example, not issuing the October 9 statement or by seeming more sympathetic toward the issues raised by the Arab ambassadors. But did these events actually cause the embargo? Also, it is difficult to imagine any foreign minister acting much differently—once the embargo had been imposed—either in resisting the initial demands or in orchestrating a gradual policy change. Thus, individual variables help explain the immediate reasons why the Dutch got themselves into a difficult situation, but not the acts they took to extricate themselves.

One other actor should be mentioned here—the prime minister, J. M. den Uyl. Because of the constitution and the multiparty system, the prime minister would not normally be involved in this sort of problem. This issue seems to have been something of an exception for a number of reasons. An unusually intelligent and impressive man, den Uyl was interested in foreign affairs. Like many of his generation, he had been deeply influenced by the annihilation of Dutch Jews, which had claimed many of his friends. Also, as minister of economic affairs, he had been largely responsible for developing the Dutch natural-gas export policy. The mechanism of European Political Cooperation required den Uyl's involvement because it was based on periodic meetings of prime ministers as well as foreign ministers and political directors. According to a strict reading of the Dutch constitution, he should have been accompanied by his foreign minister if foreign policy was to be discussed. Indeed, this was formally accepted by the EPC at one point, but in fact it was never done (Deboutte and van Staden 1978, 81). Den Uyl was therefore required to act within the EPC. He was also involved because he was more popular within his own party than was van der Stoel, and he was thus important in giving van der Stoel some protection from his party critics, particularly those on the left (Fred Grünfeld, personal interview, February 1983). Den Uyl's role in the crisis seems to have been to allow van der Stoel to work out the issue as he thought best, but this itself was a choice on his part.

Impact of the Embargo on Dutch Middle East Policy

How did the oil weapon influence Dutch policy toward the Arab-Israeli conflict? The single most important factor in explaining Dutch policy toward the Middle East has been consistent public support for Israel. This support reached its zenith during the 1967 war and declined thereafter. Dutch governmental policy toward the Middle East began to reflect this change to some extent before the 1973–1974 embargo, but the Palestinian issue was not a major consideration in Dutch policy at that time. Particular individuals did things that probably increased the Arab dislike for the Dutch position.

The Dutch never met most of the Arab demands. They asked the Israelis to withdraw from occupied territories (not "all" territories, as the Arabs demanded), and they abandoned their independent Middle Eastern policy in favor of one developed by the European Community (but never publicly admitted it during the embargo). Israel was not condemned for aggression, the right of Palestinian self-determination was not recognized, military assistance was not extended to the Arab states, and diplomatic relations with Israel were not broken. On balance the Arabs got relatively little of what they demanded, despite having created a real sense of crisis in the Netherlands. Grünfeld notes that "domestic consensus, finally, turned out to be an effective counter-weight to outside pressures" in the Netherlands (1985, 170).

In the longer term, over the ten years or so since the embargo, Dutch policy has become more pro-Arab within the EPC framework. In particular, the right of Palestinian self-determination has been recognized, although other Arab demands have still not been met. Although it is difficult to establish empirically, these changes do not appear to have been caused primarily by the threat of another embargo, with consequent oil-supply shortages and economic damage. The general sense among the people with whom I spoke was that, in the event of a renewed embargo, the Dutch government would not be likely to make more concessions than in the past. One indirect result of the embargo, as mentioned earlier, was that the Arab states became very good customers of the Dutch, and the Dutch, for whom international trade has long been central in shaping their foreign policy (Voorhoeve 1979; Grünfeld 1985, 48–51; Soetendorp 1984, 45), are inclined not to cause unnecessary trouble for good customers. This reluctance still seems to have real limits, however, presumably because of a consensus among the Dutch public and policymakers that Israel's continued existence is a moral imperative that the Dutch government must support. As a result of this attitude, the Dutch are likely to be more sympathetic to the side that is seen as the underdog at the moment. Any Arab successes in the region will increase Dutch concern for the continued survival of Israel, thus making it harder for the Arabs to influence Dutch policy. Conversely, and for precisely the same reasons, greater Israeli success in the area may well encourage Holland to express increased concern for the plight of the Palestinians.

3

Great Britain: "Friendly" with Reservations

During the 1973–1974 oil embargo Great Britain was one of the countries classified by the Arab oil producers as "friendly." As a result of the oil companies' behavior, this designation did not gain much oil for Britain but it did distinguish this country as a state the Arabs themselves had singled out as unusually supportive. This makes it particularly significant for our study. If Great Britain, which was viewed by the Arab states themselves as initially sympathetic to their position, did not alter its policy as the Arabs wished, the oil weapon clearly had only a limited political impact on its targets.

Historical Background

Of all the countries examined in this study, Great Britain has the longest history of serious involvement in the Middle East. This has occurred for three quite disparate reasons: geographic access to India, oil supplies, and foreign markets. It is interesting that, despite these changes in motivation, British policy has remained essentially constant over time.

In the nineteenth century British involvement in the Middle East stemmed from a desire to protect access to India. The two major routes were either through Egypt, Suez, and the Red Sea or through Syria, the Euphrates, and the Persian Gulf (Hollis 1987, 179); Britain's central concerns thus tended to be Egypt and the Gulf. Although this interest changed over time, it did not diminish because "at the same time as India was lost to the Empire, Britain was becoming dependent on Middle Eastern oil" (Edwards 1984, 47).

Britain has had a long-standing interest in Middle Eastern oil: From 1912 to 1914, Winston Churchill, as First Lord of the Admiralty, masterminded the government's purchase of control over the Anglo-Persian Oil Company (now British Petroleum) in order to secure fuel oil for the Royal Navy. Indeed, until 1971 the Foreign Office maintained its own separate oil department (Frankel 1975, 275–277; and Byrd 1973, 177). The concern for oil supplies also explains Britain's close ties with the small sheikdoms in the Persian Gulf, legal protectorates of Britain as recently as 1971 and still linked to Britain by a variety of ties. (For differing views of the relationship, see Mangold 1978, 83–88; Halliday 1977.) A British observer with both academic and governmental experience has suggested that Britain's long history of involvement in the Middle East rendered the government irrational vis-à-vis the country's vulnerability to oil cutoffs (personal interview, June 1983); he cited Hugh Thomas's remark: "Ever since Churchill converted the Navy to the use of oil in 1911, British politicians have seemed indeed to have had a phobia about oil supplies being cut off, comparable to the fear of castration" (Thomas 1970, 39). Another researcher disagreed that Britain's "phobia" was stronger than fears in other countries, noting in particular the 1973 public panic in Japan (personal interview, June 1983). In addition, Britain has become nearly self-sufficient in oil as a result of its North Sea fields. In 1973, 75 percent of the oil Britain consumed came from the Gulf; this figure dropped to 14 percent in 1983, when Britain became a net oil exporter, although no one is sure how long this new supply of oil will last (Hollis 1987, 195).

The Middle East is still an important market, however, for British goods and services, for domestic investment, and as a source of support for the international role of the pound sterling (Edwards 1984, 47–48; Prittie, 1983b). The region thus remains important for Britain, albeit for a third reason. Indeed, the worse the British economy gets, the more Britain seems to need the Arabs. Therefore, the end of British dependence on Middle Eastern oil has essentially had no impact on British policy toward the Middle East.

All three of these motives have caused British Middle East policy to focus on the Arabs. Perhaps we need to recall that, as one relatively minor part of this involvement, Palestine was a British responsibility. The Balfour Declaration, which marks the initial recognition of a Jewish national home in Palestine, was, in fact, one aspect of the British campaign in the area during World War I. Britain held the mandate for Palestine from the League of Nations between the world wars, while tensions between the Arabs and Jews mounted. Nonetheless, Palestine was never central to British Middle East policy. It did not control the Suez Canal, as did Egypt; it had no oil, as did Iraq or the Gulf States; and after 1973 it was much less important as a market than the Arab oil producers. Thus, Israel has always been seen as a distraction from Britain's primary interest in the area.

British control of Palestine lasted until 1948. After World War II the area was torn by a three-cornered war in which Arabs and Jews fought the British and each other. This was a nasty conflict, one of the first "wars of national liberation." The British were left with bitter memories of failure and of terrorism, particularly that practiced by certain Jewish groups.

The next traumatic event in the region was the Suez crisis of 1956, when Britain allied itself with France and Israel in an attempt to regain control of the Suez Canal and to oust Gamal Abdel Nasser from power in Egypt. The result was catastrophic, with the United States and the Soviet Union teaming up to humiliate Britain. The Suez crisis created a deep fissure in the Western alliance, a fundamental reappraisal of Britain's role in the world, and a domestic political debate that split the nation and led to the retirement of its popular prime minister, Anthony Eden. It was also the catalyst for perhaps the single most important postwar shift in British Middle East policy: the recognition that British troops would not be able to guarantee British interests in the region and "the apparent belief within the Foreign Office that Britain's interests . . . can be protected and furthered by the establishment of a fund of goodwill for Britain in the major countries of the region" (T. Taylor 1973, 12). It is worth noting, however, that this newly circumscribed role did not rule out lower levels of British military advice and assistance.

As a result of the Suez crisis, diplomatic relations between Britain and

Egypt were broken off. These were not restored until George Brown, foreign secretary of Harold Wilson's Labour government, reestablished ties in 1966—a particularly interesting policy in view of Wilson's personal sympathy for Israel. Brown's overtures were interrupted briefly by the events leading up to the 1967 Arab-Israeli war. President Nasser expelled U.N. forces from the Egyptian-Israeli border and closed the Straits of Tiran to Israeli shipping. Brown and Wilson proposed that Britain set up and join an international naval force in an effort to reopen the straits. This proposal was not strongly supported either within the cabinet or by other naval powers, and the ensuing Israeli attack made it academic in any event. (For differing views of this episode, see Wilson 1981, 332–341; G. Brown 1971, 136–137; Crossman 1976, 2 : 355, 358, 381–382; Gordon Walker 1974, 45; and Shlaim, Jones, and Sainsbury 1977, 214–215.) James Cable argues that the main effect of the proposal was to lend credibility to the rumor that British and American planes had aided the Israelis in the ensuing war, thus leading to "the June explosion of wrecked Embassies, broken relations, oil boycotts and injury to the economies, the interests, the influence and the Middle Eastern residents of Britain and the United States" (1971, 145–146).

At any rate, the proposed use of naval force proved to be an aberration in British policy toward the Middle East. A much more important development was the formulation of U.N. Security Council Resolution 242. A good many individuals and governments claim credit for it, but it was sponsored by Britain and seems to have been drafted primarily by George Brown and Lord Caradon, British ambassador to the United Nations (Shlaim, Jones, and Sainsbury 1977, 215; personal interviews, Shlaim, June 1983, and former senior British diplomats, June 1983). The resolution called for Israel to withdraw from territories occupied during the 1967 war in return for peace. Its adoption and, perhaps more important, the negotiation process that led both superpowers, Israel, and all the Arab states except Syria to support it, can fairly be called a British diplomatic triumph. (See Caradon et al. 1981 and S. Bailey 1985 for more detailed analyses of the formulation of Resolution 242.)

The Suez Canal was closed to shipping as a consequence of the 1967 war, and this hurt Britain economically. Wilson told the House of Commons it was costing the country about 56 million pounds a month (*New York Times,* October 22, 1967). To make matters worse, a short-lived Arab oil embargo was declared. Devaluation of the pound followed five months later (Mangold 1978, 69), which, in turn, influenced Britain's decision to withdraw from "east of Suez" (Gordon Walker 1970, 129–130). This "fairly traumatic" experience influenced subsequent policy toward Israel, particularly with regard to arms sales (personal interview, former British diplomat, June 1983). Indeed, if the criterion for success is the alteration of a target's policy, one can argue that the Arab oil weapon had a greater impact on Great Britain in 1967 than in 1973.

Crossman discusses at some length a protracted battle within the government after 1967 over the sale of Chieftain tanks to Israel to replace the aging British Centurions. The initial debate apparently occurred in November 1968, with the foreign minister opposing the sale and the defense minister supporting it. The cabinet then decided to sell the tanks in 1970–1971. But in 1969 the defense and foreign ministers jointly proposed that the sale be delayed because of the "war of attrition" then going on between Israel and Egypt. A former senior diplomat suggests that this shift took place in response to fear of Arab retaliation, particularly with oil, which might have forced another devaluation of the pound (personal interview, June 1983). Crossman was concerned that the British government was assuring the Israelis that the weapons sale would go through when, in fact, it would not. The situation was further complicated when the government agreed to sell Chieftains to Libya on the mistaken assumption that tanks would also be sold to Israel. This sale was aborted when a military junta deposed King Idris in 1969. The tanks never were sold, although Wilson later claimed in debate that "there was an understanding between the Labour Government and Israel that if at any time the Soviet Government started to supply T62s to Israel's neighbours we would make Chieftains available in strength" (United Kingdom 1972–1973, 861: 439). Wilson also rejected Foreign Office pressure for a formal statement of the British position on the Arab-Israeli conflict (Crossman 1976, 3: 251, 253, 466–467, 513–514, 516, 685–686, 873; Jenkins 1970).

Britain also seems to have opposed Israel's demands for direct peace talks between Israel and the Arabs and for a formal written agreement (Shlaim, personal interview, June 1983). Thus, British policy, while neutral, tended to be somewhat pro-Arab during this period, at least as far as the Israelis were concerned.

A landmark in Middle East negotiations, Resolution 242 also became one of Britain's central policy statements vis-à-vis the Arab-Israeli conflict. Another such statement was made in October 1970 by Sir Alec Douglas-Home, foreign minister in Edward Heath's Conservative government, in a speech at Harrogate ("an obscure party gathering in Yorkshire" [Adams and Mayhew 1975, 90]), presumably in response to the same pressures that Wilson had resisted (Jenkins 1970). Interestingly, Douglas-Home delivered this speech just after he had attended Nasser's funeral, an act that some saw as a significant gesture toward the Arabs and others regarded as the end of the Conservatives' anti-Nasser attitudes dating from the Suez crisis (personal interviews with former British officials, July 1983).

The Harrogate speech reiterated the central bargain of Resolution 242—land for peace—"but also spoke of the Palestinians' 'legitimate aspirations' and their desire for 'a means of self-expression'" (United Kingdom 1983, 2). Some have argued that the Harrogate speech marked a significant change of British policy (Maull 1976, 123). It certainly caused concern in Israel, not only because it specified a set of territorial solutions that the Israeli government preferred to leave to negotiations, but also because it immediately preceded Prime Minister Golda Meir's arrival in Britain (Prittie and Jones 1970, 1–3). In general, however, the speech marked only a fairly small change. As one journalist phrased it: "Sir Alec, it is true, has not said very much. The important fact, however, is that he has said it publicly. . . . The practical significance of what Sir Alec said is that, without making a bad business worse, it can gain us a little stock in the Arab world" (Jenkins 1970).

As British influence declined in the Middle East, the European Political Cooperation process offered another way to play a role there. Britain was not a charter member of EPC but became formally involved in January 1973, when Britain joined the European Community. Its prior position, as laid out in Resolution 242 and Harrogate, however, meant there was little trouble integrating its Middle East policy with that of its EC partners. As we shall see, Britain also saw the EPC process as an important part of its participation in European affairs.

During this period, the British government was becoming increasingly aware of the need to help guarantee the flow of oil and mineral resources by making political gestures. Historically, Britain has often tailored its foreign policy to assure economic goals. It was not much of a jump, therefore, to adopt discreetly pro-Arab positions to assure oil supplies. Examples of such gestures are seen in Britain's willingness to receive the head of Nigeria's new military government, General Yakubu Gowon, and President Sese Soko Mobutu of Zaire in 1973 (Wallace 1975, 246).

Britain's long history in the Middle East thus seems to have affected British policy in several ways. First, the British government, particularly the Foreign Office, had a considerable number of people with knowledge and expertise in the Middle East. These Arabists were not only important in regional departments but also in many cases were promoted to senior positions within the diplomatic service.

Second, the British think they understand the Middle East, or at least the Arabs, better than practically anybody else (they rarely talk about understanding the Israelis). In their view, the Americans have power but no understanding of the Arab mind, whereas the Europeans have neither (David Watt 1973a). (Interestingly enough, the Suez debacle actually strengthened this perception inasmuch as the Foreign Office had no hand in the decision to regain control of the canal [Monroe 1981, 198; Wallace 1975, 14–15, 21–22; Barber 1976, 29].) This perception is shared by people in other countries as well. This self-perception is an important factor in British policy, whether or not it is correct.

Third, British policy toward the Middle East is widely believed to have been consistent over time. One former official argues that it has not changed materially since Winston Churchill's virtual creation of the Anglo-Persian Oil Company in 1912. Another suggests that, while the policies themselves have changed, the basic goal has not; it remains "to try and ensure that conditions exist whereby oil will flow freely and on reasonable terms to this country" (personal interviews, July 1983).

Finally, the Middle East has been an important British foreign policy concern for a long time. One former senior official has observed that, on anyone's list of important issues for British foreign policy, the Middle East would be in the top three (personal interview, July 1983). British policy toward the Middle East is thus seen as having been thought through at great length by the professionals at Whitehall. As such, it is unlikely to change except under very unusual circumstances.

The Oil Crisis

For the British, the events of 1973 confirmed the wisdom of British policy. Unlike the Americans and Israelis, the British argued before the 1973 war that the stalemate in the region was unstable (Shlaim, personal interview,

July 1983); the war confirmed their view. The British do not seem to have anticipated the link between another Arab-Israeli conflict and threats to oil supplies for the West. When the Arab oil producers decided to employ the oil weapon, however, they classified Britain as a "friendly" country, thus in theory allowing it to have all the Arab oil it wanted; this also was seen as vindication of British policy.

But the British situation was not without its difficulties. At least three decisions had to be made during the period of the embargo. The first decision was probably the easiest: Britain declared an arms embargo on the combatants as soon as the war broke out. This was not a symbolic gesture: About half of Israel's tanks were British (Insight Team 1974, 238), and, given the high level of attrition in the early weeks of the war, Israel would need spare parts, replacements, and ammunition. Britain had also imposed an arms embargo during the 1967 war but had made it conditional on similar acts by other powers; the embargo was lifted after twenty-four hours when such acts were not forthcoming (Brecher 1980, 273). These conditions were not attached to the 1973 decision.

Douglas-Home argued that because both Arabs and Israelis had purchased arms from Britain, the arms embargo was really an "even-handed" action, unlike the 1967 embargo; thus, Jordan as well as Israel used Centurion tanks and presumably needed ammunition and spare parts. Indeed, he stated that more materials were in the pipeline for the Arabs than the Israelis when the embargo was imposed; he added that the Israelis wanted 4000 shells while the Jordanians wanted 10,000 (United Kingdom 1972–1973, 861:30–42, 421–424). Jordan was not involved in the war, however, and the embargo was clearly more important to the Israelis; it is hard not to see this as an anti-Israeli act (see also International Institute for Strategic Studies 1974, 33). The policy was challenged in Parliament and upheld as a result of many cross-party votes and abstentions. One informed British observer reports that although he had no definite information, he is "not 100 percent sure" that the embargo was rigidly enforced (personal interview, July 1983). We should also note that Douglas-Home specifically stated at the time that the destruction of Israel "could not be tolerated" (Mangold 1978, 30; see also United Kingdom 1972–1973, 861:424). The arms embargo remained in place until the disengagement agreement between Israel and Egypt was concluded in late January 1974 (United Kingdom 1974, 867:1203).

Britain's second decision during the embargo was to deny the United States the right to use British bases, both in the U.K. and in Cyprus, for its airlift to resupply Israel. (Interestingly, Crossman claims that in 1967 "the locals" had refused to allow the British to use their Cyprus base, presumably for the same reason [1976, 2:381].) The British government denied that it was refusing permission, apparently by making its position so clear in advance that the Americans never asked (Kissinger 1982, 709; Schmidt 1974, 225; Lieber [1976, 12] has a somewhat different version). Britain's decision flowed logically from the arms embargo (Shlaim, personal interview, July 1983), but, in fact, the stakes were a good deal higher because its major ally would be alienated. The government apparently felt, however, that "U.S. irritation" could be borne and would not be a major long-term problem (David Watt 1973b). In any event, all of the other EC nations had done the same thing (Maull 1976, 117). The British decision could therefore be seen as a choice between conflicting loyalties rather than as simple defiance of the United States.

Britain's third decision during the embargo "caused far more heart-searching within the government" (David Watt 1973b): Should oil be diverted to help the Dutch, who were under Arab embargo? The Arab oil producers reportedly had made it clear to the British that if they allowed any oil at all to go to Holland (or the United States), they would be penalized (*Sunday Times,* November 4, 1973); one source cites a threat of a 25 percent cut (Kelly 1980a, 402). Domestic difficulties only compounded the Arab threat: The Heath government was engaged in a major confrontation with the coal miners' union that resulted in a strike and curtailments in home heating. Threats from abroad to cut off oil were therefore very potent. As a potential oil producer, too, Britain was not anxious to set an oil-sharing precedent within the EC that might cost it dearly in the future (Venn 1986, 148). Finally, the Dutch had not followed British advice on Middle East policy; why, then, it was argued, should Britain have to pay for the consequences?

At the same time, however, "the argument for assisting the Dutch [was] admitted to be very strong" (David Watt 1973b). Both the government as a whole and Edward Heath in particular were dedicated to the European connection; indeed, the question of EC membership "dominated everything" in foreign policy during this period (Shlaim, Jones, and Sainsbury 1977, 167). Thus, it was hard to "join in what amount[ed] to a blockade of an EEC country" (Shonfeld 1973) or, if the EC were seen as an international actor, to allow its internal economic policy to be determined by outsiders (Shlaim, personal interview, July 1983). Moreover, the Dutch had been one of the foremost supporters of British entry into the EC a few months before. One justification for the policy was that if, by supplying the Dutch, the British population had to suffer energy shortages, there might be popular rejection of the EC, an interesting use of a pro-European argument to justify an apparently anti-European policy. In addition, the Dutch situation did not appear to be desperate (David Watt 1973b).

The result was a two-part policy. Diplomatically, the British refused to send oil to Holland, regardless of source, and refused to allow either the EC or the OECD to implement oil sharing (Turner 1974, 410–411). Heath denied that the Arabs had made any specific demands (United Kingdom 1972–1973, 864:660), but one authority asserts that Britain had assured Saudi Arabia, Kuwait, and Abu Dhabi that they would not reexport oil elsewhere, "least of all to the Netherlands" (Kelly 1980a, 402; see also Laqueur 1974, 241). Britain seems to have taken a leading role in constructing the Brussels statement of November 6, perhaps reflecting Douglas-Home's remark of the previous day: "In the light of suggestions by Arab oil producers that they would impose an embargo on those countries who agreed to furnish the Netherlands with oil, it would be much better to see how Europe could influence a political settlement in the Middle East" (Kelly 1980a, 403).

The second part of British policy was to oppose the efforts of the oil companies to spread the production cuts evenly by shipping non-Arab oil to embargoed countries. (Indeed, because Britain was classified as a "friendly" country by the Arabs, presumably the only institutions restricting British oil imports were the oil companies.) At one point, Heath called the directors of Shell and British Petroleum to Chequers and demanded that Britain receive as much oil as normal, regardless of the problems of any other countries. Because the British government held the majority of shares in British Petroleum, it should have been able to enforce such demands; in fact, it was unable to do so. For obvious reasons, the text of the discussion has not been made public, but, reportedly, after a certain amount of discussion, the companies said that if Heath wanted them to break contracts with other countries, he would have to pass legislation to free them from the resulting legal liability. The prime minister was unwilling to do so, and the issue lapsed (Turner 1978a, 178; Lieber 1976, 17; Stobaugh 1976,

189–190). Indeed, when the British press reported that some unnamed government officials had accused the companies of diverting oil from Britain, the government apparently apologized to the companies (*Times,* November 23, 1973, 1; Vielvoye 1973).

Britain appears not to have considered that its Middle East policy should change as a result of the Arab use of the oil weapon. At one level this is not surprising, in view of Britain's classification as "friendly" at the beginning of the crisis. There were reports, however, that Britain was exempt only "for the time being" and that Kuwait had wanted to exclude Britain from the "friendly" category because of the money sent to Israel by British Jews (*Times,* November 6, 1973; Kelly 1980a, 408). There are also suggestions that the British government was pressed by the Arabs to make further concessions but refused to accede. At any rate, there seems to have been no debate in Britain as to whether the country should move in a more pro-Arab direction. The debate in the public press tended to focus on whether Britain should "yield to blackmail" by the Arabs (for examples, see Shonfeld [1973] and the resulting letters to the editor). Interestingly, the press reported King Faisal's decision to increase the amount of oil for Britain, but this concession was traced to Lord Aldington, who pleaded with the king while on a government mission that without the additional oil the Labour party would win the approaching election, presumably introducing communism into Britain (*Guardian,* February 23, 1974). Despite this unusual support, Heath lost the election because of his clash with the coal miners.

British Middle East Policy Since the Crisis

"Since Britain's entry into the European Community in 1973, British Middle East policy has been developed jointly with her EC partners" (United Kingdom 1983, 1). Such reliance on the EPC process has made it more difficult to trace the intrinsic movement of British policy; for example, a useful background paper (quoted above) from the Foreign and Commonwealth Office (FCO) relies heavily on EC statements to illustrate British policy. The Brussels statement of November 6, 1973, is thus another important British policy statement, building on Resolution 242 while also referring to the Palestinians' "legitimate rights." The British foreign secretary, when addressing the United Nations in 1976, spoke of a "land for the Palestinians." That same year the nine EC countries stated:

"We have . . . made clear that the exercise of the right of the Palestinian people to the effective expression of its national identity could involve a territorial basis in the framework of a negotiated settlement." In 1977 they called for a "homeland for the Palestinian people" (United Kingdom, 1983). Speaking for the Nine at the United Nations, the Irish foreign minister in 1979 said the PLO was one party whose assent was necessary for a settlement. The process culminated in the 1980 Venice Declaration affirming that the Palestinian people should be able to "exercise fully their right to self-determination within the framework of a comprehensive peace."

> Britain has subsequently made it clear that this logically includes the right to a State if that is what the Palestinians decide for themselves, but it has emphasized that the Palestinians should make their decision in the light of all the circumstances, and bearing in mind what is likely to be negotiable. . . . The question of formal recognition of the PLO does not arise for the UK, nor for most members of the EC, who recognize only states. (United Kingdom 1983, 2)

The United Kingdom maintains contact "at an official, but not Cabinet, level" with the PLO (Hollis 1987, 203).

Britain also strongly opposed Israel's 1982 invasion of Lebanon, joining with its EC partners to condemn Israel and to impose an arms embargo on that country. Britain subsequently sent a token number of troops to join the abortive Multinational Force in Beirut in 1983, until the United States decided to withdraw its troops after the loss of 241 marines; at that point Britain also withdrew. Britain has not sold arms to Israel since the 1982 invasion (Hollis 1987, 205–206, 214).

British policy attempts essentially to meet Arab demands as much as possible while continuing support for the integrity of Israel and not alienating the United States. Edwards explains its rationale by noting the "essential duality of purpose irrespective of the political complexion of the British government: to influence the Arabs as far as possible to take up a more conciliatory attitude and to influence the Americans to press the Israelis to the same ends" (1984, 3). British policy is thus a continuation of pre-embargo policy. Indeed, most observers argue that the oil weapon had no notable effect on British policy toward the Arab-Israeli conflict except perhaps to accelerate its pro-Arab direction, and, in general, this seems a fair conclusion.

Some observers contend that the process was slowed by the Labour government, which lasted from 1974 to 1977 (Shlaim, personal interview,

July 1983). But most argue that, in fact, the last Wilson government really made no impact on policy in this area (Tapsell 1975, 1–2). The first Thatcher government (1979–1983) maintained British policy but was "prepared to adopt a somewhat higher profile" in the Middle East, primarily because of Lord Carrington, the foreign minister (Edwards 1984, 52). Carrington's resignation in 1982 seems to have reduced the level of Britain's interest in Middle Eastern political affairs (Ismael 1986, 121).

In voting at the United Nations, Britain has been in the center of the EC group (Stein 1982, 57–58). It is difficult to determine whether such agreement represents EPC pressure to restrain British pro-Arab tendencies or, instead, demonstrates the British ability to get most of its partners to support its preferred position (Edwards 1984, 48–49). This pattern of British-EPC linkage is not unique to the Middle East. As one British observer notes: "In general it is difficult to think of major stances of EPC which do not harmonize with the lines that British foreign policy would be likely to take in any case" (Hill 1983, 24).

Historically, Britain has been a major arms supplier for the Middle East. After its 1973 arms embargo, it announced in June 1974 it was resuming sales to the region. Such sales have become more important to Britain as the capacity of the Arab states to purchase advanced weapons has increased; since 1966 roughly half of all British arms exports has gone to the Middle East. Although this has caused some concern, the British have not sold arms to simply anyone with the money: In 1975 they refused to sell fighters, tanks, and submarines to Libya, and a sale of aircraft to Saudi Arabia fell through when the Saudis were unwilling to provide assurances that the arms would not be used against Israel. British military expertise and training are also highly valued in many parts of the Middle East, giving Britain a significant, although probably diminishing, advantage over some of its competitors (Pajak 1979, 149–151; Hollis 1987, 213–215).

Britain has traditionally been a major trading partner of the Arab states. The Middle East is now Britain's third-largest market (after Western Europe and North America) because of the quantum jump in Arab wealth during this period. Also, since 1980 the decline in oil imports (owing to the North Sea production) means Britain has run a trade surplus with the region, a very important consideration given Britain's general problems with trade balances. Arab tourists and students are also increasingly important economically in the United Kingdom (Hollis 1987, 222–223).

One result of this trade activity has been increased pressure to observe the Arab economic boycott. Such pressure is not new: In 1963 Lord Mancroft, a Jewish director of the Norwich Union Insurance Company, was dismissed because of Arab pressure (Nelson and Prittie 1977, 60–63). More recently there have been reports of demands that investment houses or merchant banks with Jewish connections not be involved in transactions using Arab money.

At various times the British government has responded with strong statements by notables opposing the principle of the boycott, including foreign secretaries David Owen and Lord Carrington and Prime Minister Harold Wilson. At the same time, it has resisted antiboycott legislation, arguing that each firm must be free to decide whether or not to comply to satisfy its own commercial interests (Nelson 1978; Phillips 1979, 28–34, 74–91; and Prittie 1982a). Thus, in 1974 a British official said that American antiboycott legislation was harmful to British trading interests (Stanislawski 1981, 146). In 1977 Zionist groups started pressuring for an antiboycott law. Lord Byers, Liberal leader of the House of Lords, introduced the Foreign Boycotts Bill—patterned after the 1977 U.S. Export Administration Act—as a private bill in 1977 and tabled it for a second reading in 1978 (Nelson 1978; Phillips 1979, 74). (In Britain, a private bill is submitted without government support and is therefore unlikely to pass.)

A select committee was finally appointed by the House of Lords, and it investigated and published a report unanimously condemning the bill, arguing that it would make trade with the Middle East more difficult. The report did, however, call for the government to oppose the boycott by administrative measures and to raise the issue before the EC council to try to establish common antiboycott legislation among the Nine (Prittie and Nelson 1978). But none of these steps seems to have been carried out. The government was "stubbornly unresponsive to the select committee's suggestions" (Phillips 1979, 79) because of opposition from government and business.

Several observers have suggested that Britain has been more responsive to the boycott than other European countries. Charlotte A. Phillips (1979, 86) argues that West German banks, perhaps because they had been given considerable informal support by their own government, resisted the boycott demands more than British banks. A pro-Israeli source asserted in 1982 that "Britain is the only European country which expedites 'negative certificates of origin'" (Prittie 1982a, 1). The French case is particularly interesting. The French parliament passed antiboycott legislation in 1976, but the government made it inoperative by employing several different interpretations (Phillips 1979, 41–52). The Mitterrand government, however, has apparently been prepared to enforce it (Prittie 1982b, 1–2). In any event, Britain has not passed an antiboycott law, and the issue has not really been important in British politics (Prittie, personal letter to the author, July 1983; personal interviews, July 1983).

Essentially, Britain seems to have accepted the boycott with equanimity (Nelson and Prittie 1977, 107, 112, 210–213), apparently thinking the situation is well in hand. One former official sees a kind of tacit bargain at work, whereby the British agree not to violate the boycott overtly and the Arabs agree not to enforce it rigorously. A middle-level official knew of an instance in which a ruler of a Persian Gulf country bought a big earthmover powered by a British Leyland motor, even though British Leyland was on the boycott list. The motor name was filed off, but one Kuwaiti said that although he didn't know who made the motor, he did know where to get spare parts (personal interview, July 1983).

British concern for Arab sensibilities had also been shown in other ways, such as the furor over the television film *Death of a Princess,* an unflattering portrait of contemporary Saudi Arabia made by an independent producer for the BBC. Saudi Arabia reportedly threatened economic sanctions if the film were shown on television. After some dispute the film was shown, and apologies were extended by the British government (Edwards 1984, 48; Shlaim, personal interview, July 1983). It must be noted, however, that Margaret Thatcher refused to receive an Arab League delegation in 1983 headed by King Hussein because it included a PLO representative; in retaliation, the Arabs canceled a trip to the Gulf by British Foreign Secretary Francis Pym (Edwards 1984, 54). Despite support for a Jordanian-PLO peace initiative in 1985, Thatcher again refused to meet with PLO representatives in that year (Ismael 1986, 121).

External Influences on British Policy

The British see their Middle East policy as more or less impervious to external pressures other than those applied by the Arabs, and there seems little reason to doubt this judgment. Two possible sources of pressure are other European Community members and the United States. Because of their greater regional expertise and their deeper commitment to the EPC process, however, British government officials see themselves as leading the Europeans in this area. The relationship with the Americans is seen as more complicated, wherein Britain's role alternates between subtle leadership and occasional deference. Britain's general sense of superior competence, however, makes it invulnerable to any real external direction (Shlaim, personal interview, July 1983). Even British vulnerability to U.S. financial pressures, graphically revealed by the 1956 Suez crisis, has not given the United States much leverage.

The EPC process has become an important component of British participation in the European Community, both externally and internally. Externally, the British think the EPC uses their strength in political affairs, as opposed to economics, where they are weaker. Internally, EPC is increasingly seen as an important way to sell EC membership to a skeptical British public. Moreover, if Britain could be influential in EPC, it could gain more influence by speaking for Europe as a whole (Edwards 1984, 48).

The EPC statements have a number of justifications: First, regardless of their content, they show the European Community as an actor on the international stage. Second, in the absence of an Arab-Israeli settlement—seen as dependent on sufficient U.S. pressure on Israel—European declarations may reward moderate Arab regimes and behavior, make the Arabs more willing to grant concessions to Israel in the eventual negotiations, and render them less amenable to Soviet influence in the meantime. Third, such statements may encourage the Arabs to allow the Europeans to benefit economically (oil supplies, markets, and so forth). Finally, the statements exert pressure on the United States by showing an alternative Middle East strategy. Indeed, the Reagan plan, a proposal for Palestinian self-rule on the West Bank in association with Jordan, was widely seen as an attempt to move closer to the European position and thus was an important endorsement of EC strategy.

The precise role of Britain in EPC is impossible to determine (either in general or on particular issues such as the Middle East) in large part because of the ability of those involved to keep the proceedings secret for more than a decade. This is an impressive achievement to an American accustomed to leaks. Despite occasional lapses—such as the 1973 leak to the Arabs of Holland's role in preventing the British and French from speaking for Europe at the United Nations—the EPC process is a mystery. (One former senior official notes that this very secrecy has made it hard to show the British public how useful the process has been; another has argued that secrecy allows for policy shifts by member countries so compromises can be achieved [personal interviews, July 1983].)

British officials generally see their role in the EPC as crucial. Indeed, two officials said that EPC had really not accomplished much until the British joined (personal interviews, July 1983). A third official argues that Britain had been particularly influential on Middle East issues: "I think the story on European Political Cooperation in the Middle East is that we are the people who have turned up with the propositions carefully worked out, with the answers to all the questions, and have given the lead. . . . This is an area where our Community partners have been very happy to give us the lead" (personal interview, July 1983). British confidence appears to be of relatively recent vintage, however; in a study of the FCO in the mid-1970s, Geoffrey Morehouse found little confidence among diplomats that Britain could play a leading role in this area (1977, 39–40). British academics remain somewhat more skeptical, noting that everyone claims credit for successes (and EPC is generally regarded as a success in Britain) but that such claims cannot be substantiated (personal interview, July 1983).

There is a potential conflict here. If EPC is seen as important in its own right, and if British Middle East policy is seen as both important and essentially correct, what happens if the other EPC members decide to adopt a divergent Middle East policy? This has apparently not yet happened; most observers agree that British policy toward the Middle East would not be significantly different if EPC did not exist (Edwards 1984, 57). But at least one observer has suggested that some of Britain's pro-Arab acts, particularly U.N. votes, might result more from European policy than Middle Eastern pressures (personal interview, July 1983). So far the British have been skillful and lucky enough to avoid this conflict, presumably by encouraging their EPC partners to adopt positions similar to their own. The international reputation of the FCO for Middle East expertise has undoubtedly helped; it was constantly cited by respondents in the other countries studied here.

France is the obvious competitor in Middle East expertise, but Great Britain is emphatically unwilling to learn anything from the French. Indeed, the level of anti-French feeling among my British respondents was startling. This attitude can be justified in realpolitik terms, in view of their competition for political leadership of the EPC, but Britain's distaste for France appears to stem more from perceptions of national character, at least among the British elite. The besetting sin of the French is that they are believed not to play by the rules. The British see the French as unreliable EPC partners who will act in their own self-interest, regardless of previous arguments. In commerce they are seen as unprincipled because they directly link political and economic activities; thus, their successes are more threatening than those of West Germany, Japan, and the United States—more significant competitors of Britain than is France, in economic terms. (The British apparently ignore or are unaware of the new French government's willingness to enforce legislation against the Arab economic boycott [Prittie 1982b, 1–2].) French willingness to sell arms to practically anyone, particularly Libya and Iraq, is often cited as an example of irresponsibility. British officials also tend to see France as a strong, unified, and efficient state. Coupled with this perception of a powerful opponent (which is essential to make any fight worthwhile) is more than a hint of envy (personal interviews, July 1983; D. C. Watt 1982, 8, 12; Morehouse 1977, 158–159).

Relations with the United States are necessarily more complex. The U.S. government, in British eyes, is obviously powerful and fundamentally well-meaning, albeit naive. However, its diplomacy is clumsy, domestic pressures force it to support Israel, and it does not understand the Arabs. It also cannot maintain a consistent policy for very long or keep secrets from anyone; thus, it is a frustrating partner. The British have therefore adopted an attitude of benevolent tutor, aimed at inculcating Americans with the more enlightened view that Arab governments should be granted substantial concessions and that Israel should be pressured into altering its policy. Given the ignorance of so many American administrations in foreign affairs, the British see such an education as vital. Incidentally, the Reagan plan was believed to be a significant, although limited, success in this process (Shlaim, personal interview, July 1983). The notion that Britain should play Greece to America's Rome was an important aspect of British foreign policy after World War II but has largely been abandoned. Its dominance of Middle East policy is a tribute to the FCO's special expertise in this region and perhaps to a certain amount of wishful thinking.

On balance, then, the British see themselves as having a special position in the Middle East, stemming from a unique historical experience and particular competence. As Douglas-Home expressed it in the House of Commons in 1973: "The Arabs . . . could get themselves, in the heat of emotion, so involved as to do most serious damage to the Europeans and the Western world who are essentially their friends. We must preserve our ability as a country to persuade the Arabs that this is totally against their interests. *We are perhaps the country of all in the world best able to do this*" (United Kingdom 1972–1973, 861:426, emphasis added). This sense of uniqueness makes the British more difficult to influence than other target nations with less self-confidence.

Societal Influences on British Policy

Public Opinion

Any serious discussion of British public opinion about the Arab-Israeli conflict must distinguish between two issues: (1) which side, if any, is perceived sympathetically, and (2) what the public wants the government to do about it. British public sympathy vis-à-vis the Middle East tends to be divided between neutrality and support of Israel; there is little pro-Arab sentiment. The use of the oil weapon did not alter sentiments much, at least in the short run. In any event, British sympathies have never been translated into strong support for intervention or even low-level involvement in support of the Israelis.

We have unusually good data on trends in British sympathies toward the Arab-Israeli conflict because the Gallup Poll has asked the same question of the British public for a number of years (table 2). If we measure from our earliest poll data in 1955, the British public started with a relatively "evenhanded" opinion but has become more sympathetic toward Israel over time. It now clearly has much more sympathy for the Israelis than for the Arabs; the important division is between sympathy for Israel and sympathy for neither side. This opinion shift, which occurred between the 1956 Suez crisis and the 1967 war, has not changed appreciably. A temporary increase in sympathy for Israel is recorded after the 1967 war, and no significant change took place after the 1973 Arab oil embargo. There is a recent secular decline in support for Israel, resulting in less sym-

Table 2.   British Public Opinion: Sympathies in the Arab-Israeli Dispute

	Favor Israel (%)	Favor Egypt/ Arabs (%)	Favor Neither (%)	No Opinion (%)
December 1955	13	7	41	39
April 1956	19	6	75	a
Suez Crisis				
November–December 1956	31	5	48	16
May 1967[b]	46	4	50	a
1967 War				
June 5–11, 1967	55	2	27	16
June 12–18, 1967	59	4	22	15
January 1969	41	8	29	21
March 1969	53	6	21	20
August 1969	36	4	31	29
September 1969	41	5	35	20
October 1969	34	5	32	29
February 1970	41	5	27	27
March 1970	46	6	24	24
April–May 1970	43	5	52	a
October 1970	33	8	37	22
October 1972	40	5	30	25
May 1973	34	7	35	24
1973 War				
Arab Oil Embargo				
December 1973	35	10	36	19
June 1974	39	6	34	21
July 1975	33	8	31	28
August 1982	25	16	40	19

SOURCES: Gallup 1976, 376, 932, 1032, 1045, 1063, 1079, 1090, 1092–1093, 1097, 1110, 1209, 1248, 1290, 1331, 1425; data for December 1955 and November–December 1956 from U.S. Information Agency polls reported in Erskine 1969–1970, 628; first set of data for June 1967 and for September 1969 are Gallup polls reported in Erskine 1969–1970, 630; figures for August 1982 from Market & Opinion Research International poll, data from the Roper Center, University of Connecticut, obtained through the Center for Computer and Information Services, Rutgers University.

NOTE: Some rows may not add up to 100 percent because of rounding errors.

[a]"Favor Neither" and "No Opinion" responses not separated.

[b]This poll was reported in the press ("Gallup Poll" 1973) but is not listed with others in the published compilation (Gallup 1976). Its date is given as "just before the 1967 War"; May seems a reasonable guess.

pathy for both sides rather than in any increase in pro-Arab sentiment, although opinion does not approach the low levels of support of the 1950s. This change is difficult to date precisely, but it seems to have occurred in 1969 with reversions in early 1970 and 1972; at any rate, it does not seem to have been produced by the embargo.

Although the British public's sympathies clearly lie with the Israelis, it also wants its government to stay out of the conflict, regardless of the impact on Israel. In June 1967, 60 percent approved of "the way the Government has handled the Middle East situation" (presumably including the arms embargo to both sides), and 71 percent opposed giving armed support to Israel even if the United States did so. If fighting broke out again, 52 percent felt Britain should have nothing to do with the situation, 33 percent favored urging the United Nations to get involved, 10 percent said Britain should help Israel, and 1 percent said it should help Egypt. The closest thing to activism was that 41 percent said Britain should work with other countries to stop the fighting; 30 percent preferred to leave Israel and the Arabs to fight it out, 19 percent favored supporting Israel, and 1 percent felt Britain should support the Arabs (Gallup 1976, 932–933). Similarly, in 1969, the poll asked whether Britain should supply arms to the Middle East "in view of the fact that Russia is continuing to supply arms to the Arabs"; 22 percent supported supplying arms to Israel (presumably by purchase, although this was not clarified in the question), and 45 percent favored sales to neither side, despite the suggestive wording of the question (Gallup 1976, 1067).

This reluctance to get involved does not appear to be simple realpolitik: During the embargo 34 percent said Britain should support Israel even if the Arab countries delivered less oil, 24 percent said Britain should cease its support of Israel in order to get more oil, and 42 percent were undecided (Gallup 1976, 1290–1291). The question implied that Britain supported Israel, and the response indicated a reluctance to be pressured by anyone. Economic sanctions generally translate the issue at hand (regardless of its substance) into resistance to external pressure. The reluctance of the British public to intervene probably reflects preexisting feelings rather than fear of Arab retaliation.

Overall, relatively few people seem to care very much about the issue, and pro-Israeli feeling, although dominant, is offset by a profound reluctance to get involved. This gives the government room for maneuvering. As one observer notes, "A Government which wished to be more explicitly pro-Arab would suffer attack and inconvenience—and that is one reason why most British Governments, who like a quiet life, will try and prevaricate. But it would be unlikely to fall to the assault of pro-Israeli sentiment alone" (David Watt 1973a).

Given the lack of impact of the public's opinions, much of the discussion about British "public opinion" is really about elite opinion. Unfortunately, we do not have good data on elite opinion and must therefore rely on impressions. Some observers note an intellectual tradition that has been important historically: "There is also in Britain a contrary tradition that is unknown in the United States—namely, a romantic attachment to the Arabs. Only a relatively few of the British are affected by this second tradition, and it is found mostly in intellectual circles. . . . The tradition has helped to sustain a school of Arabist sympathizers in a section of the British foreign service" (Smith and Polsby 1981, 65–67; see also Cobham 1983, 8–9).

Insofar as elite opinion has changed at all, it seems to have become more pro-Israeli just after the 1967 war (owing to admiration for Israel's accomplishments) and more anti-Arab during the early 1970s (in response to terrorism directed against commercial aircraft). As a result of its anti-Israel policy, for example, the British Communist party lost a significant part of its already small membership (*Sunday Telegraph*, June 18, 1967, 91), and a Labour M.P., Margaret McKay, who supported the Arab position while on a trip to Egypt, was not renominated by her local party group (*Daily Telegraph*, February 11, 1970; *Sunday Telegraph*, February 22, 1970; *Times*, March 8, 1970).

More recently, however, representatives for the Arab cause in Britain have become more sophisticated and are beginning to make headway. Admiration for the Arabs increased after Egypt's successful attack on Israel in 1973. Public unhappiness with the Arabs' use of the oil weapon also increased a sense of vulnerability among the elites, while the war in Lebanon further increased support for the Arab cause. The plight of the Palestinians is also seen as important, particularly among educated younger people in Britain (personal interviews, June 1983). Michael Adams and Christopher Mayhew (1975, 12–15, 66) argue that this shift in opinion really dates from 1973 and is the direct result of the Arabs' demonstration of power. Although difficult to measure, the shift seems to have begun even earlier and to have been more gradual than Adams and Mayhew allow. The shift

Table 3.   British Public Opinion: Sympathies in the Arab-Israeli Dispute,
            Compared to Canadian and American Attitudes

		Favor Israel (%)	Favor Arabs (%)	Favor Neither (%)	No Opinion (%)
United Kingdom	*April 1956*	*19*	*6*	*75*	a
Canada	December 1957	15	5	80	a
United Kingdom	*June 1967*	*59*	*4*	*22*	*15*
United States	June 1967	56	4	25	15
United Kingdom	*January 1969*	*41*	*8*	*29*	*21*
United States	January 1969	50	5	28	17
United Kingdom	*August 1969*	*36*	*4*	*31*	*29*
Canada	July 1969	26	9	66	a
United Kingdom	*March 1970*	*46*	*6*	*24*	*24*
United States	March 1970	44	3	32	21
United Kingdom	*December 1973*	*35*	*10*	*36*	*19*
United States	October 1973	47	6	22	25
Canada	November 1973	22	5	42	31
United States	December 1973	50	7	25	18
United Kingdom	*July 1975*	*33*	*8*	*31*	*28*
United States	April 1975	37	8	24	31
United Kingdom	*August 1982*	*25*	*16*	*40*	*19*
United States	July 1982	41	12	31	16
Canada	October 1982	17	13	70	a

SOURCES: Data for the United Kingdom from Gallup 1976, 376, 932, 1032, 1063, 1092–1093, 1290, and 1425, respectively; 1982 figures for the United Kingdom and Canada from the Roper Center, University of Connecticut, obtained through the Center for Computer and Information Services, Rutgers University. Remaining data for Canada taken from Benesh 1979, 24, 63. Data for the United States for the years 1967, 1969, 1970, and 1973 found in *Gallup Opinion Index* 1979a, 28; 1975 data from Gallup 1978, 458; 1982 data from *Gallup Report* 1982, 4.

NOTE: Some rows may not add up to 100 percent because of rounding errors.
    [a]"Favor Neither" and "No Opinion" responses not separated.

has not notably affected policy directly, but it has made it easier for the government to act as it wishes.

Another way to illustrate the impact of public opinion on foreign policy is to compare opinions in different countries. As it happens, there are some data available on the sympathy question in both the United States and (to a lesser extent) in Canada (table 3).

These data reveal that American and British public sympathies on the

Arab-Israeli issue were practically identical, at least until the 1970s. These governments followed significantly different policies toward the conflict, however, resulting in the Arab oil-producers' embargo of the Americans and the classification of the British as "friendly." This is confusing in view of the conventional analysis of public opinion, which posits that public opinion sets limits within which a government may work. But the existence of these significant policy differences between the United States and Great Britain, despite very similar patterns of public opinion, suggests that the latitude given to governments by their publics is very great indeed. In terms of the 1973 embargo, this analysis suggests that policy on the Middle East is formed by the governments and elites with little connection to the opinions of the mass public.

Mass Media

At present an intense debate is taking place among a few people in Britain as to whether the British media are pro-Arab or pro-Israel. Terence C. F. Prittie (1980, 67), for instance, has argued that the British press has become more pro-Arab because of its increasing dependence on the FCO for information, citing the onerous costs and hazards of maintaining reporters in the Middle East (e.g., the murder of David Holden of the *Sunday Times*), massive advertisements from Arab governments, and skillful propaganda issued by the FCO itself. He also sees a shift in opinion in favor of the Palestinians and attributes this to the traditional British support for the underdog, more sophistication by Arab lobbying groups such as the Council for the Advancement of Arab-British Understanding, and adverse reactions to Israeli policies under the Begin government. After the 1967 war the Labour cabinet discussed the BBC's pro-Arab bias and sent it a letter of complaint (Crossman 1976, 2:370). Chafets (1985, 59–62, 90–91) discusses the shooting of Berndt Debusmann, Reuters' Beirut bureau chief, and the departure of BBC's two-man Beirut staff under threat of assassination in 1980; interestingly, these events were not extensively reported by the parent news organizations.

In contrast, Adams and Mayhew argue that reader pressure keeps the British press pro-Israeli, asserting that the *Manchester Guardian* lost so many Jewish subscribers because of its opposition to the Suez invasion that its financial stability was threatened; they also cite direct pressure from advertisers and the Zionist tendencies of British reporters. In particular, the authors contend that BBC coverage reveals a significant Zionist bias, although they think the press has become considerably more balanced since 1973, attributing the change explicitly to the use of the oil weapon (1975, 66–105, 121–139, esp. 134). The common thread in all these arguments seems to be that the British press has moved in concert with British opinion (whether it led or followed is hard to say) and has become significantly less pro-Israeli than it was before 1970. Whether it was "fairer" before or after must be a personal judgment.

Aside from systematic biases, what resources do the British media deploy in the Middle East? Helena Cobham (1983, 9–11, 28), formerly the Beirut correspondent for the *Sunday Times*, sees few policy divisions among different papers, at least in the "quality press" (the *Times*, the *Daily Telegraph*, *The Guardian*, the *Financial Times*, the *Sunday Times*, the *Sunday Telegraph*, the *Observer*, and *The Economist*). She also argues that the BBC has been able to maintain editorial independence from the government on this issue. Like the press in many other Western countries—regardless of their editorial policies toward the Arab-Israeli conflict—British papers base a disproportionate number of their newspaper correspondents in Israel, mostly, Cobham says, because it is so much easier to work there. Overall, the numbers are quite small; Cobham estimates that the British media maintain perhaps a dozen full-time people in Israel, some of whom also work for American media, with perhaps six full-time British correspondents and about a dozen regular contributors and stringers spread among Beirut, Cairo, and the Gulf States. These are severe resource limitations with which to contend while reporting on a large and diverse area.

Interest Groups

The Zionist lobby is clearly weaker in Britain than in the United States or Canada, and pro-Arab groups seem a bit stronger. There are no objective measures of such differences, of course, and they remain subject to interpretation, with each side naturally seeing itself overmatched by powerful opposition. One indicator, however, is that the Arab economic boycott has not become a major political question in Britain (Prittie, letter to the author, July 11, 1983). This contrasts sharply with experiences in the United States and Canada, where interest groups have at least been able to make it a major issue despite their failure, in Canada, to obtain a federal antiboycott law. The Zionist lobby in Britain has also lost considerable influence since 1948 (personal interview, June 1983) whereas the pro-Arab groups are on the ascendancy. Only about 0.5 percent of the British population is Jewish, as opposed to about 3 percent in the United States (M. Freedman 1974, 169). Nevertheless, in 1975 an acute observer listed the Zionist lobby as "the strongest such group" pressuring British foreign policy on controversial issues (Wallace 1975, 110).

The Arab lobby is led by the Council for the Advancement of Arab-British Understanding (CAABU), founded after the 1967 war (Adams and Mayhew 1975, 70). It is linked to the Parliamentary Association for Euro-Arab Cooperation, to which about fifty members of Parliament belong (Phillips 1979, 90). The CAABU has become a sophisticated counterpart to the Zionist groups in Britain and is probably the strongest such group in any of the countries included in this study.

The presence of an active Arab lobby in Britain can perhaps be attributed to the widely shared perception that Britain is in serious economic trouble and that Arab goodwill and financial investment are therefore essential if it is to survive, much less prosper. Initially, of course, this concern centered on oil supplies; Britain was seen as vulnerable to any cutoff. Now, of course, Britain is perhaps the least dependent among OECD countries on Middle Eastern oil; its North Sea fields have given it energy independence and have even linked its economic fortunes with OPEC to some extent.

But, as noted earlier, the British became dependent on Arab purchases of British exports and support of sterling just as their dependence on foreign oil declined. Britain, after all, has historically been dependent on foreign trade for its economic health; the experiences of two world wars have driven home the lesson of its vulnerability, making it more similar to Japan than to the United States and Canada. In addition, a good deal of Arab money is now invested in Britain; in 1982 it was estimated that about $15 billion of OPEC assets were invested there and another $55 billion was in nonsterling U.K. bank deposits (Ismael 1986, 120–121). This has created, as Prittie puts it, a "new 'Arab lobby,' which is not concerned with oil as such" (letter to the author, July 11, 1983). This dependence probably influenced both British publics and elites to grant concessions to the Arab oil-producers (Nelson and Prittie 1977, 213). Of course, the idea of linking foreign policy to commercial concerns is not new. A former senior official observed that "British governments have always seen British foreign policy towards the area as designed to contribute not only to the flow of oil on reasonable terms from the Middle East to Britain, but also as creating the atmosphere in which British exporters can operate. I don't think there is anything new there" (personal interview, July 1983).

It is also worth noting that public opinion and special interest lobbies do not usually have much impact on British foreign policy because of the parliamentary system, the lack of coalition governments, party discipline, and a strong tradition of deference to the FCO (Maull 1976, 123; Shlaim, Jones, and Sainsbury 1977, 22; see also Barber 1976, 66–75, and 89–113). Such political characteristics tend to strengthen the role of interest groups that have direct access to the bureaucracy; in practice this means big business.

Business has privileged access for a number of reasons. British business recruits the same kind of people sought by the civil and diplomatic services. Moreover, corporations are the only organizations that the government needs to carry out its policy. Speaking of the access of various pressure groups to the FCO, William Wallace says, "For commercial and business organizations of course the relationship is far closer; they after all represent major British external interests in themselves" (1973, 265–266, 272).

These business interests tend to oppose pro-Israeli positions not only because they include British exporters but also because they include representatives from the financial pillars of the City of London, which remains a particularly important financial center for Arab money. Thus, during the oil crisis, a study by Alpha Market Research showed that about 60 percent of "top City businessmen . . . felt that Britain should be gradually adopting a policy of closer relations with the Arab block to protect the country's oil supplies" (*Guardian*, October 20, 1973). Indeed, one critic has pointed out that the sons of two recent British foreign ministers, Sir Alec Douglas-Home and Lord Carrington, worked for Morgan Grenfell, a London merchant bank deeply involved in loans with Arab states (Kelly 1980b, 16).

The oil companies have traditionally occupied a special place within this group. As mentioned earlier, the FCO had a separate oil department until 1971

> to inform companies on current government policy as it may affect overseas operations, and to advise on developments in oil-producing states. . . . The needs of the companies are of national importance. Thus the companies were consulted on the question of military withdrawal from the Gulf—which does not mean that their views were in fact implemented. (Byrd 1973, 177)

Presumably, the oil companies also encourage the British government to adopt pro-Arab positions, although it is worth remembering that the oil companies essentially undercut the oil weapon and refused to comply with intense pressure from the British government during the crisis. Their rejection of these pressures made them very unpopular with some senior members of Heath's Conservative government, although Louis Turner suggests that at the time they also had support within the government for their actions (1978a, 178–179).

The oil companies' actions may have contributed to the decision by Wilson's Labour government to create the British National Oil Company (BNOC) to handle British oil from the North Sea. Indeed, the establishment of BNOC has been seen by some as a way to avoid sharing oil with British allies in the event of another embargo. Conversely, a former senior official argues that any Labour government probably would have created a BNOC and that it was essential to reassure the British public that its interests were being looked after. He also suggests that the close relations with the oil companies over development of North Sea oil had erased any lingering resentment of their role within the government in 1973 (personal interview, June 1983).

There are, then, all sorts of realpolitik reasons why any British government should support the Arabs, and, in general, the governments have. Yet there remains a tension between what is and what ought to be. This tension is best described by David Watt, who argues that British policy is not evenhanded and that everyone knows that although British policy may not be anti-Israel it is "discreetly pro-Arab" (1973a). Although Britain has a moral commitment to the survival of Israel, it believes the United States would not allow Israel to perish under any circumstances. For the cynical observer, this belief leaves Britain free to pursue its own best interests more freely. Officials and academics alike are quite clear on Britain's perception of national interests that leads to such a policy, but somehow they cannot quite bring themselves to act on their conclusions.

That being the case, why can we not continue to have our cake and eat it undisturbed, as the French so blatantly succeed in doing? Our basic interest is precisely the same as theirs and in practice our policy is very similar. But instead of ruthlessly displaying the bones of it to the Arab world, we insist upon clothing it in the language of high-minded impartiality. The French do not seem to suffer for their frankness, rather the reverse. We, on the other hand, are lambasted by both sides—by the Israelis for perfidy and by the Arabs for hypocrisy. Why on earth do we do it? (David Watt 1973a)

The reason may be found, perhaps, in the significant numbers of the British public and elite who are uneasy about simply following the dictates of national interest; as one observer put it, "British public life has a certain lack of tolerance for overt cynicism" (personal interview, June 1983). Moreover, Israel still has a peculiar status, perhaps stemming from the Holocaust, perhaps from the Palestinian mandate, which makes it seem a particularly inappropriate object of power politics. "There is a peculiar taboo against going too far [toward the Arab position]" (personal interview, June 1983). A former senior official made an interesting analogy in arguing that Britain's pecular sense of obligation is not felt toward Africa, for example, because the British feel that they handed over power in African countries at the right time, mostly to the right people; if problems have emerged later, it isn't their fault. They have more of a bad conscience about the Palestinian mandate. Another former official noted a sense of obligation and concern for Israel, although a middle-level official interpreted this as an obligation to seek peace in the Middle East, not necessarily to favor Israel (personal interviews, June 1983). This comment suggests that there may be limits beyond which British public and elite opinion will not tolerate anti-Israeli policies; unfortunately it is extremely difficult to discern in advance what those limits might be. The limits may be particularly difficult for the government to see; "I think perhaps the Foreign Office thinks too much in rational terms" (personal interview with a former senior official, June 1983).

British officials talk readily in terms of British interests, and Palmerston's famous remark about Britain having no permanent allies or enemies, only permanent interests, was quoted during interviews a number of times.

There remains, however, an apparent unwillingness to follow this dictum through to its logical conclusion. Thus, when asked about the possibility of eventually breaking diplomatic relations with Israel in response to Arab pressure, former senior British officials strongly opposed the idea (personal interviews, June 1983).

Governmental Influences on British Policy

Parliament and Parties

One of the peculiar difficulties of researching recent British foreign policy is the high level of secrecy that surrounds it. Ostensibly, this grows out of the tradition of cabinet government, under which the cabinet is responsible as a whole for any government decisions. In practice, a system of committees within the cabinet formulates policy, with the cabinet as a whole maintaining ultimate responsibility. In order to maintain that responsibility, in fact, even the names and composition of cabinet committees are secret (for a not particularly helpful partial list, see "Close Up on the Cabinet" 1982). Other parliamentary systems, however (including those of Canada and the Netherlands), seem to survive without such restrictions. The problem this secrecy poses has been partially relieved by some good studies of the making of British foreign policy. An author of one of these studies regretted, however, that he did not focus on the *management* of policy toward the Arab states—a regret I share (Wallace 1975, vii; see also Boardman and Groom 1973).

As noted above, Parliament generally does not have an important role in foreign policy. There are exceptions, of course, such as the Suez crisis, and perhaps they should not be casually dismissed. Nonetheless, the generalization that Parliament has little or no impact on foreign policy issues seems to hold (Shlaim, Jones, and Sainsbury 1977, 19; Wallace 1975, 90–100; Richards 1973, 245–261; Maclennan 1974). David Watt argues that many members of Parliament, and not just those who supported Israel strongly, were concerned about the moral aspects of decisions taken during the embargo: "Not only is our posture unheroic but it appears to be going back on the spirit in which we sold arms to the Israelis and possibly even the spirit in which we allowed the idea of a national home for the

Jews to be propagated" (1973b). These opinions do not appear to have greatly influenced the decision makers at the time, although Watt argues that they markedly increased the political risk for the government in the event the policy failed.

The literature is divided on whether the Labour party is more pro-Israeli than the Conservatives or whether the difference is negligible. Zionism has a traditional connection to socialism, and the Labour parties of Israel and Great Britain have been linked through the Socialist International. Jewish intellectuals and financial backers also have been important for the Labour party. The Conservatives, given their close links with business, have been more interested in the Arabs than in Israel and have been more willing to support monarchist Arab regimes because of greater concern for the Soviet threat. Some older members of the Conservative party are also more prone both to paternalistic feelings toward the Arabs, particularly Jordan's King Hussein with his British background, and to bitter memories of Jewish terrorism during the British mandate (Edwards 1984, 53; Lieber 1986, 62; David Watt 1973a).

But contrary tendencies also exist. Britain's decision to abandon its Palestinian mandate was made by a Labour government, and some commentators trace a pro-Arab shift in British policy to George Brown's role in the Labour government before 1967 (Tapsell 1975, 1; Shlaim, Jones, and Sainsbury 1977, 209). Because Labour had devalued the pound in 1967, it may have been particularly vulnerable to Arab threats to force another devaluation. Like other Western European Socialist parties, Labour was split by a pro-Palestinian movement in the 1970s that advocated Third World anti-imperialism and a dislike for Israeli militancy and ruthless efficiency. Also, the Begin government was not Socialist, which further weakened the ties between Israel and Labour. At its 1982 party conference Labour called for a Palestinian state and invited the PLO to participate in the negotiations as the sole representative of the Palestinians, explicitly rejecting a condition that the PLO should first accept Israel's right to exist (Edwards 1984, 50; David Watt 1973a).

The Conservative party has always contained Zionist sympathizers (Lord Balfour and Winston Churchill among them). Moreover, anti-Nasser feelings ran deep in the party, particularly during the Suez crisis. The way Watt describes it, the Israelis might have been especially created as the type of modern British Conservative. The efficiency, nationalism, and the touch of bloody-mindedness strike resonant chords among the Right wing of the party, while the sufferings and gallantry of the Jewish race continue to move many Conservative liberals, many of whom in any case wish to make amends for their class' reputation of anti-semitism. (1973a)

Whatever the differences may have been, since 1967 they appear to be mostly a matter of tone (Maull 1976, 123; Richards 1973, 258; M. Freedman 1974, 168–169). As Edwards explains it:

In general . . . the Arab-Israeli dispute cuts across party lines; each side has its own pressure group within all parties. To some extent therefore they tend to cancel out each other's influence. Nor are there particularly significant pressures which might alter party positions. In crude political terms, few, if any, seats actually depend on the Jewish vote. . . . The Jewish vote is not particularly homogeneous; it is perhaps too diffuse and too integrated. (Edwards 1984, 51)

Bureaucratic Organizations

We have noted earlier that for most industrial countries Arab-Israeli policy has both economic and political components, complicating the question of which government organization should make and manage policy. In Britain the economic component is especially strong, but despite this the Foreign and Commonwealth Office (FCO) seems to dominate policymaking, at least publicly. There are probably a number of reasons for this. Traditionally other government organizations defer to the FCO and the Treasury, and Treasury seems to have little interest in moving into this policy area, perhaps because of a certain conservatism stemming from its role of having constantly to say no to the rest of the government. Thus, for example, the FCO coordinates British international economic policy and tightly controls British policy within the European Community (Shlaim, personal interview, June 1983).

Because of the intense secrecy, the process of coordination between departments is practically impossible to research in detail; much of it takes place between the FCO and Treasury permanent under-secretaries, the highest-ranking civil servants, presumably in connection with the cabinet secretary whose role seems to have been strengthened in recent years

(Wallace 1975, 49; see also Gordon Walker 1970, 48–56). As one observer put it:

> Whitehall is not Washington; the open conflicts between sections of the administration which characterize bureaucratic politics in America have no exact parallel in Britain. . . . It is part of the style of Whitehall that differences are muted and as far as possible concealed from the public eye, and that interdepartmental disputes are subject to the acceptance of an overriding common interest. Even so, there are clear departmental interests and clear clashes of interest arising from the very nature of large-scale organizations and of human behavior. (Wallace 1975, 9; see also Nossal 1984)

As discussed earlier, the Foreign Office's dominance of Middle East policy is explained by its unique regional expertise. The FCO puts a good deal of emphasis on skills in "hard languages" (the principal languages of Asia, Africa, and Eastern Europe), and about half of its incoming personnel are given intensive training in one. Oppenheim and Smart observe that "the significance of this hard language training is not only that it provides a pool of valuable talent but also that it constitutes the Diplomatic Service's most important element of internal specialization" (1973, 90).

The Arabists hold pride of place among the FCO's sizable pool of talented specialists. The FCO's largest such group, they are also one of the most cohesive. Arabists get greater financial rewards for their competence than their colleagues and can even boast of a separate training center, the Middle East Center for Arab Studies in Lebanon (now in Jordan because of recent violence in Lebanon). The size of the group is justified by the fact that Arabic is the native language of more foreign capitals than any other hard language. In 1975 182 fluent Arabists worked in the FCO, as compared to 159 in Russian and 35 in Chinese (Oppenheim and Smart 1973, 91–92; Wallace 1975, 32; and Morehouse 1977, 77–84). There were so many Arabists that, according to a former senior official, when most Arab states broke relations with Britain in 1967, a number of people had to be retired because they could not be used at other posts (personal interview, June 1983). Most observers agree that these Arabists are responsible in large part for the FCO's unique role in Britain and, indeed, for the country's consistent Middle East policy.

Like most other foreign offices in this study, the FCO has a pro-Arab reputation. In a sophisticated explanation of similar tendencies within the

United States, Steven Spiegel argues that one can view the Arab-Israeli dispute within one of three contexts or levels of analysis: (a) as part of some global issue, (b) as a regional problem, or (c) as a particular conflict (1985, 3–10). Those who see it as a global issue or a unique conflict may wind up supporting either the Arabs or the Israelis. Those who are primarily concerned with the Middle East region, however, almost necessarily tend toward the Arab position, inasmuch as support of Israel makes relations with all other countries in the region more difficult. Most foreign ministries are organized on a regional basis and thus tend, as a group, to be more pro-Arab than other members of the elite and concerned public.

In more concrete terms, Harold Wilson, perhaps the most pro-Israeli British politician of recent times, contends the FCO is not pro-Arab because of ideology but because many more of its representatives have been stationed in Arab countries than in Israel (1981, 129; Prittie 1980, 68). One former official was unable to recall if any British ambassadors to Israel had returned to work on Middle East policy in the FCO (personal interview, June 1983), despite Foreign Minister James Callaghan's decision in 1974 to move Britain's Tel Aviv–based diplomats to Arab capitals (Morehouse 1977, 160). In contrast, many people with experience in Arab countries have risen within the diplomatic service. Another former official has observed that "some of the best brains in the Foreign Office have started as Arabists" (personal interview, June 1983). Morehouse confirms this observation: "At one point in 1975, Arabists were the Private Secretaries [to], the diplomats closest to, the Foreign Secretary and three of his four subordinate Ministers" (1977, 83).

Officials explain the FCO's "pro-Arab" tendency as the result of British interests in the Arab world, including oil, trade, and currency stability. Support for Arab governments is seen as necessary for British national interests, and there is really no countervailing interest in Israel other than a general moral commitment (personal interviews, June 1983). As one protagonist of Israel puts it: "Whether or not an Arab lobby exists [within the FCO], a Jewish one does not" (Prittie 1980, 68).

The long institutional memory of the FCO has also worked against good relations with Israel after Begin came to power. A former FCO official recalls a briefing at which several very senior officials in the new Begin government were named specifically as having been involved in terrorist acts in the 1940s, including the assassination of Count Bernadotte (personal interview, June 1983). More recently, the British refused to accept an Israeli ambassador, apparently because of his terrorist background, and a former senior official suggested in an interview that the Israeli foreign minister was similar to the ambassador in this regard (personal interview, June 1983).

The degree of the FCO's influence presumably depends on the strength of the political leadership, as Wallace notes: "Where political direction is clear, it is able to carry the administrative machine with it; in the absence of firm political pressures, however, administrative policies prevail" (1975, 9). Given the weight the Arabists have within the FCO, Crossman argued that bureaucratic influence is likely to be more pronounced vis-à-vis Middle East policy than in other areas of foreign affairs (*Times*, October 17, 1973). A former senior official, an Arabist, contends this is not true, however, because all specialists are regarded with suspicion by ministers, and their advice is usually rejected (personal interview, June 1983). It nevertheless is likely that the large number of former Arabists who now occupy high positions within the FCO exert influence in favor of Arab positions.

The FCO's impact on the decisions taken during the embargo seems to have been considerable. One critical step was to evaluate the Arab threat to cut off oil supplies:

> On this point ministers have to rely upon the judgment of Arab experts in the Middle Eastern departments of the Foreign Office and the oil companies. This appears to be that the producers have limited aims but within those aims they are in deadly earnest. . . . Threats to punish those who transgress this code will certainly be carried out and threats to punish those who ship Arab oil to others already being punished will almost certainly be carried out as well. (David Watt 1973b)

The FCO has also taken notable pro-Arab positions in various interdepartmental disputes, arguing strongly for British financing of the abortive Egyptian pipeline between the Gulf of Suez and the Mediterranean (SUMED) from 1967 to 1973, for instance (Wallace 1975, 183–187)

Given the research problems already noted, it is difficult to determine how the British government ultimately reaches decisions. It is clear, however, that the cabinet committee—a group of ministers that "has parallel and equal powers with the Cabinet itself"—is important (Gordon Walker

1974, 44). Such committees, wherein department views are discussed and reconciled, are at once the most important and least known part of the British foreign policy machinery. David Watt asserts that the critical decisions during the embargo were made by Prime Minister Heath; Foreign Minister Douglas-Home; Lord Carrington, who chaired the cabinet committee dealing with the oil crisis; Lord Rothschild; and, "to a lesser extent," Peter Walker of the Department of Trade and Industry (1973b). (Trade and Industry was involved, not to make policy, but because it would have to carry out any rationing or allocation systems.) Except for Lord Rothschild, none of these men had any connections with Israel, and the more pro-Israeli cabinet ministers (among whom Watt included Margaret Thatcher) were not involved in the decisions. As Watt describes the situation, "the traditional discretion given to the Foreign Secretary has virtually cut them out. There seems to have been a certain amount of general discussion in the Cabinet throughout the crisis but the key decisions—to apply an arms embargo and to subscribe to the European statement [of November 6]—were apparently taken without previous reference to the Cabinet" (1973b).

The Influence of Individuals on British Policy

If British Middle East policy is dominated by the Foreign and Commonwealth Office, and particularly its Middle East experts, the role of personality is presumably fairly limited. We can test this hypothesis by examining the impact over time of several prominent individuals on these policies.

   Harold Wilson was prime minister from 1964 to 1970 and again from 1974 to 1976. He was generally regarded as pro-Israel, an attitude demonstrated by events ranging from publicly kissing Golda Meir in 1974, which stirred up the Arabs and the FCO about equally (Adams and Mayhew 1975, 37–40; personal interviews, Shlaim and others, June 1983; Morehouse 1977, 160–161), to writing a history of the relationship between Britain, America, and Israel after his retirement (Wilson 1981). Indeed, Israeli leaders seem to have found him sympathetic (Brecher 1980, 98). He was also associated with the idea of using naval forces to open the Straits of Tiran for Israel in 1967, probably the most pro-Israeli proposal of any British government during this period. (It is notable that even though he and his foreign secretary, George Brown, were for once united in support of the naval force, it was nevertheless rejected by the cabinet [Gordon Walker, 1974, 45].)

Crossman, a member of the Wilson cabinet and a fervent although not uncritical supporter of Israel, argued that Wilson was hypocritical, telling the Israelis he supported them but then being unwilling to translate this into action when opposed by the Foreign Office (1976, 3:513–514). Crossman is also our source, however, for the assertion that in 1970 Wilson rejected a Foreign Office initiative that the British government clearly state its (presumably pro-Arab) position on the Arab-Israeli conflict, "in this election year, at least" (1976, 3:873).

Sir Alec Douglas-Home, foreign secretary in the Conservative government that succeeded Wilson in 1970, evidently had a different reaction to these ideas and consequently decided to make the Harrogate speech. Home phrases the difference in terms of parties rather than individuals: "For the Middle East I felt that the Conservatives must proclaim a policy more definite than that which the Socialists had felt able to pursue" (1976, 258). The policy difference, however, was more likely the result of Wilson's personal interest in Israel, inasmuch as the Labour cabinet was divided on the issue.

The personal sympathies of high-level officials still do not add up to much. A prime minister with pro-Israeli sympathies was able to propose only an aborted naval initiative and to delay for perhaps a year or two a formal statement of evidently accepted British policy (Wilson does not criticize the Harrogate speech).

George Brown, foreign minister under Wilson from 1967 to 1968, is undoubtedly one of the more fascinating personalities in British politics during this period. Brown had visited the Middle East several times during the 1950s and had met a number of important Arab leaders, including Hussein and Nasser. At the same time, he had close connections with Israel, and his wife was Jewish (G. Brown 1983, 228–232; personal interviews, Shlaim and others, June 1983). When he became foreign minister, Brown said he wanted to do three things: direct foreign policy, move Britain toward Europe, and "change British policy towards the Middle East on to 'more sensible lines'" (Shlaim, Jones, and Sainsbury 1977, 209). In particular, he wanted to reestablish diplomatic ties between Britain and Egypt, which had been broken after Suez; this was done in 1966. At one point he transferred a British ambassador out of Israel because he had become too attached to his host country (Morehouse 1977, 159). When the 1967 war broke out, Brown reversed course, supporting the use of naval forces in the Straits of Tiran. After the war began, however, he reverted to a more neutralist position, culminating in his drafting of U.N. Resolution 242.

How did Brown's personality influence British policy? The 1967 naval force initiative was the product of Brown and Wilson—it is hard to imagine the FCO proposing such an idea, and in fact the rest of the British cabinet rejected it (Mangold 1978, 182; Gordon Walker 1974, 45). Wilson's stance vis-à-vis the region can presumably be explained by his sympathy for Israel. Brown's attitude is more difficult to explain.

Although Resolution 242 did not mark a change in British policy, Brown's deep concern and his willingness to put an extraordinary amount of work into it (he reportedly saw about thirty foreign ministers in a single day at the United Nations [personal interview, June 1983]) undoubtedly contributed to its success. Brown was fired by Wilson not long after the 1967 war, but his concern for the Middle East continued. Some time after leaving the government, he undertook a private peace mission to the Middle East. In retrospect the visit was a long shot indeed, but his willingness to try certainly demonstrates his deep personal concern for the Middle East (G. Brown 1983, 234–240; personal interview, June 1983). Again, however, Brown's deep concern seems to have had relatively little impact on British foreign policy.

Edward Heath served as prime minister from 1970 to 1974. His background made him a fervent supporter of European unity (David Watt 1973b; Barber 1976, 18–19). Despite this position, his government opposed helping the Dutch during the oil embargo, apparently even covertly. The fact that Heath could make such a decision suggests that any other prime minister, who presumably would have been less committed to Europe, would have done the same. In addition, Heath deliberately established and maintained a distance between himself and the United States which probably made it easier for him in 1973 to refuse bases to the United States to resupply Israel (D. C. Watt 1984, 154–155; personal interview, June 1983).

Sir Alec Douglas-Home does not seem to have been particularly interested in the Middle East, although he delivered the Harrogate speech. His personal commitment to European Political Cooperation, however, reinforced by his stature as a former prime minister, may have helped the British raise the image of EPC within the European Community after it joined in 1973.

James Callaghan was foreign minister under Harold Wilson from 1974 until 1976, when he became prime minister. He saw himself as taking a quite different tack from Douglas-Home (and, presumably, from Brown as well). He started his tenure by declaring his expectation that the FCO staff would be pro-Arab, pro-Catholic, and pro-Europe—the clear implication being that he was not. Despite the oil embargo, Morehouse writes, Callaghan

> none the less reacted in the classic Labour party manner. The diplomats were at once given to understand that a high priority of the new government would be an effort to recover the middle ground between Israel and the Arabs; they most certainly were not to expect any repetition of . . . Harrogate . . . which had inclined Britain heavily toward the Arabs. . . . Callaghan [was] quite unwilling to make those gestures which—sometimes much more than substance—come as manna to the Arab soul. The diplomats were equally insistent that the overwhelming priority was to avoid another oil embargo, and that this meant a very clear pro-Arab policy. (1977, 155, 160)

Morehouse argues that this fight lasted about three months and ended with a compromise that left British policy, in effect, unchanged.

Margaret Thatcher and Lord Carrington have been unusually strong personalities in the recent Conservative governments. Thatcher came to office in 1979 without much expertise in foreign policy in general or on the Middle East in particular, although she represents a constituency with "the largest synagogue in the United Kingdom" (personal interview, June 1983). She had been one of the more pro-Israeli members of the Heath cabinet during the oil embargo. In general, she was quite willing to delegate foreign policy to Carrington while she concentrated on domestic issues. Her particular concern with economic issues prompted her visit to the Middle East early in her term of office. She was inevitably involved in political questions, however. Two examples are useful: her refusal to receive an Arab League delegation in 1982 that included PLO members, and her support of Reagan's Rapid Deployment Force (RDF) (Edwards 1984, 52–54). Her decision vis-à-vis the Arab League was linked to the PLO's use of terrorism, a touchy issue in Britain because of the Irish Republican Army (IRA). Friends and colleagues of Thatcher's have been killed by terrorist bombs, and she is understandably sensitive on the issue. The terrorism issue cuts both ways, however. She apparently "has an absolutely gut distrust of . . . Begin," based, it appears, on his own terrorist background (personal interview, June 1983). Her support for the RDF was inspired more by her vision of Britain as a significant military power than by serious analysis of the RDF's utility in the region.

Carrington, in contrast, possesses a serious personal interest in the Middle East. As noted earlier, he played a key role in the embargo decisions as chairman of the cabinet committee on the issue (David Watt 1973b). In contrast to Douglas-Home, for example, he was more knowledgeable about the Middle East and more concerned about the economic problems, a central issue for the Thatcher government (personal interviews, Shlaim and others, June 1983). He was eager to increase the role of Britain within Europe in foreign policy, particularly in view of the Labour government's previous policies, and the Middle East seems to have been selected as the obvious issue on which to do this. Carrington also had unusual personal leverage with Prime Minister Thatcher in part because, as a member of the House of Lords, he was not eligible to become prime minister and therefore did not pose a political threat to her (personal interviews, June 1983; "Britain's Foreign Office" 1982, 19).

Carrington's first priority as foreign minister was to settle the Rhodesian problem, which he accomplished. World attention was shifting to the Middle East again as the Camp David negotiations lagged and the Russians invaded Afghanistan. At the same time, Carrington, by national rotation, was about to become president of the Council of Ministers of the European Community. The time briefly seemed right for a British initiative in the Middle East (Edwards 1984, 52), connected, it appears, to intensive British efforts to shape the 1980 Venice Declaration and to Britain's reception of the Fahd plan in 1981. Carrington, as president of the Council of Ministers, entered into talks on the Fahd plan in Riyadh, even though the plan was not acceptable as it stood.

A British initiative would also have justified British membership in the Sinai Multilateral Force & Observers (MFO), formed to help implement the Egyptian-Israeli peace agreement. A former, very senior EC diplomat said he had expected Carrington to propose a diplomatic initiative, even though this diplomat was skeptical of its chances for success (personal interview, February 1983). Some Americans were concerned that Carrington was prepared to support a Palestinian state, headed de facto by the PLO, and that he opposed the MFO (discussion following the Colloquium on European Foreign Policy-Making and the Arab-Israeli Conflict, Amsterdam, February 1983). Strong private American opposition to any new initiative led Carrington himself to state in July 1981 that Europe by itself could not do much (Garfinkle 1983, 71–72). Carrington's resignation over the Falklands war, however, made the entire discussion academic, despite the considerable degree of continuity provided by Douglas Hurd, minister of state in charge of the Middle East, who shared many of Carrington's views (Shlaim, personal interview, June 1983). Again it is hard to know whether this loss of momentum was caused by Carrington's departure or by the need to shift attention and resources to the Falklands war; the subsequent changes in the Middle East as a result of the Israeli invasion of Lebanon also must be taken into account.

In general, then, individuals do not seem to have exerted tremendous influence over the course of British policy toward the Middle East; the bureaucracy played a considerably more important role. This contrasts with other countries such as the United States, Canada, and Japan, where the personalities of leading politicians are important determinants of Middle East policy at certain times. The British political system is different because it insulates civil servants from public pressure and permits the FCO special policymaking powers on the Middle East.

Impact of the Oil Weapon on British Middle East Policy

British policy toward the Middle East did not change much during or just after the use of the oil weapon. The priority given to this issue over other important aspects of British foreign policy, however, highlights certain features of that policy. The Heath government bowed to Arab pressures at the expense of its alliance with the United States in refusing permission to use British bases to resupply Israel. Similarly, European unity suffered when Britain refused to implement an oil-sharing agreement for the benefit of the Netherlands. The latter was particularly surprising given Heath's commitment to the European idea, although one can argue that it was not really anti-European because all the other EC countries did the same thing.

It is impossible to trace the precise degree of British influence over the

European Community's policy shift in the decade following the use of the oil weapon, but it seems to have been considerable. Janice Stein points out that much of this policy movement is really cosmetic and that the Western governments agree on most basic issues (1982, 54). Nonetheless, the shift toward acceptance of the Palestinians as a national group represented by the PLO and the increasing rhetorical pressure on Israel to withdraw from the occupied territories—common threads in European declarations— presumably derive in large part from Britain. On its own, Britain has expanded arms sales to Arab countries (Pfaltzgraff 1978, 163–164), curtailed arms sales to Israel, and acquiesced in the Arab economic boycott more than many of its EC partners, including "immoral" France.

The reasons for these shifts seem straightforward. The FCO has argued persuasively that British economic interests will not be served by upsetting the Arabs. Despite some concern in the press and Parliament about the moral implications of this policy, it has been unchallenged. There is no bureaucratic opposition; indeed, government groups concerned with promotion of foreign trade have supported the FCO position. Unlike pro-Israeli lobbies in some countries, particularly the United States and Canada, Jewish groups in Britain have been unable to mount a sustained campaign in favor of Israel, even on the issue of the Arab boycott. The most impressive indicator of FCO dominance is that the debate is conducted in realpolitik terms; almost no one admits to preferring a pro-Arab position, but it is a commonplace that such policies are in the British national interests and should therefore be adopted.

Why hasn't British policy become even more pro-Arab? Some argue it is lack of pressure, that Britain would have recognized the PLO, for example, if the Arabs had demanded it (Shlaim, personal interview, June 1983), despite the official demurral that Britain recognizes only governments, not organizations (United Kingdom 1983, 2). A decision to deny diplomatic recognition to Israel would violate the British tradition of recognizing effective governments, regardless of their behavior, which is perhaps why it seems unthinkable. Moreover, British policy is not mindlessly pro-Arab. Stein points out that when the European countries divide on Arab-Israeli issues in the United Nations, Britain tends to be in the moderate camp (1982, 57–58). Similarly, Britain has been fairly restrained in arms sales to the Arab frontline states and Libya, despite an almost desperate need for exports (Pajak 1979, 149–151).

Given the dominance of pro-Arab exponents within the British government, this restraint requires explanation. I attribute it to an underlying, shared sense of morality within the foreign policy elite, a general unwillingness to push realpolitik arguments to their logical extreme (as opposed, according to the British, to the French). I do not know enough about British foreign policy in general to judge whether this is typical. Why this moral sensibility does not apply to the Arab economic boycott is also unclear. Moreover, the fact that this consensus is unspoken makes it difficult to judge its real impact if it comes into conflict with more tangible national interests. Nonetheless, if it is true that "Britain has no permanent allies or permanent enemies, only permanent interests," one of these interests seems to be the belief that its foreign policy behavior should fall within some vaguely defined sense of moral correctness. (Tillman [1982, 43, 49] advocates a similar interpretation of national interest for the United States.)

It is notable that Britain's policy has become more pro-Arab since the use of the oil weapon even though British dependence on Middle Eastern oil has drastically declined because of its North Sea oil resources. Although Britain still imports oil from the Middle East (because it is technically easier to work with), this is more a convenience than a necessity—a radical shift from 1973. The fact that this change has not been reflected in British policy toward the Middle East suggests that the supply theory of economic sanctions is not a useful predictor of British foreign policy.

Why has there been so little change? One reason is that British policymakers have not been sure how long the North Sea oil will last; estimates during the 1970s were as low as two to three years, although the combination of further discoveries and decreased consumption has increased estimates to twenty or thirty years. More important, however, has been the impact of increased Arab wealth and its consequences: increased British exports to the Arab world, higher Arab consumption of British services such as banking and insurance, and Arab support of the pound in international money markets. If the oil weapon had left the Arab states poorer instead of richer, it seems likely that British policy would have moved less rapidly in a pro-Arab direction.

4

Canada: Low Dependence and High Responsiveness

$\mathbf{C}$anada presents an interesting case for a study of the oil weapon because it was the OECD country least dependent on Middle Eastern oil at the time of the embargo. According to the conventional theory of economic sanctions, which stresses supply vulnerability, it should be the Western country least likely to alter policy as a result of the oil weapon.

Historical Background

"The formulation and implementation of an independent Canadian foreign policy is a twentieth-century development" (Stanislawski 1981, 1). Within this larger context, Canada's involvement in the Middle East has been intense but episodic. Canada may fairly be said to have had no interest in the Middle East until the end of World War II. It had no economic interest in the region. Nor, unlike Australia, for example, did Canada have troops in the Middle East during either world war, at least in part because Canada did not perceive Suez as a lifeline and, while quite prepared to defend Britain itself, was generally reluctant to defend other portions of the empire (personal interview, former senior official, March 1983).

Quite suddenly, however, the Middle East became important to Canadian foreign policy just after World War II because of Canada's commitment to the United Nations and to the Western alliance—both of which faced a major test over Palestine. Indeed Palestine was seen as the first non-Western test of the postwar international system (Dewitt and Kirton 1982, 13), and Canadian leaders played a prominent role in the complex negotiations that led to the partition of Palestine and the establishment of the state of Israel. Justice Ivan C. Rand, Canadian representative to the U.N. Special Committee on Palestine, was a central figure in arguing for partition. Lester Pearson, who later dominated Canadian foreign policy for two decades as foreign minister and prime minister, was then chairman of the United Nations First (Political) Committee, which sponsored the Special Committee. He was motivated both by concern for the United Nations and by a desire to reduce conflict between Britain and the United States (Stanislawski 1981, 36–41; Canada 1985, 49–50).

We should note that Canada did not recognize Israel immediately upon Israel's independence in May 1948. In November of that year, however, the St. Laurent government was inaugurated, with Pearson serving as secretary of state for external affairs, and by December Canada had recognized Israel de facto. In May 1949 Canada voted for Israel's admission to the United Nations, implying full recognition (Dewitt and Kirton 1982, 14–17).

Canadian involvement in the Middle East was thus focused closely on the Arab-Israeli dispute from the beginning, although it grew out of concern for the United Nations and the alliance rather than out of interest in the dispute itself. As Janice Stein puts it, Canada had a policy on the United Nations in the Middle East rather than a policy on the Middle East (1976b, 272). This commitment probably reached a high point in the Suez crisis, arguably the greatest threat the Western alliance has seen (S. Spiegel 1982, 1–2). In addition, Pearson was largely responsible for developing the idea of U.N.-sponsored peacekeeping forces, and the Canadian military became central to the resulting force in the Middle East. Eventually, however, the Canadian government developed a gnawing dissatisfaction with its peacekeeping duties during the 1960s, which had grown increasingly thankless, frustrating, and expensive.

Canadian involvement in the Middle East was so dependent on the United Nations that Canada did not even have diplomatic representation in the area until the mid-1950s. By the time of the Suez crisis, however, Canada had embassies in Israel, Egypt, and Lebanon. Ambassadors to Egypt and Lebanon were also accredited to other countries—including Sudan, Syria, and Jordan—but Canada had no representation in the Persian Gulf or on the Arabian Peninsula. A second set of embassies was opened during the 1960s for essentially domestic reasons. Relations with Francophone Third World states were to be encouraged (Delvoie 1976, 29), and, as a result, embassies were opened in Tunisia, Algeria, and Morocco.

An important part of the complex events leading up to the Arab-Israeli war of 1967 was Egypt's demand that the U.N. Emergency Forces (UNEF) be withdrawn from the Egyptian-Israeli border. The Canadian troops were an important part of this force. Secretary-General U Thant agreed to the withdrawal in spite of Canadian government protests that this would fatally weaken peacekeeping forces in the future. This conflict is said to have "ended the two decades-long special relationship between Canada and the United Nations as an entity" (Dewitt and Kirton 1982, 60).

Canada remained an important actor within the United Nations, however, as shown by its participation in the 1967 negotiations that produced Security Council Resolution 242. That same year the Palestinians re-emerged as a political issue. Canada's response, consonant with Resolution 242, was that the Palestinians posed a refugee problem, not a nationalist one, a stance consistently demonstrated by its opposition to yearly U.N. resolutions on the rights of the Palestinian people from 1969 to 1972 (Noble 1985a, 85–106). One analysis shows that, from 1967 to 1972, Canada's votes on Middle East issues coincided with the position of the United States 81 percent of the time, while agreeing with that of the European Community in only 44 percent of the instances (Grossman 1978, 11).

In 1968 Pierre Trudeau became prime minister, and a new era opened in Canadian foreign policy. This change of government had a good deal more impact than the earlier accession to power of John Diefenbaker's Conservative government, which had essentially continued Pearson's Middle East policy (Dewitt and Kirton 1982, 45–55). Even though he was Pearson's successor as leader of the Liberal party, Trudeau came to power committed to changing the country's foreign policy by stressing Canadian national interests—especially the federation problem, his first priority—rather than trying to solve international problems. (We will examine Trudeau's policies and his impact on Canadian Middle East policy in more detail later.) The net result was a deemphasis of the Middle East from 1968 to 1973 not only because of Trudeau, but also because "the two traditional stimuli to Canadian foreign policy in the Middle East—disagreement between her major allies and a U.N. presence in the field—were no longer present" (Stein 1976b, 276).

The Oil Crisis

Canada's first reaction to the outbreak of the October war was to have its foreign minister restate its Middle East policy. "Sharp's statement [to the House of Commons, October 16, 1973] reflected traditional Canadian preoccupations: the preservation of international peace and security, support for the resolutions of the United Nations, and emphasis on the necessity for negotiations between the parties" (Stein 1976a, 97). Essentially, Canada continued to behave as if it had no interests at stake other than Middle Eastern peace. But this stance was shaken by the use of the oil weapon.

Although Canada was the OECD country least vulnerable to an Arab oil embargo (Canada exported considerably more than it imported), it did import some Middle Eastern oil. Canada's problem was basically regional, because its oil was in the West, and there were no pipelines to carry it economically to the East, particularly to Quebec and the maritime provinces. This was not an insoluble problem: Oil could be shipped by tanker from the West Coast through the Panama Canal to eastern Canada. Alternatively, it could be piped from Alberta to the Great Lakes and sent by tanker through the St. Lawrence Seaway. It was also possible to ship by rail. But all of these methods were inconvenient and expensive, and they would all take time to arrange. Moreover, Canadian oil was already committed by contract, mostly to the United States; breaking those contracts raised other political problems. Canada's oil problem was basically one of time, getting the one to two months necessary to divert enough oil to guarantee that eastern Canada would not suffer. "In December 1973, Prime Minister Trudeau announced his proposals for a complete change in

Canadian energy policy. He decided to scrap the national oil policy in favour of a new policy in which western oil would be joined to eastern markets by extending the Interprovincial Pipeline east from Sarnia into Montreal" (Turner 1978a, 197). It was not until the 1979 oil crisis, however, that the pipeline was extended to Quebec.

I interviewed several people who suggested that Quebec's vulnerability to the oil weapon made the issue much more politically sensitive, especially to a government desperately trying to avoid giving Quebec separatists any issues to use politically. It is interesting that the economic-sanctions literature often argues that sanctions serve to unify the target state. The Canadian government feared the opposite, in part because rising oil prices meant resources would be transferred from eastern to western Canada, much as happened in the United States at the same time. Nevertheless, a former cabinet minister has said the French-English issue was "not particularly" important in making decisions (personal interview, March 1983).

Given eastern Canada's vulnerability, the government was understandably worried about an embargo. Leaving aside Canada's Middle East policies, which, at least in public, all government spokespersons assumed were clearly neutral, two technical factors made an embargo likely. The first was Canada's open border with the United States. If the Arabs were serious about cutting the United States off, Canada was an obvious target. More immediately, Canada's Middle Eastern oil was in fact imported from a pipeline that started in Portland, Maine. The line itself was a remnant of World War II, built by Standard Oil of New Jersey (now Exxon) in 1941, before the war broke out, to allow tankers going to Canada to remain under U.S. naval protection.

In Ottawa a good deal of confusion arose, in public and private, after the embargo was declared. The government was unable to determine for some time whether or not it was on the embargo list. This uncertainty is understandable—a number of other countries seem to have been similarly in the dark. But the Canadian case is unusual because ten years later very well informed observers of Canadian foreign policy still did not know for certain if Canada had been embargoed, and a number of them suspected the government itself never really found out. (The only comparable case I know of is Denmark, which apparently was subjected to a partial embargo early in 1974, also shrouded in mystery.)

On October 25, the *Toronto Globe and Mail* reported that Abu Dhabi had stopped a Gulf oil shipment en route to Canada (Ismael 1976, 263). There were also rumors that a British Petroleum (BP) shipment had been stopped. In response to questions in Parliament,

> Sharp reported to the House on October 26 that officials were still unable to discern the intentions of Saudi Arabia and Abu Dhabi. As an official in the Department of External Affairs described the situation, "There was the feeling that we didn't know what the hell had hit us. Some Arabs said that we were on the embargo list and some said the opposite." Confusion on this issue was to persist well into November. (Taras 1983, 18)

Ismael suggests that the political aspects of the shutdown were not all that difficult to understand:

> In the House, Mr. Sharp expressed his inability to comprehend the Arab states' action. The following day the Arab Information Agency, in response to queries by reporters, replied that the Canadian government had taken the same stand as that of Holland which had been cut off from oil shipments. Further, on November 1, the Egyptian Ambassador to the U.N., Mr. Neguid, said that unless Canada took positive steps to change its policy it could expect continuing disruption of oil shipments. Mr. Sharp initially thought that the suspension of oil shipments was a mistake because of the destination of Portland where the Canadian pipeline is located. . . . Abu Dhabi and Saudi Arabia suspended oil shipments to Canada. (1976, 263–264)

Part of the confusion stemmed from disagreement among the Arab states themselves over how to treat Canada (Taras 1983, 18). A former member of the Canadian Department of External Affairs (DEA) observed, for example, that the Algerians, who played a leading role in the embargo, were not able to clarify the status of the Gulf shipment because, apparently, they simply did not know what had happened to it (personal interview, March 1983). The evidence I have gathered suggests there were two distinct embargo periods. Initially, Canada was indeed embargoed, not because of its politics but because of its Portland-based pipeline and because it supplied oil to the United States (Lenczowski 1976b, 65; personal interview, June 1983). This was a difficult period because the Saudi government seems to have been internally divided, stating that Canada was not embargoed, while Saudi oil companies claimed that their orders were not to send oil. This problem was finally resolved only by King Faisal himself in December when he classified Canada as "neutral" (Byers 1974, 257; personal interview with a senior official, June 1983).

During this second stage, Canada was not embargoed but was presumably subject to production and export cutbacks (Canada 1985, 51). (Kelly [1980a, 407] asserts that Canada was able to avoid the cutbacks by offering to stop shipping its domestic oil to the United States, but I have no confirmation of this.)

If the embargo was a shock to Canada, a second shock immediately followed, centered on the question of peacekeeping. We have already noted that Canada was not happy about its participation in United Nations peacekeeping, feelings that had been exacerbated by its experience in Vietnam. Nevertheless, DEA continued to regard peacekeeping as an appropriate role, in part to avoid being subject to the oil embargo (Byers 1974, 257), and as early as October 10 a departmental representative announced that Canada was willing to participate in a U.N. peacekeeping force in the Middle East if certain conditions were met.

No one seemed to have considered the idea that Canada might not be acceptable to others. It was therefore a distinct shock when, on October 15, Egypt condemned Canada for its pro-Israeli press statements. "In November it became apparent that Canada was not wanted by Egypt" (Ismael 1976, 259–262), in large part because of Canada's opposition in 1967 to President Nasser's request for the withdrawal of UNEF troops (Canada 1985, 50–51). (As an aside, it seems unfair for Egypt to have penalized Canada for its opposition to a move that triggered the disastrous 1967 war.) Egypt finally allowed Canada to participate in the subsequent peacekeeping force only on condition that Poland also be included, clearly implying that Canada was pro-Israel and had to be balanced by a more pro-Arab, Warsaw Pact government.

Neither the embargo nor the peacekeeping question was a major issue in Canada at the time, however (Taras 1983, 18–19). Trudeau was leading a minority government, and political attention was focused elsewhere. Public alarm over oil shortages was negligible; Canada regarded itself with good reason as a rich country with ample resources, and short-term energy needs were met by purchases on the spot market (with the minister for energy and natural resources dramatically directing them in person). The initial response to the embargo was that it was not a Canadian problem.

Noble argues that, despite this rather relaxed public attitude, the embargo produced almost immediate diplomatic movement on the Palestinian issue in order to demonstrate Canadian neutrality (1985b, 113–116). This issue was selected because it would not alienate Israel too much and because the European Community's November 6 statement had already blazed the trail. The shift became apparent in several ways. First, Minister Sharp began to use the term "Palestinian" in public statements while the embargo was still in force. Second, he agreed in principle that the Palestinians should participate in Middle Eastern peace negotiations, although remaining vague on how this might be done. Finally, Canada joined the European Community on December 7 and abstained on the annual U.N. Palestinian resolution (Resolution 3089D), which was essentially identical to the one it had opposed the year before. By doing so, Canada acknowledged for the first time that Palestinians had collective rights as Palestinians as well as individual rights as refugees, a clear departure from Resolution 242. Indeed, Canada's official explanation of its abstention stressed precisely this point (Stein 1983, 154).

This was, in fact, the central change of Canada's Middle East position during the embargo period, although the rest of the decade was spent working out its implications. The problem of how the Palestinians would be represented and, in particular, the appropriate Canadian attitude toward the PLO continued to be an issue, but the basic decision to acknowledge the collective rights of the Palestinians outside the restraints of Resolution 242 had been taken.

No one seems to have noticed. Trudeau himself said, "There has been no change in the Canadian Government's policies regarding the Middle East conflict as a result of the Arab oil boycott or for any other reason" (*Toronto Star,* June 17, 1974, cited in Stein 1976b, 293). But the policy changes themselves did occur during the 1973 embargo.

By the end of the crisis, then, things looked fairly normal. Oil supplies had not been drastically cut, and Canada was once more participating in a U.N. peacekeeping force in the Middle East. But this appearance was deceptive. Canada for the first time had concrete issues at stake in the Middle East, first oil and then trade, and the government understood that many Arab governments thought its position was not neutral but pro-Israel. Canada responded to this situation by significantly altering its policy on the Palestinian issue.

Canadian Middle East Policy Since the Crisis

We can divide Canadian policy toward the Arab-Israeli conflict after 1974 into several different parts, including the Palestinian issue in general, the PLO's role in two U.N. conferences in Canada, the Arab economic boycott, and the location of Canada's embassy in Israel.

The Palestinian Issue

Canada's policy changes on the Palestinian question produced a voting pattern on the PLO that one leading analyst has aptly labeled "confusing": "When the PLO alone is invited, Canada abstains since it does not accept *a priori* the claim of the PLO to represent the Palestinians; when such an invitation is issued by consensus, Canada does not dissent. Canada distinguishes, however, the specialized agencies and conferences from the General Assembly of the United Nations" (Stein 1976b, 281–282).

Canadian rhetoric on the Palestinian issue is similarly vague. Because of the touchy issue of Quebec separatism, the term "self-determination" has been avoided by Canadian governments. In 1975 Allen MacEachen, the new minister of external affairs, supported a "Palestinian entity" in a speech to the United Nations. In 1976 the phrase "political self-expression within a limited type of political arrangement" was used repeatedly. In 1981 Robert Stanfield recommended that Canada support the Palestinian right to a "homeland" negotiated with "its neighbors," including Israel (as distinguished from self-determination, in which Israel would not be involved). Stanfield's report itself had no real effect on the government, however. In the same year the minister of external affairs referred to the Palestinians' "right to a homeland within a clearly defined territory and by that I mean the West Bank and the Gaza Strip," which seems to have become the new Canadian formula (Noble 1985b, 116–149; Canada 1985, 51–52).

Grossman maintains that Canada shifted from support of the United States on such questions to agreement with the European Community (1978, 11). (K. J. Holsti found that Canada's Index of Agreement with the United States on U.N. voting declined somewhat after 1968 [1982, 99]. Holsti does not break the figure down by issue, however, so we cannot determine if Canada's change vis-à-vis the Middle East was mirrored in other issues or was unique.) In a more sophisticated analysis, Janice Stein (1982, 57–69) reaches a more complex conclusion, arguing persuasively that the differences between the United States, Canada, Japan, and the European Community are more apparent than real: They agree on the right of Israel to exist; the need for Israel to withdraw from occupied territories; opposition to unilateral change in the status of the occupied territories, such as Israeli settlements; and the need for a political and territorial solution to the Palestinian problem. The differences concern means rather than ends and have been deliberately exaggerated by the Europeans (and presumably the Canadians) in order to appear different from the United States.

With regard to the United Nations, Stein's analysis does not show a simple division between Europe and the United States. The pattern, instead, is one of shifting coalitions on different issues. The United States and its allies, including the EC, do seem to disagree on various issues, such as general recognition in the mid-1970s of the Palestinians' collective rights. On other issues, however, such as granting the PLO formal-observer status, the alliance split differently, with the United States, Canada, and several European countries (typically Belgium, West Germany, the Netherlands, Britain, and Denmark) often opposing France, Italy, and Japan. On other issues, such as the Zionism-as-racism resolution, the United States, Canada, and the EC voted together, with only Japan abstaining.

What does Canada do when the United States disagrees with a majority of the EC? Stein's analysis shows that on the few issues where this is true, Canada supported the EC in 1975 and 1976, split evenly in 1977, and supported the United States more in 1978 and 1979 (Stein 1982, 57–69; Stein 1983, 153–156). Canada thus continues to hold a middle position within the Western alliance. This analysis is basically supported by a more detailed historical study by Paul Noble (1985b, 115–116, 122–127, 134–141). As Stanislawski explains it:

> Canada began to apply what Canadian diplomats call the "company principle," a voting posture which takes into critical account the anticipated actions of like-minded states. As the Middle East situation at the U.N. evolved, American company ceased to be regarded by Canadian decision-makers as "good or acceptable company." Instead, Canadians began to take their cues from the West European and Scandinavian states. (1981, 67)

U.N. Conferences and the PLO

Canadian policy toward the Arab-Israeli conflict was not confined to voting in New York, however. When the PLO was granted observer status by the United Nations (over Canadian opposition), it automatically was invited to conferences sponsored by the General Assembly. Two of these had already been scheduled in Canada, a conference on crime in 1975 in Toronto and another on housing, called Habitat, held in 1976 in Vancouver. Part of Canada's agreement with the United Nations about these conferences included a stipulation that Canada could not refuse admission to delegates or observers. Under Canadian immigration law, however, members of the PLO could not be admitted to Canada unless the government specifically requested an exemption, which had in fact been done before (Stanislawski 1981, 69).

The idea of representatives of the PLO, widely viewed as a terrorist organization, participating in an international conference on crime upset a lot of people in Canada, and a major political brouhaha resulted. Intense domestic pressure, led by an organization called Canadians Against PLO Terrorism (CAPLOT) and generally supported by the Canadian press, developed to prevent PLO representation at the conference. Both the provincial government of Ontario (a cohost of the conference) and the Toronto city government opposed it. Within the government, the departments of External Affairs and Immigration led opposing factions in a "highly volatile process. Officials produced over a dozen draft memoranda for cabinet, cabinet itself debated the issue on six occasions, and finally produced a decision which bore no resemblance to the recommendation forwarded by officials through [the Department of External Affairs]" (Dewitt and Kirton 1983, 188; see also Bones 1985, 157–159). The result was a formal decision to request postponement for a year. The United Nations responded by moving the conference.

Trudeau reportedly was outraged by the decision process and made sure it would not be repeated for the question of PLO representation at Habitat (Dewitt and Kirton 1983, 189). Housing was a less incendiary topic than crime, and perhaps for this reason much less controversy surrounded the question of PLO involvement in the second conference. After some discussion, the Canada-Israel Committee actually participated in the housing conference in return for the government's commitment to avoid a "politicization" of the conference. Habitat proceeded on schedule, honoring Canadian policy to admit PLO delegates who were not themselves terrorists or who were not members of PLO terrorist subgroups. But the government was so unsuccessful in keeping the conference nonpolitical that it voted against the final conference communiqué precisely because it was anti-Israel (Bones 1985, 161–163).

The Arab Boycott

This controversy surrounding the housing conference, however, was only preliminary to a more sustained public discussion over the appropriate Canadian reaction to the Arab economic boycott of Israel. Boycotts have been a part of Arab strategy since before the founding of Israel. They consist of several levels. A primary boycott is a refusal to buy from or sell to Israel. A secondary boycott is a refusal to buy from or sell to anyone who does business with Israel or Israelis. Thus, a tertiary boycott is a refusal to deal with anyone who in turn deals with anyone who does business with Israel or Israelis. In order to enforce these more indirect aspects of the boycott, the Arabs often require companies with which they do business to sign documents stating they abide by the terms of the boycott. To add to the complexity, Zionists are often included as well as Israelis. In practice, this often serves as a synonym for Jews, regardless of their nationality (Stanislawski 1981, 107–122; Nelson and Prittie 1977, 25–50).

In his definitive study of the boycott in Canada, Stanislawski argues persuasively that it violated many values widely held in Canadian society: support of Israel's security, freedom of Canadian trade, lowering of trade barriers, and Canadian economic autonomy (1981, 4; summaries are in Stanislawski 1983, 200–220; Stanislawski 1987, 223–254). The problem for analysts, then, is to explain why no federal antiboycott legislation was ultimately enacted. The boycott had gone on for many years, but it did not become an issue in Canada until after 1973, when Canada was encouraging exports and when Arab wealth had made the Middle East an attractive market. Stanislawski says the boycott emerged as a public issue in 1975 when Herb Gray, a prominent Jewish M.P. and former cabinet minister, revealed that the Export Development Corporation, a Canadian government corporation, had insured transactions with boycott clauses (1981, 177). This in turn meant Canadian companies would be forced not to do business with Israelis and possibly not with Canadian Jews either. On May 8 Trudeau said that such acts were "alien to everything the government stands for."

Despite Trudeau's flat condemnation, the government proved unwilling to introduce legislation to require companies to report boycott demands and to forbid them to violate the civil rights of Canadian citizens; those who advocated such legislation often pointed to the example of the United States. On October 21, 1976, the new minister of external affairs, Don Jamieson, announced a new boycott policy. There would be no new legislation, but no government facilities or support could be used for transactions with boycott clauses; in addition, a mandatory reporting system would be introduced.

In practice, however, the departments of External Affairs and Industry, Trade, and Commerce proved unwilling to implement these policies when, for example, Bell Canada received a billion-dollar contract from Saudi Arabia with boycott clauses (Stanislawski 1981, 198–232, 266–277). Indeed, government spokesmen said the government did not have the power to enforce reporting (Canada 1985, 101). A report by the ad hoc Commission on Economic Coercion and Discrimination, headed by Professor Irwin Cotler of McGill law school, heightened the controversy by revealing the extent to which Canadian companies had in fact been involved in the Arab boycott (Commission on Economic Coercion and Discrimination 1977).

By 1978 the boycott had become a major political issue. An election was approaching, and in 1974 the boundaries of parliamentary constituencies had been changed. The previous borders gave Toronto only two districts in which the Jewish vote would be crucial. But under the new plan, six or seven districts out of about twenty-five in the whole city fell into this category.

Most political commentators, as well as the opinion polls, considered the 1978 election to be very close. . . . It was widely believed that Ontario, and especially southern Ontario, held the key to the election. In metropolitan Toronto, Jews were seen as playing a significant electoral role in four to eight key constituencies. Thus, for the first time in Canadian electoral history, the question of the "Jewish vote" arose and was seen by many as having a major role to play in the election's outcome. (Stanislawski 1981, 326)

In a move widely regarded as linked to the coming election, the government announced at a press conference on August 21, 1978, that legislation would be introduced in the next session to force reporting of boycott requests, although the minister of industry, trade and commerce—in charge of the department that, presumably, would administer such a bill—was notably absent. The bill was introduced in December but was not brought forward by the government for a vote, with the government and the opposition blaming one another for the failure (Stanislawski 1981, 262– 264, 310–329). Thus, the issue was held over until after the election. Meanwhile, impatient with the delay at the federal level, the Province of Ontario passed its own antiboycott law in November 1978 (Stanislawski 1981, 284–308). Because the province includes the major financial center of Toronto, this was not a trivial gesture, although corporations can bypass the law by signing agreements outside of the province (Canada 1985, 101–102). The Arab states responded to this furor by quietly dropping the secondary and tertiary boycotts and focusing almost exclusively on the primary boycott (Canada 1985, 100).

The Jerusalem Embassy Affair

In April 1979 Joe Clark, the Conservative candidate for prime minister, caused a political storm by promising that, if elected, he would move the Canadian embassy in Israel from Tel Aviv to Jerusalem. Begin had asked Trudeau to do this in 1978, but the idea was not well received. Clark, while in Israel in January 1979, was also asked if he would do it; he said he would make no decision until the Israeli-Egyptian treaty was signed. This was done on March 26, and Clark made his pledge on April 25 at a meeting with leaders of the Canada-Israel Committee. He also reaffirmed a pledge to support antiboycott legislation (Stanislawski 1981, 326–328).

The embassy pledge seems to have been made on the advice of a few candidates for the Toronto parliamentary seats rather than the regular Conservative party strategists; even the Jewish community was divided on the issue (Takach 1980, 22–28; Dewitt and Kirton 1982, 87; Bones 1985, 163–165). An alternate theory is that Clark simply felt he needed "a foreign policy initiative to round out his campaign" (Bruce Garvey, *Montreal Star*, June 16, 1979, cited in Leyton-Brown 1981, 229); indeed,

Clark seems to have planned to move toward recognition of the PLO as well, with the two gestures presumably creating a new Canadian Middle East initiative and distinguishing him from the Liberal party (Takach 1980, 39, 44; Dewitt and Kirton 1983, 191–192).

As an electoral strategy, it seemed to work. On May 22, 1979, the Conservative party (P.C.) won the election: "As anticipated, southern Ontario had proved to be the key to a P.C. victory; the margin, however, was larger than had been anticipated. . . . With only one exception, P.C. candidates were elected in the constituencies with the most substantial Jewish populations" (Stanislawski 1981, 331). We should note that although the Conservatives' boycott position may have been influential among Jewish voters, some Toronto Conservatives attributed the electoral victory instead to the mortgage-deductibility scheme (Stanislawski, personal letter to the author, August 1983; Takach 1980, 43).

The 1979 elections mark the high point of apparent Jewish electoral influence on Canadian Arab-Israeli policy, but the limits of such influence were quickly demonstrated. The Conservative government proved unable to move its embassy to Jerusalem, and the antiboycott legislation, ironically, became an almost incidental casualty of the same political fight. The fact that Clark had made his pledge during the election did not, of course, obligate him to carry it out afterward; in 1976 Jimmy Carter had made a similar promise that was quietly forgotten (Takach 1980, 19). Indeed, some observers argue that Clark never really wanted to move the embassy in the first place. But at his first press conference after the election, Clark was asked whether he would carry out his campaign pledge to move the embassy. The issue became entangled in whether or not he was going to be ruled by the civil servants in Ottawa, and Clark declared that he was going to move the embassy (Takach 1980, 45–46; Stanislawski 1981, 376; 1983, 2).

Clark's declared intention turned out to be remarkably unpopular. Indeed, it is hard to find any Canadian who will admit to having supported it. For the first time in the history of the department, career officials at External Affairs threatened to resign over an issue (Stanislawski 1981, 376). Even Clark's own minister of external affairs, Flora MacDonald, was reportedly opposed (personal interviews, March and May 1983). The Jewish community was divided. The press, which had strongly supported antiboycott legislation, generally opposed the embassy move.

Arab reaction was quite strong—Iraq went so far as to embargo oil to Canada again in 1979, although this was not formally communicated to the Canadians until the following year (Takach 1980, 56–57). More important, the embassy issue marked the entry of the Canadian business community into the public debate over the Middle East. Arab governments effectively communicated to the business community that there was a link between Canadian foreign policy and economic access to their countries, as they had not over issues such as antiboycott legislation or U.N. resolutions. In turn, Canadian corporations overcame their usual reluctance to become publicly involved in political discussions and played an effective role (Takach 1980, 94–98; Stanislawski 1981, 331–345, 397–398, 402–409).

In order to reduce the political temperature, Clark appointed Robert Stanfield, former leader of the Conservative party, as a one-man commission to study the embassy issue and Canadian Middle East policy in general. (Takach discusses the debate over Stanfield's instructions [1980, 84–85].) Strictly speaking, Stanfield's report had no impact on the government; by the time he completed it, Trudeau and the Liberal party had won the 1980 election and were back in office. The Stanfield study, however, served as a convenient excuse for the Clark government to shelve not only the embassy issue but the antiboycott legislation as well. On October 29, 1979, to no one's surprise, Stanfield, in an interim report to Clark, recommended that the embassy not be moved. His final report also recommended against antiboycott legislation (Stanislawski 1981, 345–353). The Trudeau government showed no interest in introducing such legislation after the election.

Dimensions of Change in Canadian Middle East Policy

How has Canadian policy changed since 1973? Before that time, Canada followed the United States' lead quite closely. As a coauthor of Security Council Resolution 242, Canada supported linking Israeli withdrawals from occupied territory with guarantees of Israeli security. Officially, it had viewed the Palestinian issue as a refugee problem, the heart of which was to guarantee the individual rights of people in distress rather than the collective rights of a national group. Just after the oil weapon was used, these policies all changed, so quietly that no one seemed to notice. Canada is now prepared to contemplate dealing with Palestinian representation separately from other issues in Resolution 242. It also acknowledges that the Palestinian issue involves collective as well as individual rights. It has thus moved significantly away from the United States, particularly on U.N. votes.

Canada presents an interesting contrast to the European states. On the one hand, Canada moved more quickly toward the Arab position after the embargo. On the other hand, it has not gone as far as the Europeans. It has supported the step-by-step ("Camp David") initiative of the United States rather than the comprehensive ("Geneva") solution preferred by the Europeans (Stein 1982, 69–70; Stein 1983, 161), and it was cool to the Venice Declaration, perhaps because of the phrase "self-determination" for the Palestinians, which suggests the problem of Quebec to many Canadians (Noble 1985b, 130–131). Canada has thus remained in an intermediate position within the Western alliance on the Arab-Israeli conflict.

Canada in 1973 quite correctly did not see itself as dependent on Middle Eastern oil, so much so that, ten years after the event, knowledgeable observers still professed not to know if Canada was embargoed. Despite Canada's relatively low vulnerability, Ottawa made at least rhetorical concessions to soothe Arab sensibilities, some offered very quickly indeed. Canada thus presents another challenge to the supply-shortage theory of economic sanctions: Why did it alter policy at all, certainly more than the United States, which was significantly more dependent on Arab oil?

External Influences on Canadian Policy

The United States is a constant, looming presence for Canada, in foreign policy as well as in other areas. Moreover, American policymakers have invested a good deal of time and prestige in the Middle East (in contrast to areas such as sub-Saharan Africa, for example). In addition, Canadians and Americans tend to share moral perspectives on the region, with a common commitment to the continued survival of Israel and to a negotiated settlement. It is not surprising that the United States has greatly influenced Canadian Middle East policy, although this dominance may be waning.

Indeed, one indication that the year 1973 was a turning point in this regard has been Canada's increased willingness since the embargo to take positions independent of the United States—albeit not *too* independent (Benesh 1979, 2–3). Part of Canada's tendency to move away from the United States can be explained by Trudeau and his desire for "counterweights." But the idea of counterweights is hardly new in Canadian foreign policy (Tucker 1980, 4), and it is worth noting that although Trudeau came to office in 1968, Canada's Middle East policy did not change until *after* the oil weapon had been used.

I found little evidence of overt American pressure on Canada over Middle East issues. Perhaps the American government has such good access to Ottawa that it does not need to air its complaints publicly. Alternatively, the Canadians may simply be doing what the Americans want anyway, although Dewitt and Kirton note, for example, that Canada's reluctance to participate in the U.N. force in Lebanon had surprised the American government (1982, 72–73). A third possibility is that the Canadians are so unimportant that the Americans do not really care, but this is unlikely: The Canadians have played a significant role in the Middle East through the United Nations, where American allies have been scarce lately. Perhaps more likely is that various American administrations have gotten more sophisticated about tolerating dissent in the north, particularly when the Canadians sound so much more responsible than some of the Europeans' more extreme pronouncements. At any rate, American influence was no doubt important, but it was not felt directly, and Canada demonstrated a clear tendency to move away from direct linkage with U.S. policy.

One interesting exception to this general tendency took place at the Tokyo summit in 1980, when Jimmy Carter reportedly told Joe Clark that if Canada moved its embassy to Jerusalem, the Camp David peace process would suffer. Another version holds that the same message was communicated only through Secretary of State Cyrus Vance to Foreign Minister Flora MacDonald in a conversation at the United Nations because of the administration's concern for Canadian sensitivities about U.S. pressure (Dewitt and Kirton 1983, 184). One informed observer regards this communication as the real turning point for the Jerusalem embassy affair, although this position is by no means unanimous (Takach 1980, 100; Stanislawski, personal letter to the author, August 1983). Thus, the one clear example of American pressure actually involved the promotion of a less pro-Israeli policy.

A second external influence on Canadian policy is the United Nations, which clearly has been central to the formation of Canadian foreign policy, despite Trudeau's efforts to introduce a narrower definition of Canadian national interests. The United Nations' consensus position on the Arab-Israeli dispute has shifted as that organization has become more dominated by Arab and Third World states. Canada seems to have reacted to this change in two ways. On the one hand, it has edged away from the United Nations; the kind of simple support Canada gave the international organization in the 1960s became harder to maintain. On the other hand, it is hard to escape the sense that to some extent the U.N. consensus has pressured Canada to alter its policies, particularly on the Palestinian issue. Some of this external pressure comes through the Department of External Affairs (we will discuss the department in more detail later), which has generally encouraged Canadian politicians to adopt a less pro-Israeli posture. This may be partly because of the department's desire to stay roughly in the middle of this issue in the United Nations in order to preserve Canadian effectiveness (Stein 1976b, 273; Benesh 1979, 2, 14).

A third external force at work on Canadian policy toward the Middle East has been the example of Western Europe. Trudeau's intellectual predilections in favor of the use of counterweights were greatly strengthened by the "Nixon shocks" of 1971 and by his general concern that, in a world increasingly dominated by international trading blocks, Canada was without equal partners. The result was a concerted push for a "contractual link" with the European Community. This link was to involve an agreement in principle rather than any change of existing treaties, a symbol of willingness to diversify. The agreement was finally signed in 1976, and the negotiating process increased Western Europe's weight vis-à-vis a number of foreign policy decisions, including Middle East policy and the Canadian contribution to European defense.

The Canadians thus began to pay more attention to the shifts in EPC statements on the Middle East (Noble 1985b, 111). In addition, the existence of differences within the Western alliance may have encouraged the Canadians to take a middle ground in order to act as negotiators, which they see as one of their traditional roles. In practice, since 1973 the Canadians have not supported any of their traditional allies completely, except on the right of Israel to exist.

The French-English split in Canada, an internal problem, probably amplified Europe's influence. As one observer argued: "No Canadian government over the long haul can afford to be too out of step with France." France is one of the most pro-Arab members of the European Community, so Canada was pressured to keep its public statements cautious and vague and to adopt positions somewhere between the United States and France, winding up fairly close to the Germans and the British: "That's a nice position for us," a Canadian academic declared (personal interview, May 1983). Quebec was Trudeau's main concern; thus, Quebec separatism made the European position more attractive. In contrast, Stanislawski argues that Britain has more influence than France on Canadian foreign policy (letter to the author, August 1983). The terrorism of Quebec separatists may also have made it more difficult for the government to side with the PLO (Noble 1985a, 91).

A fourth external factor stems from the Arab states themselves. One result of the embargo was increased contacts with the Arab states as trade grew. Dewitt and Kirton may overstate its relative weight, but it was clearly noteworthy:

> Of greatest importance, however, was the more sensitive understanding Canada developed of the views of individual Arab states, through its expanding network of diplomatic posts, visits, and joint organizations, and its policy of forging relationships with particular states. To be sure, the representations Arab states made on the anti-boycott issue through the Arab League Information Office in Ottawa were insufficiently focused and too late to affect Canadian actions. Yet the vigorous, widespread criticisms made by Arab states on the proposed transfer of the embassy to Jerusalem helped secure a rapid shift in the government's policy. (1983, 184)

It is important to note, however, that the Canadian government did not think the Arab oil states were pressing hard for PLO recognition and indeed found them divided on the issue. During his 1976 trip to the Middle East, Allen MacEachen, the external affairs minister perhaps most sympathetic to such a change, found that Saudi Arabia urged recognition, Egypt did not push the point, Jordan did not raise it, and Iraq was opposed. "Analysts concluded that recognition of the PLO was not a prerequisite to the improvement of Canadian-Arab relations" (Stein 1976b, 282).

Similarly, in recent hearings on Middle East policy by the foreign affairs committee of Parliament, speakers from all points of view have agreed that the Arabs do not link Canadian foreign policy positions to economic relations (personal interview, March 1983). Stanislawski argues that Arab governments were also influential in the antiboycott legislation issue, but most of this seems to have been indirect (1981, 434, 436). The only time this leverage was used directly by Arab governments was in the Jerusalem embassy affair, where it had considerable direct and indirect impact (Takach 1980, 93–94). It is not yet clear whether the Arabs' success in this instance indicates a shift of tactics or was simply an exception to the rule.

Societal Influences on Canadian Policy

Public Opinion

It is generally believed that Canadian policy toward the Arab-Israeli conflict has deeply interested only a fairly small number of Canadians (Benesh 1979, 20–21, 39–40, 77–86; personal interview, March 1983). The relatively few public opinion polls suggest that most Canadians, before and after the embargo, either supported neither side or did not know. But among the minority that did support one side, Israel had a clear advantage, with about 21 percent of the sample supporting Israel and 6 percent supporting the Arabs, producing an image of general public support for Israel (Benesh 1979, 23–25; Cohn 1979, 38).

We do not have much data comparing public opinion among countries. Some material is available from the various Gallup organizations, however (table 4). The data must be used with caution because the wording of the question varies a bit over time, and we only have four data points for Canada, some of which cannot be matched precisely in the other countries. Despite these constraints, the pattern is fairly clear. Since 1967 Canadian public opinion has been notably less supportive of Israel than public opinion in either the United States or Britain. This finding is particularly interesting because most observers believe Canadian foreign policy has been more supportive of Israel than has British policy. If Canadian policy *is* more pro-Israel, it is not the result of public opinion.

Some differences in public opinion can be explained by the country's English-French division. In his analysis of the Canadian polls of 1969 and

Table 4.   Canadian Public Opinion: Sympathies in the Arab-Israeli Dispute,
Compared to British and American Attitudes

		Favor Israel (%)	Favor Arabs (%)	Favor Neither (%)	No Opinion (%)
Canada	*December 1957*	*15*	*5*	*80*	a
United Kingdom	April 1956	19	6	75	a
Canada	*July 1969*	*26*	*9*	*36*	*30*
English-speaking					
Canadians	*July 1969*	*32*	*7*	*39*	*22*
United States	January 1969	50	5	28	17
United Kingdom	January 1969	41	8	29	21
United Kingdom	March 1969	53	6	21	20
United Kingdom	August 1969	36	4	31	29
United States	March 1970	44	3	32	21
Canada	*November 1973*	*22*	*5*	*42*	*31*
United Kingdom	December 1973	35	10	36	19
United States	December 1973	50	7	25	18
Canada	*November 1978*	*23*	*7*	*70*	a
United States	November 1978	39	13	30	18
Canada	*October 1982*	*17*	*13*	*70*	a
United Kingdom	August 1982	25	16	40	19
United States	July 1982	41	12	31	16

SOURCES: Data for the United Kingdom from Gallup 1976, 376, 1032, 1045, 1063, and 1290, respectively; August 1982 figures for the United Kingdom and Canada from a Gallup poll, data from the Roper Center, University of Connecticut, obtained through the Center for Computer and Information Services, Rutgers University. Data for Canada for the years 1957, 1969, and 1973 taken from Benesh 1979, 24, 63, and 24, respectively; 1978 data from Gallup 1980, 131. Data for the United States for the years 1969, 1970, 1973, and 1978 found in *Gallup Opinion Index* 1979a, 28; 1982 data from *Gallup Report* 1982, 4.

NOTE: Some rows may not add up to 100 percent because of rounding errors.
a "Favor Neither" and "No Opinion" responses not separated.

1973—using fairly generous criteria (significance levels of .05 with asymmetric lambdas of .030 or greater)—Benesh found language was significant in 1969 but not in 1973, whereas religion was significant in 1973 but not 1969 (1979, 52–53). Even English Canadians were much less supportive of Israel in 1969 than Americans and perhaps a bit less so than the British as well (table 4).

Nonetheless, it is clear that more Canadians favored Israel than favored the Arabs. Noble attributes this to the perception of Israel's legitimacy, which in turn was based on sympathy for the plight of European Jews during World War II (1985a, 89). In a sophisticated analysis of Canadian public opinion data, Benesh found that concern for Jewish suffering was the single most important factor in explaining why pro-Israeli Canadians felt as they did (1979, 32). Noble contends that this sympathy has translated into a general sense that Israel is the aggrieved party in the region, trying only to protect itself. In contrast, the Arabs are seen as trying to destroy Israel by aggression. Interestingly, Benesh found that Arab supporters simply reversed this image; Arabs were seen as the underdog protecting themselves against Israeli aggression (1979, 32). Thus, the rather small Canadian Jewish groups had a disproportionate impact on Canadian policy, but only when their policies were supported, at least passively, by most Canadians (Bones 1985, 166–170).

These attitudes have changed over time, as in many other Western countries. French Canadians have always been less sympathetic toward Israel than English Canadians (Cohn 1979, 38–45; Taras 1983, 10; Benesh 1979, 62–64), something attributed variously to traditional anti-Semitism in Quebec (Cohn 1979, 32–34) and to a sense of identity with the Palestinians as a national group seeking self-determination. French Canadians saw Israel less as a valiant underdog and more as a dominant regional power. Over time, concern for the Holocaust had less impact on policy, some European states changed their policy, and the Palestinians were increasingly seen as a people with a legitimate grievance rather than as simply a collection of terrorists. Sadat and Begin also made the Arabs appear more desirous of peace and the Israelis less so.

> This shift in the perception of legitimacy was important. Although it might not have constituted the main stimulus or motivation for the initial moves toward a more even-handed policy in the fall of 1973, it was a basic prerequisite for these moves and for any subsequent evolution of policy. In particular, it tended to remove some of the constraints on government actions arising from the perceptions and attitudes of members of the media, community leaders, and the public at large. (Noble 1985b, 110)

In 1979 a poll found that 69 percent of Canadians felt the government should not move its embassy to Jerusalem because some Arab nations had threatened retaliation (Gallup 1981, 89).

Interest Groups

Interest groups have not played much of a role in Canadian foreign policy (Stanislawski 1981, 12, 102). Legislatures are traditionally better targets of such activity than are bureaucrats, but the Canadian Parliament has a much less influential role in foreign policy than, say, the U.S. Congress. This notwithstanding, the Canadian Jewish community operated as an effective pressure group, even though it was relatively small (250,000–300,000 [Canada 1985, 98; Stanislawski 1983, 201]) and its influence diminished over time.

The lead lobbying organization was the Canada-Israel Committee (CIC), which was founded after the 1967 war by the Canadian Jewish Congress, the Canadian Zionist Federation, and B'nai B'rith "to represent them to the government and the non-Jewish public of Canada on all issues of Canada-Israel public affairs" (Stanislawski 1981, 411). The committee could boast of small, highly professional staffs in Toronto, Montreal, and Ottawa; 10,000 members of Canadian Zionist organizations; 150,000 Jews in Ontario; "immediate access . . . to M.P.s, media, and influential citizens across Canada"; and the ability "to reach virtually every practicing Jew in Canada" (Dewitt and Kirton 1983, 186; Bones 1985, 153). The CIC interacted directly with parliamentarians and bureaucrats, and its ability to mobilize significant numbers of people for voting and protests, especially in Toronto, gave it direct political clout.

The CIC made efficient use of its resources by focusing on key decision makers and the media, spending much less time with backbench members of Parliament or the general public (Bones 1985, 167). A DEA official, speaking at a 1982 conference on pressure groups, said the CIC was the most influential lobby in shaping Canadian foreign policy and was a model of professionalism and skill (Lyon 1982, 4–5). This kind of reputation can sometimes be transformed into a political resource. Conversely, the CIC was often seen as too aggressive (Taras 1983, 4–6).

The Jewish community had one other unusual electoral resource. During the middle 1970s the two most heavily Jewish parliamentary districts in Canada were represented by Prime Minister Trudeau and Minister of External Affairs Sharp. As we will see, however, it was not easy to translate this situation into effective leverage, particularly over Trudeau.

The CIC's successes, however, often set the stage for subsequent defeats. The U.N. conference issue of 1975 was the committee's first public test (Stanislawski 1981, 411). It was clearly successful in that the conference was not held in Canada. Its success caused Trudeau to alter the decision-making process, however, in part to reduce the proclivity of some of his cabinet ministers to allow their desire to please Jewish voters to influence policy (personal interview, March 1983).

The CIC's major public fight concerned antiboycott legislation. Stanislawski argues that the committee's battle was a new tactic in Canadian foreign policy:

> Canadian Jewish organizations determined to embark upon a large-scale, long-term lobbying effort on this issue, and, in an unprecedented manner, they began to mobilize other Canadian support for their position. . . . The Arab boycott case enlisted a wide array of Canadian groups and individuals in the Canadian foreign decision-making process, breaking through the wall of public apathy and the closed system of policy-making that had characterized previous foreign-policy issues. (1981, 28)

Of course, the fact that the Jewish groups had to go public in this manner indicates that they had been unsuccessful in earlier, more subtle attempts at influence because of the opposition of government officials. They succeeded in getting an antiboycott statute enacted in Ontario and almost got one at the federal level as well. Opposition groups were able to block the antiboycott law only by using the furor over the Jerusalem embassy affair. In any event, antiboycott legislation probably would have been easier to enact than to enforce.

The CIC was divided on the embassy issue. Some members argued that it should be supported because Begin had requested it, but some Jewish businessmen actually urged Clark to retract his decision (Takach 1980, 22–28, 99n).

The Arabs never developed much of a domestic pressure group to offset the CIC. There are probably only about 19,000 Canadians of Arab descent, and about 100,000 Canadian Muslims (Dewitt and Kirton 1983, 187), and they are not politically united. The Arabs, for example, include a rather large contingent of Lebanese Christians. The Arab Information Center in Ottawa is run by the Arab League rather than by a domestic organization. Moreover, some of their efforts lacked sophistication. Bones attributes Trudeau's decision to postpone the crime conference in part to statements by the Canadian Arab Federation calling Israel "racialist [*sic*], exclusivist and fascist" and threatening to destroy Canadian-Arab trade (1985, 159).

As for Christian opposition, one influential member of Canada's United church, Rev. A. C. Forrest, editor of the journal *United Church Observer*, focused on Palestinian grievances in the late 1960s and early 1970s. The church spoke with different voices, however: Several former ministers in Parliament were pro-Israel, and the church moderator, Dr. Bruce Macleod, wanted to avoid a battle with the Jews (Taras 1983, 6–9).

The strongest counterbalancing interest group to the Jewish community was clearly Canadian business. Its interest in the issue increased as the Arab states became a more important market after 1973; Canadian-Arab trade increased twenty times between 1970 and 1980 (Ismael 1985, 15). This recent upsurge in interest makes it difficult to determine the real impact of business groups such as the Canadian Export Association and the Canadian Manufacturers' Association. They seem to have played no real role in Palestinian policy or the crime conference and Habitat issues. They were important, however, in the antiboycott legislation and in the Jerusalem embassy debate.

Ostensibly, business opposed the antiboycott legislation as government interference, but the central issue clearly was Canadian competitiveness with corporations from more permissive countries (presumably not the U.S., which enacted some fairly stringent laws). The corporations exerting pressure were almost entirely Canadian—American subsidiaries stayed out of the controversy for the most part (Stanislawski 1981, 430–431)—and they clearly influenced government policy. Not only did the Department of External Affairs accept their arguments, but also the economic aspects of the issue, combined with Trudeau's preference for collegial decision making, served to involve government economic agencies in the decision. Business was thus very strong as long as the decision was being made within the government.

When the decision was moved into the public arena, however, business groups had a much more difficult time with the antiboycott issue. They were quite unable to prevent, for example, the adoption of an antiboycott law in Ontario (Dewitt and Kirton 1983, 188). Stanislawski attributes this to the lack of strong bureaucratic opposition to such legislation; such opposition at DEA and ITC worked at the national level (1981, 441–442). The personal role of Trudeau may have been the decisive element, although Stanislawski himself does not agree (letter to the author, August 1983). In any event, business by itself could not carry the day in Ontario, and the antiboycott position seems to have been popular with the public.

The Jerusalem embassy affair evoked the greatest involvement of business groups in the Canadian debate over Middle Eastern policy. As one of Clark's advisors said: "They were coming through the windows" (Takach 1980, 94). This was an easy victory, however, because there was no real opposition. It is therefore difficult to draw many conclusions from it.

Governmental Influences on Canadian Policy

Parliament and Parties

Canadian foreign policy on the Middle East is an unusual issue because Parliament competes with the Department of External Affairs. In general, DEA personnel believe Canada has been too pro-Israel, whereas parliamentarians tend to support Israel. The motivations and depth of this support in Parliament range from a small minority, particularly among the older generation, who see support of Israel as a central moral issue, to the majority for whom the Middle East is a marginal issue with no electoral risk (Taras 1983, 1–2, 9–12).

Parliament, preoccupied with the existing minority government and forthcoming elections, did not devote much attention to the October 1973 war and the resulting oil embargo. When the war broke out the House of Commons was not in session, which may have been just as well, because the Jewish community immediately pressed the government for a statement condemning the Arabs for starting the war. The government was not prepared to issue such a statement, however, and neither of the opposition parties pursued the matter, perhaps because by the time Parliament reconvened, it was clear that the war was going to be very costly and that intense superpower activity would be involved (Taras 1983, 12–21). Such a statement by the Dutch may have been one reason they were embargoed by the Arabs.

Generally, Parliament's role in foreign affairs has not been particularly significant. Committees are crucial, because M.P.s cannot all be expected to keep up with developments in every area. The Standing Committee on External Relations and National Defense (SCEAND) would be a key institution, but it cannot set its own agenda and has only vague and limited powers to obtain witnesses and records (Tucker 1980, 49–52, 150–151). Its hearings on Canadian Middle East policy in 1982–1983 were noteworthy not because of their content so much as because hearings had never been held before on this topic. Similarly, when the Senate Committee on Foreign Affairs issued a report in 1985 on Canada's relations with the Middle East, it noted that such a report had never before been prepared (Canada 1985, 1–2). It is probably true, however, that the general parliamentary consensus in favor of Israel has imposed some limits on government policy (Dewitt and Kirton 1983, 185).

In fact, one aspect of Parliament did exert some influence on Canadian policy toward the Arab-Israeli conflict. Given party discipline, the caucus of the majority party is crucial. During this period, except for a short Conservative interregnum, this has meant the Liberal party caucus. A study of Canadian elites found Liberals significantly more sympathetic to Israel than Conservatives (Benesh 1979, 115–116, 140). We might assume, therefore, that Liberal dominance would strengthen Israel's position. In fact, however, the Liberal caucus is divided into Anglophone and Francophone caucuses. Pierre Trudeau was profoundly concerned with Canadian unity and thus felt it was imperative to be sympathetic to the Francophone caucus of his own party. One 1983 estimate holds that almost all of the Francophone Liberals were pro-Palestinian (Ismael 1985, 21). When Israel invaded Lebanon in 1982, for example, the Francophone Liberal caucus agitated for a policy change toward Israel. In particular, members wanted to alter Resolution 242 to stress Palestinian rights more strongly, call for Palestinian self-determination, and perhaps recognize the PLO. Their initiative was headed off only when the government agreed to reevaluate its Middle East policy (personal interviews, March 1983). In fact, the Canadian government opposed Israel's invasion of Lebanon and supported its subsequent withdrawal, as did a Senate report (Canada 1985, 71–72).

Bureaucratic Organizations

Within the executive branch of the government, the single most important organization vis-à-vis foreign affairs is clearly the Department of External Affairs (DEA). Indeed, until Trudeau came to power in 1968, the DEA was really the only government agency concerned with the Arab-Israeli conflict; the Department of National Defense was only occasionally involved in peacekeeping issues. The DEA's monopoly did not mean that all of its recommendations were accepted by the political leadership, but it did mean it had no bureaucratic rivals.

Trudeau ended the DEA's monopoly for a number of reasons. He distrusted DEA, thinking it was too conservative for the innovative foreign policy he wished to pursue. He wanted to orient Canadian foreign policy more toward Canada's economic interests. Naturally, this meant departments concerned with economics, such as Finance and particularly Industry, Trade, and Commerce, were now seen as relevant to foreign policy. Trudeau preferred a multiple-advocacy decision-making system, and so he encouraged other departments to become involved. Moreover, as foreign policy became more politicized, the decisions moved to the cabinet level. At the same time, Trudeau greatly augmented the role of the Privy Council office and the prime minister's office. Thus, the effective power of the Department of External Affairs was severely curtailed (Tucker 1980, 61–68).

Trudeau's reorganization notwithstanding, DEA remains the most influential department in most decisions relating to the Arab-Israeli conflict. It is the lead agency for any issue involving the United Nations (Dewitt and Kirton 1982, 83); thus, it was a prominent bureaucratic combatant over PLO involvement in U.N. conferences in Canada, as well as over the U.N. votes that determine so much of Canada's Palestinian policy. Also, Trudeau was forced to allow his departments a good deal more independence during the minority government from 1972 to 1974 (C. Campbell 1980, 87–88); this presumably strengthened DEA's position during the period of the oil weapon and the policy shift on the Palestinian issue. In contrast, DEA probably had less influence on the Arab boycott issue.

As noted earlier, the Department of External Affairs has pressured the government to move toward a less pro-Israeli policy. We have unusually good data on the department's basic assumptions because of a 1975 study of the Canadian foreign policymaking elite, which asked several hundred individuals to rank twenty international actors as to whether they had a positive impact on the international system. Israel ranked below Egypt, OPEC, and the Soviet Union, although still above the PLO, which was at the bottom (Byers and Leyton-Brown 1977, 625, 628–629).

It is not altogether clear why DEA encourages policy movement away from Israel, but the following explanations should be considered. We have already noted Steven Spiegel's argument that foreign offices in general tend to be pro-Arab because they regard the Arab-Israeli dispute as a regional problem, and the regional predominance of Arabs makes a pro-Israeli position look unreasonable (1985, 5–6). We can also trace some considerations that are more particular to DEA. The Middle East division is well regarded and has attracted competent people, despite Canada's lack of historic involvement in the area. Depending on its financial condition, the Canadian government has periodically supported DEA people who were willing to learn Arabic, for example, and has sometimes sent them to the Middle East Center for Arab Studies in Lebanon (now in Jordan), run by the British Foreign and Commonwealth Office. The Middle East division is regarded as a good, although difficult, posting within DEA (personal interviews, March 1983). Naturally, it has many more people in Arab countries than in Israel.

Another feature of the Middle East division (noted by a well-known academic observer) is that DEA has essentially two tracks, Anglophone and Francophone. The Middle East posts of North Africa are Francophone (which is why they were originally established), and the Saudi embassy also has had an unusually high number of Francophone officers. As a result, the Middle East division is disproportionately Francophone, and Francophone officers may be less pro-Israel than Anglophones. Another observer discounted this, arguing that although many Quebecois are indeed sympathetic to the Palestinians, they do not, as a rule, work for the Department of External Affairs, that perhaps 40 percent of the DEA Francophones are not from Quebec, and that particularly the older Quebecois tend to be deeply religious with classical educations and are thus sympathetic to Israel for religious and historical reasons (personal interviews, March 1983). In any event, we do not need the Francophone argument in order to explain the attitudes of the Middle East division—the same phenomenon holds true for each country in this study except Japan.

The DEA is influenced by more than its Middle East division, however. The importance of the United Nations for Canadian foreign policy means that U.N. opinion is brought back to Ottawa by DEA; as a former DEA senior official observed: "A lot of the professional thinking [about the Middle East] gets done around the U.N." (personal interview, March

1983). Similarly, European policy changes are often transmitted by DEA to the Canadian government. It is the job of any foreign ministry to communicate such external pressures to its own government; it is not surprising, therefore, that DEA tends to favor such policy changes.

Some DEA personnel may also be reacting against what they see as the politicization of foreign policy by the Canadian Jewish community. Many feel, for example, that Arab boycott cases could be handled through quiet diplomacy, given that Arab governments do not care all that much about the issue, although they cannot afford to say so publicly. Canadians pushing antiboycott legislation are thus seen by DEA as interested in public confrontation instead of actual results. We must also recall that antiboycott political pressure was a new phenomenon and, as such, presented DEA (which like most foreign ministries has little if any domestic constituency) with a novel and unwelcome situation. There was also some feeling that conservative Arab governments should be compensated for cooperating with the West in order to avoid internal political threats.

The DEA does not always recommend pro-Arab policies. Interestingly, through all this period it has retained excellent relations with Israel, with the Canadian ambassador often being extremely close to Israeli decision makers (personal interview, March 1983). Moreover, DEA recommended strong support for the Camp David peace process when most European states were refusing to do so.

When the oil weapon was used in 1973, DEA found itself in a somewhat embarrassing position because it had no embassies in any of the Gulf States, including Saudi Arabia.

> When questioned before SCEAND on May 15 [1973] about the need to have representation in oil-producing countries Mr. Sharp downplayed the necessity and noted that "we now have representation in what are considered to be the most important of the Middle East countries from the point of view of Canada and Canadian interests, for instance in Cairo, in Tel Aviv, in Beirut. These are key countries from the Canadian point of view." (Byers 1974, 256)

A number of authors have suggested that this lack of diplomatic representation made it difficult for Canada to know whether or not it was embargoed and to communicate persuasively with the Arabs (Byers 1974, 255–256; Tucker 1980, 71–72; Ismael 1976, 262; Dewitt and Kirton 1983, 179). Certainly the embargo accelerated the planned increase of embassies in the region (Stein 1976b, 287); in December 1973 it was announced that an embassy would be built in Saudi Arabia, and twelve years later Canada had resident ambassadors in ten countries in the Middle East (Canada 1985, 87). The Canadians did, however, have an embassy in Algeria in 1973, a critical country in the OAPEC decision-making process. Moreover, early in 1973 a Canadian ambassador to Saudi Arabia was officially accredited. Although resident in Beirut, he spent a good deal of time in Saudi Arabia during the embargo, was well received, and seems to have had some success in presenting the Canadian case. Canadian diplomats apparently predicted a major oil price rise but not the embargo itself, presumably because they did not foresee the Arab-Israeli war that triggered it (personal interviews, March 1983 and May 1983). Considering that Israeli and American intelligence did not foresee it either, this is not a bad record.

Although other agencies are involved in Middle East policy on particular decisions (Immigration in the crime conference and Habitat issues, Energy and Natural Resources on oil supplies, and Finance with a general concern for Canada's macroeconomic health), two others have particularly significant roles and should be discussed briefly.

The Department of National Defense (DND) is involved in Middle East issues, but only in connection with peacekeeping questions. By 1973, however, the department decided to greatly curtail Canada's peacekeeping functions. Like DEA, it was not well regarded by Trudeau and had suffered budget cuts; thus, its resources were stretched very thin. In addition, the department's experience with peacekeeping had not been encouraging. Canada's ejection from Egypt in 1967 still rankled, and its subsequent difficulties with a peacekeeping mission in Vietnam, as well as problems in Cyprus, did not help (Canada 1985, 81–82). Therefore, DND was not sympathetic to taking on another such assignment. One observer describes it as having been "hysterical" at the thought (personal interview, March 1983). It is interesting that, despite DND's lead position on peacekeeping issues, it had to accept the assignment. The unhappiness of DND may have contributed to the severe restraints on the Canadian peacekeeping commitment, however, and the department was clearly important in limiting later Canadian involvement in the U.N. peacekeeping force in Lebanon (Dewitt and Kirton 1983, 189–190).

The second government agency under discussion is Industry, Trade and

Commerce (ITC), a relatively new department with no particular interest in the Middle East until after 1973, when oil money suddenly made the Arabs look like good potential customers. The department's interest seems to have been quite simple and direct: It wanted to encourage trade. ITC joined the government committee on arms exports in 1978, and its influence is reported to have increased Canadian willingness to export weapons to sensitive areas (Dewitt and Kirton 1983, 190). The issue was not just exports, although those were always useful; it was the nature of those exports that was particularly attractive. The Canadians had no problem exporting raw materials, resources, and agricultural products, and some of those went to the Middle East. But ITC was particularly attracted by the possibility of exporting industrial goods, including high technology (Lyon 1985, 27; Kubursi 1985, 46, 57, 59–63). ITC was also a major player in the Arab boycott controversy because it would have to have administered any new laws or regulations. Because its personnel regarded such a boycott as contrary to their major mission of promoting Canadian exports, it is not surprising that they hindered rather than helped the push for antiboycott legislation.

The ITC's export drive was not a great success in any event. Several observers suggest that, in fact, Canadians have not done as well in the Arab market as hoped because their exports were mostly services and infrastructure rather than industrial goods (Canada 1985, 91–92; personal interviews, March 1983). The foreign trade component of ITC was recently integrated into External Affairs, but it is unclear if this change will have much impact on Middle East affairs because the two departments do not seem to differ substantively (Canada 1985, 94–95).

Overall, then, governmental factors seem to pressure Ottawa to move in a more pro-Arab direction. Parliament is an exception in this regard but appears hopelessly outmatched. If governmental factors favor a more pro-Arab stance, however, why has Canadian policy been slow to move in this direction? The answer may be in a hidden strength of Parliament, the fact that cabinet ministers are drawn from its ranks, combined with the shift of decision making from individual departments to the cabinet level for reasons already discussed. When the final decisions are made by cabinet ministers, who are themselves elected and whose background is in electoral politics, public opinion and pressure groups wielding electoral power are likely to have significant influence.

Canada's position on the Arab-Israeli issue has also been influenced by its federal system. The Province of Ontario played a direct role in the crime conference controversy by its opposition to PLO participation (personal letter to the author, August 1983) and in the boycott controversy by passing its own antiboycott law. Although Quebec played no such direct role, its parliamentary representatives in the Francophone caucus significantly influenced Trudeau's policy.

The Influence of Individuals on Canadian Policy

In an analysis of Canadian foreign policy toward the Arab-Israeli conflict over fifteen years, Pierre Trudeau clearly emerges as the central figure. That is not to say that others were not important at particular times, only that no one else seems to have been able to alter foreign policy because of his or her individual predispositions.

> Perhaps the dominant element of the policymaking process during this period was the increasing tendency of the prime minister to intervene in support of the economic priorities emerging at the official level, and override cabinet's un- structured processes, focused on domestic political considerations, in accor- dance with his personal philosophic values and substantive foreign policy ap- proach. (Dewitt and Kirton 1982, 81)

There are others. Within the Department of External Affairs, for ex- ample, Elizabeth MacCallum was for many years the department's only Middle East expert. Although her primary influence was felt during and after World War II (Kay 1978, 67–68, 168), it did not entirely end with her retirement (personal interview, March 1983). Another influential indi- vidual was Jack Horner, a right-wing Conservative who defected to the Liberal party, whom Trudeau appointed minister of industry, trade and commerce in 1977; "Horner's pro-corporatist inclinations . . . were to play a key role in the boycott controversy of the following eighteen months" (Stanislawski 1981, 239).

The various ministers of external affairs also were obviously impor- tant individuals, and knowledgeable observers suggest, for example, that Allen MacEachen was significantly less pro-Israel than Don Jamieson and Mitchell Sharp (Noble 1985a, 98; 1985b, 117, 129–131). There is no evidence, however, that Canadian foreign policy toward the Arab-Israeli conflict varied from minister to minister—the key shift, under Sharp, is in precisely the opposite direction one might have expected, Jamieson's impact seems to have been insignificant (although some observers disagree [personal interview, March 1983]), and MacEachen was not in office long enough to make much of a difference. Tucker puts the issue in historical context by saying they "could not and did not emulate the status held by Lester Pearson in the 1950s" (1980, 36).

Pierre Trudeau is one of the most fascinating personalities among recent political leaders, and his brilliance and complexity defy simple analysis. Nonetheless, a number of important factors seem clear: "The Cartesian Trudeau, schooled by Jesuits and trained at law," was an extremely intelligent leader who valued rationality and equilibrium highly (Eayrs 1983, 102). As prime minister, Trudeau was comfortable with multiple advocacy and long-range planning; he disliked emotionalism. His primary concern was to maintain Canadian unity by alleviating the English-French split through strong centralization. He thus disliked nationalism and interest groups.

Foreign policy was not particularly important to Trudeau for its own sake; he saw it as one of many avenues to achieve larger goals. He came into office wanting to alter foreign policy because he felt it had been dominated by "internationalism," a devotion to solving world problems with Canada as the "helpful fixer." He wanted policy to serve Canadian national interests, particularly national unity. In practice, this meant using foreign policy to encourage international trade to improve Canadian macroeconomic health, which he felt was vital for unity. He was also very concerned with the question of effectiveness: Would Canadian involvement in fact make a difference (von Riekhoff 1978, 267–270, 276; Kirton 1978, 291–293)?

Like many complex thinkers, Trudeau possessed ideas that sometimes seemed contradictory. Thus, he curbed DEA's role in foreign affairs because it was committed to internationalism while he also expected it to coordinate the activities of other departments that he had encouraged to enter foreign affairs (Tucker 1980, 62–63; Kirton 1978, 292–293). The result was that DEA's effective power varied a good deal on different issues, depending on its expertise and commitment (Kirton 1978, 293–296).

At the outset, Trudeau greatly diminished the importance of the Middle East within Canadian foreign policy. Immediately after taking office in

1968, he launched a major foreign policy review. A "laborious process of interdepartmental consultation which sought to develop the maximum degree of consensus among all concerned" produced several reports on different areas of Canadian foreign policy; the results were fairly useful for foreign policy purposes because the operative departments had been involved in their preparation (Kirton 1978, 301). The Middle East was not included in this process, however, and its omission was deliberate (personal interview, March 1983). This was not surprising. The Middle East had been the symbol of Canadian internationalist policy, even though Canada seemed to have no particular national interest there. Peacekeeping justified the military spending that Trudeau wanted to cut (Byers 1978, 321–32). Worst of all, Canada had "no cards to play," no way of significantly influencing events. The Middle East was therefore not on Trudeau's list of priority areas for foreign policy, and so Canada essentially did nothing in the region between 1968 and 1973.

The October 1973 war put the Middle East back on the foreign policy agenda by demonstrating that Canadian interests were involved, initially with oil supply and the Canadian image of neutrality, later with the possibility of significant exports. The problem of effectiveness has remained—perhaps one reason Canada keeps a relatively low profile on this issue. There is no indication that Trudeau was personally involved in the 1973 policy change on the Palestinians, but he presumably approved it. He was also not involved in the crime conference issue until it reached the cabinet level, but he was reportedly appalled at the sloppy decision process, and he structured the later decision for the U.N. housing conference much more carefully, "not because he loves the Palestinians, but because he believes in rational policymaking" (personal interview, March 1983).

Trudeau's major impact was to totally restructure the way in which the Canadian government approached Middle East issues. Regarding the particular decisions under discussion in this study, the prime minister's greatest impact was undoubtedly on the Arab oil-boycott legislation. Here, as elsewhere, he wanted to use foreign policy for economic gain, but he also developed an almost visceral dislike for what he regarded as excessive pressure from the Canadian Jewish community (Noble 1985a, 97; Dewitt and Kirton 1983, 191). Trudeau's aversion to such pressure groups probably grew out of the value he placed on equilibrium and rationality; interest groups were therefore seen to interfere with rational decision making. In addition, Trudeau personally disliked outside pressure. The fact that his own parliamentary constituency was heavily Jewish may also have made him less sympathetic, for fear of appearing susceptible to pressure; it was common for him to address Jewish audiences and not mention Israel at all (personal interview, March 1983). The Arab boycott was a hot political issue at a time when Trudeau's government was in the process of losing an election, but, aside from a few rhetorical gestures, he refused to succumb to electoral pressures. Had another individual been prime minister at this time, antiboycott legislation might have passed (personal interview, March 1983), as it did in Ontario.

One other individual had a significant impact on at least one Middle East issue. Joe Clark's promise to move the Canadian embassy to Jerusalem, in retrospect, doomed the antiboycott legislation, not to mention his own chances of remaining prime minister (Leyton-Brown 1981, 229), and this seems to have been largely a personal decision (Dewitt and Kirton 1983, 191–192); another candidate, Robert Stanfield, for example, might not have done this. But it is not clear that an antiboycott law by itself would have had much impact on other Canadian policies toward the Arab-Israeli conflict. Nonetheless, it certainly confirms Takach's observation that, from W. L. Mackenzie King to Clark and Trudeau, the prime minister has been the dominant force in Canadian Middle East policy (1980, 105).

Impact of the Oil Weapon on Canadian Middle East Policy

Why did Canada's policy toward the Middle East change in 1973, just after the oil weapon was used? Was this change a rational reaction to a threatened supply shortage, or did it take place for other reasons? We cannot dismiss out of hand the explanation of a threatened supply shortage. The shift occurred so quickly after the embargo that it is hard not to suspect some relationship. Furthermore, the degree to which Canada's oil supply was reduced and, more important, whether its renewal hinged on political concessions remain impossible to determine with confidence.

Nonetheless, I believe the oil weapon did not change Canada's policy on the Middle East for a number of reasons. First, Canada was not particularly vulnerable to cutoffs of Middle Eastern oil. More important, perhaps, its government does not seem to have perceived such a vulnerability. There is no evidence of any specific set of demands issued by the Arab governments. A former cabinet minister has volunteered that the policy shift was not brought about by the oil weapon (personal interview, March 1983). Canada was more vulnerable to trade bans by oil-rich Arab countries. But, in November 1973, it was much too early to foresee the Arabs' economic potential and carry through a policy change in anticipation. Moreover, if this were part of a coordinated policy to appease the Arabs, presumably the embarrassment caused by the crime conference affair would have been avoided by the government.

It thus seems more likely that Canada's shift on Palestinian policy was linked only indirectly to the embargo. Presumably, such a change would have been one option suggested earlier by the Department of External Affairs. The decision to implement it might well have been triggered by the realization, as a result of the embargo, that Canada's image of neutrality was threatened. The government seems to have believed this image was important for its ability to function well at the United Nations, which in turn was important to Canadian foreign policy. This reasoning remains speculative, however.

Government policies regarding PLO participation in the U.N. crime and housing conferences seem to have come from different sources. The furor over Palestinian participation in a U.N. conference on crime apparently caught the government off guard, which is perhaps not surprising because it was the Canadian Jewish community's first open demonstration of its potential political clout. Trudeau seems to have reacted by structuring the discussion for Habitat more rigidly, successfully reducing the political element in that decision. Involvement of the PLO in discussions of housing, the topic of Habitat, was also less controversial than its involvement in discussions of crime.

The antiboycott case is somewhat different; it seems to be a case where bureaucracies, particularly DEA and ITC, allied with business interests to oppose the idea. Moreover, it occurred long enough after the oil crisis so that Canadians had a sense of a large potential export market in the Middle East that might be threatened by such a law. Given its popularity with the parties, the media, and Parliament, however, the law would probably have been enacted anyway except for Trudeau's opposition, which seems to

have stemmed from personal considerations rather than from the very nature of his position. Similarly, it might have been enacted under Clark had
it not been for the Jerusalem embassy imbroglio.

In sum, then, each aspect of Canadian policy toward the Arab-Israeli
conflict contains a somewhat different balance of forces. The common patterns are clear, however, with bureaucracies in Ottawa pushing the government toward a less pro-Israeli policy, resisted by politicians and pressure
groups. The result achieves something very close to an equilibrium, with
personalities and particular events making the difference in individual
situations. The process produces a pattern of policy movement similar to
other countries in the OECD, despite the fact that Canada is much less
dependent on Arab oil than the other states. The supply-shortage theory
holds that Canadian policy would have shifted *less* than the policies of its
more vulnerable allies, when, in fact, it shifted *more*. This once again suggests that, whatever the effect of the oil weapon on Canadian policy toward
the Middle East, the supply-shortage theory does not explain it very well.

5

Japan: The "Perfect Target"

Japan was the OECD country most dependent on Middle Eastern oil in 1973, as it is today. According to the supply theory of economic sanctions, Japan should have been the most pliant of the target states examined, particularly because it has none of the other states' historic and ethnic ties to Israel and should therefore be more responsive to simple economic pressures. As such, Japan is a critical case in evaluating the impact of the oil weapon.

Historical Background

Japan has practically no historical connection with the Middle East or the Arab-Israeli dispute, although diligent researchers have found evidence of indirect contact with the Arabs as early as the sixth century (Kuroda 1985, 26–28) and evidence that a Jewish-American financier underwrote crucial Japanese naval loans prior to the 1904–1905 Russo-Japanese war, to punish Russia for its treatment of Jews (Shillony 1985, 69). Japan had no colonial links with the region, and it was not involved in the Holocaust, although there is some evidence of a long-term strain of anti-Semitism among intellectuals (Solomon 1978, 21–22; Ikeda 1986). Japan also has no Jewish community of its own. This lack of ties makes it unique among the developed democracies.

Although Japan has no historical connection with the Middle East, it has a long history of dependence on unstable foreign sources for strategic materials, including oil. Indeed, one can argue that World War II in the Pacific was caused by Japan's desire to conquer the oil fields of Southeast Asia in the face of resistance by the Allies, and the United States' oil embargo was an important reason for Japan's attack on Pearl Harbor. Japan's history has been conditioned by an almost total lack of resources, and, unlike Britain and Holland, the magnitude of the problem has been driven home by catastrophic military defeat. Takehiro Sagami,* former vice-minister for international affairs at the Ministry of Finance, said in 1980:

> Our first oil shock was in 1940, not in 1973. At that time, America stopped shipments of heavy oil because of what we had done in Manchuria and Shanghai. The U.S. embargo turned around the Navy that had opposed war with America, and forced it to move for Indonesia. From that lesson, we learned that an oil cutoff may have unforeseen political consequences . . . and may mean the difference between life and death. (Yoshitsu 1984, 42)

Valerie Yorke notes that imported oil supplied 37 percent of Japan's energy needs in 1960, 71 percent in 1970, and 76 percent by 1978 (1983, 51). The figure had dropped to 66 percent by 1980, but it was not expected to fall rapidly; the national goal for 1990 is 49 percent. Japan has had great difficulty establishing a realistic national energy program, despite its vaunted business-government cooperation (Turner 1978b, 107–108; Morse 1981). In 1983 more than 70 percent of Japan's imported oil came from the Gulf (Carvely 1985, 217), but much of this oil came from Iran, which was not a member of OAPEC. Japan is therefore less dependent on *Arab* oil than these figures would suggest (Yorke 1983, 52). Japanese dependence on Arab oil may have actually increased since 1973 as a result of the Iranian revolution and the Iran-Iraq war (Caldwell 1981a, 81–82).

---

*In order to be consistent with other chapters, personal names are in Western style, with given name first and surname second.

Japan thus had a unique dual dependency

on Middle East oil for a crucial margin (80 percent) of its indispensable oil imports and an even broader dependence on the United States as a supplier of a
variety of essential imports including food, feed, raw materials, and high technology, as Japan's most important and profitable export market, and as the ultimate guarantee of Asian-Pacific peace and the military security of the Japanese
homeland. (Saeki 1976, 258)

As a result of the traumas of World War II and its high level of economic
dependence, Japan has generally stressed economic relations in its foreign
policy and has tried to stay out of political controversies, as Okita says,
"being friendly with everybody, or at least not making serious enemies
anywhere" (1974, 723). Thus, it is perhaps not surprising that Japan is the
only Asian country with full diplomatic relations with Israel while also
maintaining stronger relations with the PLO than almost any Western
country.

Early in the 1970s a number of Japanese business and political leaders
advocated "resource diplomacy" as a result of a series of "shocks" to U.S.-
Japanese relations and the increasing power of commodity-producer states
in the international oil markets. Although this was "more a compelling
perspective than an integrated policy framework," its central assumptions
were that energy vulnerability was one of Japan's most important problems and that therefore Japan should pursue an active foreign policy to
construct multiple economic links to the oil producers (Caldwell 1981b,
159–162, 170–172). Its advocates expressed concern about delays of
proposed Japanese projects in the Middle East and tried to facilitate cooperation between business and government to expedite the process (Caldwell 1981b, 67–70). This process stressed economic rather than political
issues and thus meshed with Japanese foreign policy in other parts of the
world as well as taking into account the peculiarly strong cross-pressures
on Japan in the Middle East.

Many analysts argue that Japan had no policy on the Arab-Israeli dispute during this time, but this seems an oversimplification (personal interviews, Henry Nau and others, June 1984). A more plausible interpretation
is that it had a policy, based on the "twin pillars" of two U.N. resolutions,
but no diplomacy to implement it (Yoshitsu 1984, 1). Japan had helped
draft Resolution 242 (chairing the Security Council at that time) and re-

mained strongly committed to it (Baker 1978, 28); this resolution "set the tone for the limited role Japan took" (personal interview, Martha Caldwell Harris, June 1984).

Less well known was a 1970 General Assembly resolution (#2628) that asserted: "Respect for the rights of the Palestinians is an indispensable element in the establishment of a just and lasting peace in the Middle East." Yoshitsu correctly notes that Japan voted with the U.N. majority on Resolution 2628 but overlooks the fact that the only other industrial countries that did so were France, Greece, and Spain (1984, 1; Djonovich 1976, 221, 52). For a country that supposedly refused to take the lead in foreign policy, this vote was rather daring. On May 23, 1971, during a visit by King Faisal, the Japanese government referred to the "lawful rights" of the people of Palestine. In December of the same year Japan voted in favor of U.N. Resolution 2792D, which asserted the right of the people of Palestine to self-determination; it supported a similar resolution a year later (Carvely 1985, 42–43). (One researcher notes that all this seems to have been overlooked by the Japanese; he recalls Japanese government officials telling him after the oil crisis that their government had not taken a public position on the Palestinian issue [personal interview, June 1984].) Yorke asserts that Japan "was well in advance of other industrial states" in supporting self-determination for the Palestinians (1983, 68). Japan was so quiet about the policy, however, that it failed to reap any political benefits.

> Senior officials had constructed these twin pillars of Persian Gulf policy, believing them to be the minimal requirements for peace in the region. Nevertheless, they had refrained from fusing the two elements into one comprehensive statement. . . . Moreover, Tokyo planners had informed only one Persian Gulf leader, King Faisal of Saudi Arabia, of their views on Israeli withdrawal and the Palestinian issue. As a result, "the Arab countries were largely unaware of [Japan's] recognition of the legitimate right of Palestinian self-determination." (Yoshitsu 1984, 1, quoting from his interview with "a high-ranking Japanese official")

Another indicator of Japan's growing concern was the radical increase in Japanese contributions to the United Nations Relief and Works Agency for Palestine Refugees in the Near East. From 1957 to 1965 the Japanese donated an average of only $9,100 per year. In 1966 they gave $30,000; by 1972 the figure was more than $760,000, and from 1970 to 1973 Japan gave more than Saudi Arabia. Even though general Japanese multilateral aid was increasing at this time, this suggests a special concern for the Middle East and particularly the Palestinian problem (Carvely 1985, 41–42). Harold Solomon asserts that in the summer of 1973 the Foreign Ministry called a meeting of Japanese ambassadors posted in Arab states to discuss adopting a more pro-Arab public position. Apparently the ambassadors said it was not necessary (letter to the author, September 1986).

The Arab-Israeli conflict directly intruded into Japanese affairs in 1972 when members of the Japanese Red Army killed a number of people in Israel's Lod Airport. The government of Japan sent a mission to Israel to apologize for this act and then sent a second mission to the Arab countries to apologize for apologizing to Israel (Kuroda 1985, 28).

The Japanese government was not particularly worried about oil-supply cutoffs from the Arabs before the 1973 war; it seems to have relied on the major oil companies and its own diplomatic gestures. It was more concerned about its dependence on the major international oil companies and wanted to reduce the price of imported oil (Nau 1980b, 5; Baker 1978, 2). The Ministry of International Trade and Industry (MITI) had, by control of licenses and foreign exchange, encouraged domestic refining companies to compete with the majors, presumably as a first step in reducing this dependency. At one level the policy had been successful; competition kept domestic prices low, and MITI was able to maintain roughly a fifty-fifty balance between domestic and foreign firms (personal interview, Martha Caldwell Harris, June 1974). The resulting low profits for the Japanese companies made it difficult, however, for them to invest in more "upstream" operations in areas such as the Middle East; thus, Japan's dependence on the major international oil companies remained (Tsurumi 1976, 114–118; Caldwell 1981b, 81–149). Indeed, the Japanese felt they were allied with the producer governments against the oil companies (Baker 1978, 3–8, 20–21). Apparently the producers did not see it that way.

The Oil Crisis

Given Japan's fairly strong support for the Palestinians and its general non-involvement in the area, its expectation of being left out of the oil crisis is understandable, and at one point appeared on the way to fulfillment. On November 14 the Libyan ambassador to Japan called on MITI Minister Yasuhiro Nakasone to request Japanese economic cooperation with Libya.

Supposedly, he even expressed willingness to consider joint Japanese-Libyan oil projects. The *Oriental Economist* said at the time: "The fact that Libya made this official request for Japanese assistance even as the other OAPEC nations were applying themselves to 'disciplining' Japan can only be termed unusual" ("Using the Oil Crisis" 1973, 9).

This odd behavior interrupted a series of events that began on October 19—after the initial production cuts had been announced—when several Arab ambassadors called on Foreign Minister Masayoshi Ohira, "requesting support for the Arab position in the Middle East war." Ohira replied that Japan would continue to uphold Resolution 242 and that he hoped the war would soon be settled; there is a dispute over whether or not Ohira specifically supported Israeli withdrawal from the 1967 territories (Yoshitsu 1984, 3; Baker 1978, 29; Wu 1977, 1). If the Japanese position was diplomatically vague, the Arabs made no commitments either. Reportedly, they said Japan would "probably not be bothered by a cut in oil supply," although some said Japan should further clarify its position vis-à-vis the conflict (Baker 1978, 29).

Japan's policy was unsuccessful, however. On October 24 the Arabs demanded that the Arabian Oil Company (the major Japanese oil producer in the Middle East) reduce production by 10 percent, twice the normal cut imposed on companies (Wu 1977, 1–2). At the same time BP, Gulf, and Exxon all announced they would cut their shipments to Japan significantly to even out the shortage worldwide (Juster 1976, 52–53; Yoshitsu 1984, 2). As Harold Solomon points out, the oil company cuts actually signaled that Japan's political shifts would not result in more oil (letter to the author, September 1986), but the Japanese do not seem to have noticed this at the time. Apparently, these developments caught the Japanese by surprise.

On October 26 the Saudi ambassador returned to the Foreign Ministry and met with Vice-Foreign Minister Hogen. The ambassador was handed a note, the text of which is reportedly not publicly available. Two radically different interpretations of the note appear in the literature. Several authors see it as a *note verbale,* a diplomatic term for a note summarizing an earlier oral communication, in this case presumably the October 19 discussion between Ohira and the Arab ambassadors. Wu says it confirmed Ohira's position of October 19, which had included a call for Israel's withdrawal from all the 1967 territories (1977, 1–2); Baker (1978, 29–30) contends that it repeated the earlier conversation—he does not mention the 1967 territories issue—and that, because of the events of October 24, it added a phrase: "We sufficiently understand the various Arab nations' hope for restoration of their own territory" (*Mainichi Shimbun,* October 25, 1973, cited in Baker 1978, 29–30; see also Carvely 1985, 58).

Yoshitsu, however, argues that this document was verbatim identical, except for the title, to the later November 22 statement from Chief Cabinet Secretary Susumu Nikaido (1984, 5).

> The document stipulated (1) that Israeli forces must withdraw from all areas occupied after the October 1967 war, (2) that any settlement of the conflict must respect the legitimate right of Palestinian self-determination, and (3) that the Japanese government would reexamine policy toward Israel should that country ignore [these] principles.
> Reflecting on the document, Hogen emphasized that "nothing had really changed". . . . In a strict sense, the vice-minister was correct. The statement on the fourth Middle East war had incorporated the United Nations resolutions that Japan had supported in 1967 and 1970. Moreover, the final point on policy toward Israel had promised reexamination, not punishment. From a broader perspective, however, much had changed. For the first time, the government had drafted a complete list of views on the Middle East and then had tried to transmit it to officials in the Arab world. (Yoshitsu 1984, 4)

Yoshitsu seems to have had unusual access to senior Foreign Ministry personnel, and he may well be correct. There are several puzzling aspects to his account, however. Japanese advocacy of a complete Israeli withdrawal from the 1967 territories, backed by a threat to reconsider relations with Israel, makes sense only if it were public, or at least communicated to Israel. It would not be useful for the Japanese to tell the Arabs they supported their position but not tell anyone else. If Yoshitsu is correct, the Japanese were reluctant to make the statement in public; at least, it took them a month to do so, missing at least one obvious opportunity in the meantime.

On November 4 the Arab position became even clearer with the declaration that production would be cut by 25 percent, escalating 5 percent per month, and that countries "would have to take a more specifically positive attitude toward the Arab cause" in order to be classified as "friendly" (Baker 1978, 30). If Yoshitsu is correct, the October 26 note had not persuaded the Arabs to reduce the pressure on Japan.

The next turning point was the EC statement in Brussels on November 6. On the same day the chief secretary of the Japanese cabinet, Susumu Nikaido, issued a statement reiterating the government's position, which was very similar to that of the EC. It opposed territorial expansion by force and supported Resolution 242, supported U.S.-Soviet mediation of the dispute, and supported the right of self-determination and equality for the Palestinians as set forth in "the United Nations Resolution" (presumably 2628 and its successors, inasmuch as 242 does not address this issue). This statement did not help Japan, however, as the EC statement had helped Europe. The Arabs were apparently disappointed that Japan had not explicitly condemned Israel (Juster 1976, 55).

If Yoshitsu is correct, the earlier October 26 note, which had not been made public, was much stronger than this statement; it is therefore not surprising that the Arabs were unhappy. The Saudi and Kuwaiti governments declared Japan a "nonfriendly" country the next day (Baker 1978, 31). On November 18 the OAPEC oil ministers announced in Vienna that the EC would be exempted from the 5 percent December cuts; Japan was pointedly not exempted. "The sense of panic which some government officials had experienced by late October now became pervasive" (Juster 1976, 55).

Apparently the Arabs made further demands on Japan that, although never formally acknowledged by either side, were widely known. Japan was reportedly asked to break diplomatic relations with Israel, to sever all economic ties with Israel, and to provide military assistance to the Arabs ("Using the Oil Crisis" 1973, 10; Kelly 1980a, 409; Lenczowski 1976a, 21; Economist Intelligence Group, November 11, 1973, 5; U.S. Congress 1973–1974, 37; Ahari 1986, 125), as well as to pressure the United States to alter its policy toward the Arab-Israeli dispute (Nau 1980b, 16; Carvely 1985, 60).

These demands plunged the Japanese government into an intense internal debate. During the next two weeks, public concern mounted. Henry Kissinger came to Japan and argued against acceding to the Arab demands. Despite this special plea, the Japanese government decided to adopt a position on the Middle East independent of the U.S. government. This has been widely seen as the first independent foreign policy initiative by the Japanese since World War II (Tsurumi 1976, 124; Baker 1978, 46; Yorke 1983, 52–53; Pfaltzgraff 1980, 20).

Japan's new position was announced in another statement by Nikaido on November 22. As noted earlier, although most commentators see the November 22 statement as a turning point, Yoshitsu argues it simply repeated the October 26 *note verbale* with publicity, hoping to signal the requested policy break (1984, 5–6). In any case, the note had three parts. The first asserted that Israel should withdraw from all territories it had occupied during and after the 1967 war. Japan thus went further than the November 6 EC statement, which was unclear as to whether "all" territories were included; it also modified its support of Resolution 242, which, like the EC statement, was deliberately vague on this point. Second, it contended that Palestinian self-determination should be a precondition of any peaceful settlement in the Middle East. Finally, the Japanese government said it would reconsider its policy toward Israel if Israel refused to accept these preconditions (Wu 1977, 3–4). Another part of the same statement announced that Deputy Prime Minister Takeo Miki would travel to the Middle East. Interestingly, Kissinger calls the November 22 statement "similar to Europe's of November 6" (1982, 881).

Japan, however, did not meet any of the four Arab demands. It did not break diplomatic relations with Israel and consistently refused to clarify the vague language about when this might be done (Caldwell 1981b, 215–216). (Lenczowski reports that during the Algiers conference the Arabs issued a demand, presumably at the behest of Saudi Arabia, that Asian states break political rather than diplomatic ties with Israel, which was intended to make it easy for Japan to become a "friendly" power [1976a, 21–22].) Economic ties with Israel were also not severed. Japan used its constitutional prohibition as a pretext for not supplying the Arabs with sophisticated weapons. Apparently, this substitution of money and personal diplomacy was acceptable.

On November 28 Japan was exempted from the scheduled December 5 percent cut, but the earlier cuts remained in effect, and it was not clear if the November 22 statement would be sufficient (Juster 1977, 305). On December 10, as Miki was preparing to go to the Middle East, the Arabs announced another 5 percent export cut for January for the European Community and Japan, which was not a good sign. But on December 25 Japan was formally reclassified as "friendly" by OAPEC while Miki was still in the Middle East (Hirasawa 1976, 141), although "the importance of this was reduced by the decision of the Conference to forego the January 5 percent reduction, and to lower overall reductions from 25 percent to 15 percent—most of which would be absorbed by the embargoed nations" (Baker 1978, 32n).

While in the Middle East, Miki had promised $127 million to help rebuild the Suez Canal and $100 million in additional aid to Egypt. Sources differ as to whether this was at his own initiative (Hirasawa 1976, 141) or at the suggestion of Foreign Minister Ohira (Yoshitsu 1984, 7), but the fact that the noted international economist, and later foreign minister, Saburo Okita was included in his entourage suggests it was planned. Miki was followed to the Middle East by MITI Minister Nakasone in January and a number of other Japanese government leaders, each of whom promised more money; one source estimated the total amount of Japanese credits pledged during the crisis at $3.3 billion (*Japan Petroleum Weekly* 9 (4): January 28, 1974, 2, cited in Wu 1977, 5; see also Caldwell 1981a, 228; Yorke 1983, 53).

Japanese Middle East Policy Since the Crisis

In her analysis of U.N. voting patterns on Arab-Israeli issues, Stein (1982, 57–59) distinguishes three different coalitions just after the crisis. The United States opposed acknowledging Palestinian rights, while all of its allies, including Japan, abstained. The United States was joined by most of its European allies in opposing recognition of the PLO; only France, Italy, and Japan abstained. All voted against a resolution condemning Zionism as racism except Japan, which abstained. Thus, with the possible exception of Greece following the November 1973 military coup, Japan remained the most pro-Arab industrial country, as it had been before the oil crisis. This pattern has continued. "The level of agreement between Japan and other members of the alliance is significantly lower than among all other allies."

At the same time, Japan clearly remains aligned with the Western consensus on the Middle East, albeit at the pro-Arab end. For example, Japan abstained on the 1974 General Assembly Palestinian resolutions because they did not reaffirm Resolution 242 and did not explicitly affirm Israel's right to exist (Yoshitsu 1984, 13–14; Carvely [1985, 100–104] gives a detailed analysis of Japanese U.N. voting on Palestinian issues during this period). Thus, in relative terms, Japan continued to occupy, almost precisely, its precrisis position vis-à-vis its Western partners.

In 1975 the Japanese government invited the Palestinian Liberation Organization to begin talks on establishing an office in Tokyo, although it firmly resisted pressure to extend formal diplomatic recognition. After extensive negotiations, the office opened in 1976 (Yoshitsu 1984, 14–22).

Another aspect of Japanese foreign policy was a deliberate shift in trade and investment policy. The Japanese learned from the crisis that they had an enormous stake in the Middle East, while the Middle East had no equivalent stake in Japan. The result was to greatly increase Japanese investment in the region, particularly in Iran (personal interview, Henry Nau, June 1984) and, after the Iran-Iraq war began, in Saudi Arabia (Yoshitsu 1984, 85).

The Iranian revolution and resulting oil crisis in 1979 weakened the Japanese confidence that they could bargain money and technology for oil in a relatively nonpolitical manner (Morse 1981, 41). Iran had been important to Japan because it had not linked access to its oil to the Arab-Israeli issue; this important buffer was destroyed, as Iranian oil exports immediately dropped and, even if they were resumed, the Shi'ite Moslem theocratic regime was likely to follow policies similar to those of the Arab states. As a result, the Japanese government again altered its policy on the Arab-Israeli dispute: "In the tense international climate of 1979 Japan realized that although political gestures toward the Palestinians would not persuade oil-rich states to expand production against their interests there was nevertheless a prospect of preferential treatment to those adopting a more favourable attitude toward the PLO" (Yorke 1983, 60).

In July 1979 Masumi Esaki, minister of MITI, returned from a trip to the Middle East persuaded that Japan needed "a more realistic assessment of its regional interests" (Morse 1982, 262). In response, on August 6 Foreign Minister Sumao Sonoda presented a position paper stating: (1) Any Middle East peace must be comprehensive, with the Egyptian-Israeli treaty a "first step" in this process (an implicit criticism of the Camp David approach); (2) it must be based on resolutions 242 and 338 and involve the recognition of the legitimate rights of the Palestinians, including self-determination, in their own state if they so choose; and (3) the PLO must participate in peace talks. He also condemned Israel's occupation of the

1967 territories and its settlement policy. Thus, nine months before the EC's Venice Declaration, the Japanese had already laid out its essentials. The Japanese diplomatic position was to ask Israel for self-restraint, with a strong verbal commitment to Israeli security behind the pre-1967 borders, and to ask the PLO to accept Resolution 242 in return for self-determination.

Following a Japanese initiative in 1979 and complex negotiations, Japan in 1981 became the first major industrial country to have talks with the PLO at the prime-ministerial level. This initiative occurred for several reasons. First, the PLO had become more moderate since 1974. Second, the Europeans were moving toward Japan's position, and more movement was expected (which came in the Venice Declaration). Third, adverse international opinion was forming in response to Israeli settlements and retaliatory attacks on refugee camps in southern Lebanon. Also, the end of the autonomy talks and upcoming U.S. elections were expected to increase Palestinian frustrations. Japan, in addition, perceived a connection between the Arab-Israeli conflict and the internal stability of the Gulf states. Finally, Yasser Arafat wanted the meetings enough to make concessions. Moreover, Arafat's visit was originally to be balanced by a visit from President Anwar Sadat, a plan that ended with Sadat's assassination on October 6, 1981 (Yorke 1983, 60–62; Yoshitsu 1984, 24–37, 74–75; Caldwell 1981b, 75–78; Kuroda 1983, 29).

In 1982 Japan joined the United States and Israel in opposing a General Assembly resolution condemning Israel's annexation of the Golan Heights as an act of aggression, Japan's first clearly pro-Israeli vote since joining the United Nations in 1956 (Carvely 1985, 162–164). But, like most industrialized nations, Japan reacted negatively to the Israeli invasion of Lebanon in 1982, threatening to support Israel's expulsion from international organizations and to apply sanctions against it (Carvely 1985, 164–168). Indeed, the head of the Foreign Ministry's Middle East–Africa Bureau urged Israel to withdraw from the Golan Heights and explicitly did not make the same request of Syria on the grounds that Syrian troops were in Lebanon by invitation of the Arab League (Kuroda 1983, 23, 26–27). Yet another response was to invite the chief of the PLO's Tokyo office to the emperor's birthday party, a step one observer calls "extraordinary" (W. Campbell 1984, 149). At the same time, though, Japan gave considerable support to the Reagan peace plan in 1982 (Carvely 1985, 168–171). It is not clear if Japan has been leading the Western alliance or been pushed by it, but it continued to be just a little bit in front of its allies on the Arab-Israeli issue.

Dimensions of Change in Japanese Middle East Policy

We can trace two different kinds of policy shifts toward the Arab-Israeli conflict during the oil crisis. Rhetorically, Japan went further than any other industrial country in supporting the Arab position on Israeli withdrawal from the occupied territories and threatened to "reconsider" its policy toward Israel if this did not occur. Economically, the Japanese promised huge amounts of financial assistance to Middle Eastern countries. These acts were widely seen as Japanese capitulation to Arab pressure, readily explained by the supply theory of economic sanctions, given Japan's extensive dependence on Middle Eastern oil.

This is too simple an explanation, however. The supply theory posits that Japan would accede to practically any demand that did not threaten its basic national interests, but, in fact, the Japanese refused most of the Arab demands. Japan did not break diplomatic relations with Israel, even though such a rift would have been a mere rhetorical gesture. It also did not terminate economic relations with Israel, even though Israel was not a particularly important country to Japan. It did not supply weapons to the Arabs. It also does not seem to have tried to pressure the United States to change its Middle East policy (Kissinger 1982, 741–742).

The steps Japan did take turned out to be more shadow than substance. Israel did not withdraw from the occupied territories, but Japan has not noticeably "reconsidered" its policy as a result. Even its economic promises were in many cases not kept (Juster 1976, 81), although Japan has increased the percentage of its development assistance to the Middle East from 2.7 percent in 1975 (Morse 1981, 41f) to about 8 percent by the 1980s, concentrated primarily in Egypt, Sudan, Saudi Arabia, and Turkey (Brooks 1985, 83, 89, 91).

Japan has supported the PLO more strongly rhetorically than any other industrial country, but, in fact, it also did so before the oil crisis. As the industrial nations have shifted toward a more pro-Arab position, Japan has essentially maintained its relative position within the Western alliance.

This seems deliberate. Regarding Japan's policy shift in 1979, Yorke writes: "It also re-established Japan's lead over members of the European Community in terms of policies toward the Palestinians" (1983, 62). Thus, continuity seems more important in Japanese Middle East policy than change, even though the supply theory would have predicted major shifts as a result of the oil weapon.

External Influences on Japanese Policy

Aside from the Arabs, the United States was clearly the major external influence on the Japanese. Japan had been heavily dependent on the United States for trade and security; the United States had also been a model for Japan in both domestic democracy and foreign policy. One author points out that the Japanese-American alliance is the only American alliance in which economic cooperation is explicitly given equal billing with military security (Momoi 1974, 25–26).

By 1973 this relationship was under increasing strain, however. Differences over the Vietnam war, combined with the explosive growth of Japanese industry and trade and the rise of a new generation of Japanese with little memory of World War II, had brought about a shift in attitude. Moreover, a series of "shocks" in the Japanese-American relationship dating from 1971 had made it difficult to argue the simple pro-U.S. position within Japan (Nau 1980b, 10). In addition, the Japanese were becoming more sensitive (without necessarily being more sympathetic) to Third World views, particularly in the United Nations.

The U.S. government quite consistently entreated the Japanese to resist Arab pressures, although the real American interest in Japan's attitude toward the Arab-Israeli conflict is not entirely clear. Peter Mueller and Douglas Ross argue, for example, that it is in the United States' interest that Japan continue to get oil from the Arabs, because otherwise the United States would have to supply it (1975, 189–190). This is particularly compelling for Japan, as opposed to the European countries, because its dependence on Middle Eastern oil is greater and because it is unlikely that any Japanese action will really affect the Arab-Israeli conflict, at least in the short run. Therefore, one could argue that the United States should have encouraged the Japanese to meet any Arab demands that were primarily rhetorical. But during the crisis the U.S. government did not adopt this position. Interestingly, the Japanese themselves seem to have felt under pressure from American Jews, whom many Japanese see as particularly important in U.S.-Japanese trade (Shillony 1985–1986, 17–18).

Kissinger went to Japan to try to persuade the Japanese government not to adopt a pro-Arab position. He arrived on November 14 and met separately with Tanaka, Ohira, and Nakasone, possibly because he was uncertain where the real power lay in Tokyo. The Japanese wanted to know the U.S. plan for a Middle East settlement, how long it would take, and whether the U.S. would ensure the oil supply of the Japanese. Kissinger reportedly said that setting a timetable for settlement would be "unwise"; by implication, the crisis was not likely to end quickly. He was also not optimistic about oil supplies, reportedly saying that it might take six months for supplies to return to normal and that prices would continue to rise. He promised to ask the U.S. oil firms to "cooperate with Japan."

Whatever Kissinger felt about the meeting, the Japanese were disappointed (Caldwell 1981a, 203–204; Carvely 1985, 62–64, 68–69). Nau contends that Kissinger had missed a major opportunity, arguing that the Japanese did not know the oil companies were distributing shortages evenly among consuming countries and that Kissinger did not enlighten the Japanese about this policy (Nau 1980b, 11, 16; see also Juster 1977, 301–302). Interestingly, Kissinger's own account of this mission makes no mention of Japanese requests for guarantees or even information about oil shipments (1982, 741–745; Martha Caldwell Harris, letter to the author, July 1984). In contrast, the State Department did become a major source of information for the Japanese government on both the Arab governments and the policies of the oil companies (Juster 1977, 300).

The United States' apparent reluctance to divulge crucial intelligence probably reinforced a feeling among "some Japanese elites" that the United States was involved in a conspiracy to raise oil prices, hoping the Arabs would drain Japanese reserves and weaken its competitive position. Of course, when the U.S.-owned major oil companies began diverting oil from Japan to make up for shortages elsewhere, the conspiracy theory seemed confirmed (Tsurumi 1976, 124).

It is not surprising, then, that Kissinger's entreaties were unsuccessful. In his memoirs he seems to bear no malice, saying he found the Japanese less objectionable than the Europeans because the former were quite direct in saying what they were doing and why (1982, 741, 744). Almost alone among observers, he attributes remarkable diplomatic skill to the Japanese, citing the press coverage of his visit:

> The English-language press of Tokyo was full of stories that there had been disagreements between me and Japanese leaders (the Japanese-language press was silent on the subject). This was technically true but misstated the essential harmony and good feeling that prevailed. But it served the purpose of conveying in media most likely to be read by diplomats from the Middle East that Japan was champing at the bit. I complained to the Japanese Ambassador. In response, Ohira made a statement to the *Japanese-language* press denying any rift or disagreement. They, of course, saw no reason to print it, not having reported a disagreement in the first place. (1982, 745, emphasis in original)

According to Ohira, however, the understanding was less clear than Kissinger suggests. Ohira says he explained to Kissinger on November 14 that Japan might have to adopt a new Middle East policy, then he asked the Japanese ambassador to the United States to continue negotiations. The State Department finally said that it understood the Japanese position; only after this was the November 22 statement issued (Ohira 1979, 120). Juster disagrees, stating that the United States received no advance warning of the November 22 statement, particularly the threat to reconsider relations with Israel (1977, 304; Binnendijk 1974, 24). Even goodwill does not seem to have guaranteed successful communication between the two governments during this period. Nonetheless, despite these uncertainties, it is at least clear the Japanese knew what the Americans wanted and explicitly rejected it.

Ironically, the United States was also indirectly responsible for Japan's desire to develop contacts with the PLO after the crisis. The Japanese government had been stunned in 1971 when Richard Nixon announced his trip to China just after the Japanese had decided to maintain their policy of nonrecognition at the behest of the United States. The Japanese government took considerable domestic criticism as a result. In 1975 and 1982 senior Japanese officials worried that the United States might extend recognition to the PLO, again leaving them politically vulnerable; they thus felt it expedient to establish contacts with the PLO, although they were not prepared to extend diplomatic recognition (Yoshitsu 1984, 16–17).

Societal Influences on Japanese Policy

Public Opinion

Public opinion in Japan seems to have been particularly important in altering the process by which decisions were made, shifting control from the government bureaucrats and triggering a much more political process in which the influence of outsiders such as business leaders as well as elected politicians increased greatly (Juster 1977, 301, 310). Despite general agreement that public opinion was important, few polls specifically measuring Japanese attitudes on the Arab-Israeli dispute are available in English (Solomon says that there were some in the Japanese-language press but that the questions weren't particularly helpful [letter to the author, September 1986]). Instead, there are a series of polls from the prime minister's research office on the public's level of concern about the Middle East generally (Kuroda 1984, 4; Kuroda 1985, 31).

Of the countries studied here, the Japanese public came the closest to mass panic during the oil crisis. One Japanese observer argues that public opinion was much more inflamed in Japan than in the United States, partly because there had been previous discussions about an energy crisis in America but not in Japan (Miyoshi 1974, 144–147). At any rate, the poll results were startling. Of those polled, 69 percent said they were "greatly worried" (Caldwell 1981a, 196). One observer refers to "mass social hysteria," noting that panic buying and hoarding emptied store shelves of laundry detergent, toilet paper, and even soy sauce (Murakami 1982, 139). Another describes it as "the sort of panic not experienced since World War II" (Morse 1981, 41). A third notes this was the first postwar crisis with an immediate threat that any Japanese government had faced; Japan had not undergone a recession, for example (personal interview, Henry Nau, June 1984).

Nau even argues that the November 22 statement can be seen primarily as a response to Japanese public opinion rather than to external pressure:

Japanese officials knew that Japanese oil supplies depended more on the oil companies than on the oil-producing countries. The oil companies could offset embargoed oil to Japan with oil from non-embargoed sources. Yet the Japanese public, in the panic generated by press and other reports, demanded some action. The November [22] statement appeared decisive, even though Japanese oil supplies were cut back no more or less because of the statement. (1980b, 13)

This argument probably overstates the degree of rational calculation in a Japanese government under enormous strain, but the impact of public opinion on political leaders, particularly Tanaka and Nakasone, was apparently so great that it allowed them to fundamentally change the usual Japanese decision-making processes and produce a significant alteration in Japanese foreign policy. Indeed, Tanaka referred specifically to Japanese public opinion in explaining his actions to Kissinger at their November meeting (Kissinger 1982, 742).

Political scientist Haruhiro Fukui argues that public opinion determines the *process* by which decisions are made in Japanese politics. The *normal* Japanese decision-making process is dominated by bureaucrats: "The various ministries are, to a large extent, self-contained under the direction of the vice-minister. Outside political leaders, including the minister himself, lack a great deal of leverage within the ministry, especially in comparison with American cabinet officers" (Edwin O. Reischauer, quoted by Vogel 1975, xxiv–xxv). Not surprisingly, policy change is quite restricted in such circumstances.

Critical decision making, in contrast, is characterized by intense participation by politicians and actors outside of government (parties, pressure groups, mass media, and aroused citizens), temporarily displacing officials from control of the decision-making process. Usually the prime minister leads, working with a small group that has great freedom of action. Hierarchy replaces the usual consensual decision process. Thus, the impact of personality is greatly increased. The critical decision process is invoked when public opinion is aroused, which gets leading politicians involved, and when no satisfactory decision has emerged from the normal process (Fukui 1975, 2–6, 41). Fukui uses the Okinawa-reversion case as an example of the critical decision-making process; Juster persuasively argues that it also applies to the oil crisis (1976; 1977).

Since the oil crisis, public concern about the Middle East appears to have fluctuated with events. The Japanese have been most concerned about three regions of the world: Asia, North America, and the Middle East (or

West Asia). The Soviet Union has ranked a "poor fourth," with Western Europe fifth. The relative importance of the Middle East, however, seems to have declined in recent years. In 1980, just after the second energy crisis, 38 percent of the public was most concerned about Asia, 30 percent about the Middle East, and 29 percent about North America. But by 1983, while Asia was still the primary concern of 38 percent, 37 percent cited North America and only 13 percent the Middle East. In 1984 the Middle East percentage increased by more than ten points again, presumably because of the Iran-Iraq war. Interestingly, Japanese with higher status and who are better informed are more likely to be concerned about the Middle East (Kuroda 1984, 4–5; Kuroda 1985, 31). This erratic pattern of concern means public opinion is not likely to be an important constraint on governmental decision making unless some sort of crisis occurs. Conversely, the relatively low level of constant interest probably means that little learning has taken place and that, should such a crisis occur again, public opinion is likely to have the same sort of impact as in 1973.

Despite the lack of hard data, there is a general assumption that Japanese public opinion is more pro-Arab than is government policy. This is often assumed to be a result of Japanese dependence on Arab oil, but one informed observer argues that this is not true: "Japan supports the Arab nations because they are weak, and the Japanese are likely to favor the weak. It is true that oil produced by the Arab nations is vital to Japan. Even if there were no oil being produced in the Middle East, however, Japan would still support the Arabs and not favor the Israelis, whose behavior is based on military power" (Yasunobe 1983, 13). Yasunobe clarifies this point by saying that although the government's position is influenced by oil dependence, the public's view is based on the weakness of the Arabs and the military power of Israel (1983, 16–17). Shillony similarly argues that before the 1967 war Israel "had a positive image in Japan because it was considered an underdog" (1985–1986, 22). Solomon finds this unpersuasive, given general Japanese respect for and deference to power; he argues that the root of anti-Israeli sentiment in the Japanese population is anti-Semitism, stemming from "traditional" notions of Jewish power in Western societies, distaste for the idea of a "chosen people," and misinformation from the mass media (letter to the author, September 1986; see also Solomon 1978).

Mass Media

The Japanese press was important in creating the mood of public panic about oil shortages, paranoia about the majors withholding oil from Japan, and the resulting intense political pressures to which Japanese government leaders responded. Press reports included scenarios such as a "no-oil situation" by the end of 1973 or the major oil companies cutting oil supplies in half (Caldwell 1981a, 196). In retrospect, it seems clear that the press exaggerated the problem. The government is partially responsible for this because it allowed the press to continue to misstate the facts without trying to correct them. Nau argues that MITI did not deny early press reports that the international oil companies had diverted oil from Japan, "even though by late November it was known that supplies were higher than reported" (1980b, 7–8; see also Baker 1978, 34–37).

Certain qualities of the Japanese press itself, particularly its adversarial attitude toward the government and its tendency to play to its mass audience, reinforced this tendency toward sensationalism. Tsurumi notes that the oil shortage was not reported by business reporters but by journalists who routinely covered crime and politics; they in turn exaggerated the significance of ambiguous evidence such as some tankers arriving partially empty (cited in Nau 1980b, 7).

The press's impact was considerable. Japan's press is national, and it took a fairly united position. Solomon argues that it systematically shuts off any voices supporting Israel and has even refused to report certain events (1978, 19–20). Moreover, during the crisis it was a two-way communication device, albeit an imperfect one. Politicians saw the press as registering the "national pulse" on the issue, and the image of a country in the middle of a domestic panic was to some extent created and certainly reinforced by the press.

Before 1973 the Japanese press had relatively little coverage of the Middle East. Coverage has increased considerably since then, although its quality has been criticized as superficial. Hisao Yasunobe recently estimated that there were about twenty-four Japanese journalists in the Middle East, five of whom were from television companies, while another twenty covered the Middle East from Tokyo (1983, 26, 14). He contends that they do not remain in the Middle East long and therefore find it difficult to put individual events into perspective. In addition, they rely too much on reports from others and are reluctant to investigate for themselves, perhaps because they do not feel at home in the area. One Israeli analyst maintains that the only Japanese journalists permanently based in Israel are from the two wire services; none of the Japanese newspapers, magazines, or television networks are represented, and only one part-time reporter speaks Hebrew (Shillony 1985, 75).

The Japanese press is significantly more pro-Arab than the government (Horvat 1973). One analyst, for example, contends that the Japanese press was significantly more opposed to the Israeli invasion of Lebanon than were its American counterparts, with the exception of the government broadcasting corporation. The Japanese press, since its repression during the Meiji era and before World War II, sees itself as the watchdog of the government. This distrust makes journalists unwilling to trust government information, and the government itself is more restrained than that of the United States in providing information. The ideological attitudes of many Japanese journalists also incline them in this direction:

> Opposition to American imperialism and colonialism was considered a qualification for what was called the "progressive man of culture." Journalists are categorized as men of culture. It is probably for this reason, together with the lack of profound understanding of the Middle East, that some arguments presented by progressive men of culture supporting the Palestinian liberation movement appear in Japanese press coverage of the Middle East. (Yasunobe 1983, 13)

Solomon adds that anticolonialism buttresses notions of Asian unity and that Israel is viewed as a Western outpost on Asian soil and, as such, must be opposed (1978, 16).

Interest Groups

Unlike the other countries studied, Japan has no indigenous ethnic interest groups relevant to the Middle East, although the Japanese business community has traditionally been an important pressure group. In practice, this community is usually seen as the *zaikai*, leaders of the major business organizations: the Federation of Economic Organizations (Keidanren), the Japan Federation of Employers' Associations, the Japan Committee for Economic Development, the Japan Chamber of Commerce and Industry, and the Japan Industrial Club (Yanaga 1968, 32). Of these groups, Keidanren is generally regarded as the most powerful (Curtis 1975, 34, 54).

Interestingly, within the ranks of Japanese business leaders, executives of oil firms have occupied a secondary position. Oil was viewed as an industry that was important only because it supplied Japan's heavy industry. It consumed rather than generated foreign exchange, and many refineries were owned by foreign companies. Thus, at least before the oil crisis, the Japanese were willing, albeit reluctantly, to rely largely on the major international oil companies, allowing the domestic companies to remain relatively weak as a result of intense price competition. Therefore, the leaders of the Japanese oil companies did not play a major role in determining Japanese business priorities outside of their own field of expertise (Caldwell 1981a, 5, 18–19).

The issues of oil dependency were not novel to the Japanese business community. A small but influential group, known as the "resource faction," had looked seriously at the issues in the early 1970s and had recommended that the Japanese get more involved in overseas economic projects in oil-producing countries and establish new institutions to facilitate business-government cooperation for this purpose. The Keidanren energy committee, founded in 1960 and headed by Soichi Matsune, had recommended such a program more than a year before the oil crisis. Sohei Nakayama, chairman of the Resources Subcommittee of MITI's Economic Advisory Council (Keizai Shingikai), advocated a "resources alliance" with oil-producing nations, and, in fact, these leaders had taken concrete steps to translate these ideas into reality before the oil crisis (Caldwell 1981b, 68; Caldwell 1981a, 172–188).

These business leaders were important participants in the Japanese decision-making process during the oil crisis. Indeed, the atmosphere of crisis seems to have significantly increased their leverage. In November the Keidanren energy committee, a large group of about two hundred members, announced the formation of a subcommittee, which became the active unit of the institution. On November 14, 1973, the first day of Kissinger's visit to Japan, it held an all-day meeting of top-level business and government leaders and produced a consensus that Japan must increase its "resource diplomacy." In a highly unusual action, a group of these men called on Prime Minister Tanaka that night—interestingly, although the request for the visit was made on very short notice, it was immediately granted (Caldwell 1981a, 206–207, 238–241). "Business leaders . . . met with Tanaka the night before his meeting with Kissinger to tell him that the oil majors would not supply Japan in an emergency. They told the prime minister that to overcome the oil crisis, 'there is no other way but to adopt a clear-cut diplomatic posture inclining to the Arab side'" (*Nihon Keizai Shimbun,* November 15, 1973; cited in Nau 1980b, 12). Moreover, this process was made public—an indication of how seriously the business leaders themselves regarded the issue (Caldwell 1981a, 207).

Solomon has another interpretation of these events (1978, 15). He argues that business leaders were concerned not about a cutoff of oil supplies, but that, if the partial cutbacks were not ended, the resulting austerity would produce an anti-Arab reaction among the Japanese public. In either case, the results were the same.

This kind of pressure could reasonably have been anticipated and had parallels in other countries. More unusual is the fact that a businessman, Sohei Mizuno, president of the Arabian Oil Company, became the most important Japanese envoy to the Middle East in the early stages of the crisis. As will be discussed later, this, in part, reflected the weaknesses of the Japanese Foreign Ministry. It also is related to the special status of the Arabian Oil Company, classified by Chalmers Johnson as a "national policy company broadly defined" (1977, 57–58). Such organizations, Johnson explains,

> belong to a blurred area between enterprises launched by government initiative and purely private, even if government-regulated, businesses. These in-between firms are distinguished by having a high concentration of retired government bureaucrats on their boards of directors, strong delegations of their executives on powerful government advisory commissions, and a history of direct involvement with the government in forms such as governmental assistance at their births, administrative guidance, governmental subsidies, and governmental brokerage in effecting mergers or joint ventures. . . . Often these firms have been singled out for governmental aid or protection, as distinct from governmental efforts to foster or protect their industry as a whole. (1977, 57–58)

The Arabian Oil Company had an unusual beginning for a national policy company. It was established in 1957 as a joint venture of a number of non-oil Japanese financial interests in conjunction with the Japanese, Saudi Arabian, and Kuwaiti governments. It was actually a private initiative—MITI was basically opposed to it. Over time, however, an informal deal was worked out whereby Arabian Oil focused on exploration and production while leaving refining and distribution to other firms (Tsurumi 1976,

117, 127n; Carvely 1985, 28–29). The Japanese government subsidized the company by requiring Japanese companies to purchase specified quantities of its oil (Caldwell 1981a, 84–87). Clearly, the president of such a firm carried considerable authority in government.

Business pressures increased as the crisis went on. The fear of Japanese isolation increased when the Arabs announced on November 18 that the Europeans were being absolved from the December 5 percent cuts while the Japanese were not. But Japanese business was not united, although it is true that Europe's decision increased Japanese support for a pro-Arab shift (Caldwell 1981a, 212).

> Emotionalism became more prevalent and leading *zaikai* members such as Sohei Nakayama, former President of the Industrial Bank of Japan; Soichi Matsune, Chairman of *Keidanren*'s Energy Committee; Hiroki Imazato, Chairman of *Nihon Seiko;* and Yoshihiko Morozumi, top level advisor to MITI, echoed the sentiments of many of their colleagues by pushing for a pro-Arab announcement. Nevertheless, other businessmen such as Fumihiko Kono, Chairman of Mitsubishi Heavy Industries, continued to oppose a change in policy. The majority of this latter group, however, either lacked influence with politicians or were reluctant to speak out. (Juster 1977, 302)

Despite these disagreements, then, the balance of Japanese business opinion clearly advocated moving toward the Arab position. The president of Arabian Oil, Sohei Mizuno, epitomized this position and became an important actor when he returned from the Middle East with fresh information. During the crisis some business leaders even suggested establishing their own "submarine" ties with the Arab states if the Japanese government would not do so. Moreover, direct pressure from business leaders was largely responsible for Tanaka's decision to include the famous clause in the November 22 statement threatening to reconsider relations with Israel if the 1967 territories were not returned (Caldwell 1981a, 209, 214–215, 269).

Business leverage was greater because the major Japanese trading companies such as Mitsui and Mitsubishi were important sources of information for the government during the crisis. "Access to information and the ability to interpret it were powerful resources for those who possessed them, particularly businessmen" (Caldwell 1981a, 219). But Japanese oil companies were unhappy with the government because of the revelation of an oil scandal in MITI in September. Reportedly, they were reluctant to cooperate (personal interview, Henry Nau, June 1984; Juster 1977, 300; Juster 1976, 60–61). On balance, business groups had influence because of their intimate contacts with government officials; access rather than pressure was the key to their success (Caldwell 1981a, 218–219, 258–259).

Business involvement in Japanese Middle East policy continued after the crisis. Interestingly, although *zaikai* leaders had pushed strongly for more Japanese economic involvement in the Middle East in order to guarantee oil shipments, problems arose when the ambitious commitments made by politicians had to be translated into particular acts by corporations (Caldwell 1981a, 228–246). Projects such as the Mitsui petrochemical installation in Iran became a central feature of Japanese diplomacy, and in the process Japan became hostage to unexpected political and economic developments. In order to keep the companies involved in money-losing projects (Mitsui had invested $1.5 billion in the Iranian project), the Japanese government has become more heavily involved in coordinating and financing such projects in cooperation with groups such as the Energy Committee of Keidanren.

In a pattern we have seen in other countries, the Middle East also became increasingly important as a market and a source of investment for Japan. Caldwell notes that in 1973 the Middle East received about 4.4 percent of all Japanese exports, but that by 1979 the figure had increased to 9.8 percent (1981b, 71–75, 78–81); Kuroda states that by 1981 it had risen to 12 percent and notes that about 92 percent of these exports go to Arab oil-producing states (1984, 7). These exports were concentrated in machinery and equipment and metal and metal products (Babayan 1985, 101); thus, in 1979 31 percent of all Japan's plant exports went to the Middle East. Japan's overall trade volume with the Middle East (exports and imports combined) is roughly equivalent to the Japanese-U.S. trade. The difference is that Japan runs a large deficit with the Middle East, whereas in recent years it has consistently had a surplus with the United States (Kuroda 1984, 6–7). Given the healthy state of Japan's exports, however, the Middle Eastern market seems to have been seen as important but not critical.

Interestingly, Japan has maintained a small but significant trading relationship with Israel; from 1972 to 1983 Japanese exports to Israel increased by a factor of six, while exports to the Middle East as a whole increased by a factor of seventeen (Babayan 1985, 99). In 1983 the trade was about $500 million per year, .017 percent of Japanese trade and 3.3 percent of Israeli trade. Japan imports fruit, diamonds, chemicals, and high-technology equipment from Israel; in return, it sells a variety of products such as cars and electronic equipment (Shillony 1985, 73–74; Shillony 1985–1986, 20–21). The Japanese government does not formally support the Arab economic boycott, thus allowing this trade to take place. But it does have an informal policy of not penalizing Japanese companies that accede to the boycott demands (Kuroda 1984, 9; for an excellent, and very humorous, summary of Japanese attitudes on the subject, see Phillips 1979, 108–109). Most of the larger Japanese corporations, therefore, either do not trade with Israel or do so indirectly; Japan Air Lines and Japanese ships do not call at Israel, which means all the trade is carried in Israeli vessels (Shillony 1985–1986, 20), helping the Israeli economy somewhat.

Japanese investment in the Middle East has been relatively limited; by 1982 it constituted only about 5 percent of Japan's relatively limited total overseas investment (Kuroda 1983, 14) and was concentrated almost exclusively in a number of giant petrochemical projects (Babayan 1985, 104). Japanese investment seems unlikely to increase in the near future; indeed, the Japanese have largely abandoned such major projects, given the repercussions particularly of the Mitsui petrochemical complex in Iran. In general, Japan has abandoned the policy of trying to secure natural resources by direct investment, relying instead on the market (Brooks 1985).

Governmental Influences on Japanese Policy

Parliament and Parties

The Japanese Diet seems to have played no significant role in the oil crisis, which is not unusual. The intensity of party competition makes serious foreign policy discussion in that body almost impossible (Baerwald 1977, 38–40), and, because the Liberal Democratic party (LDP) has retained a majority in the Diet since World War II, there is no incentive for interparty discussions. Interestingly, however, the opposition parties (except for the Social Democrats) advocated a pro-Arab policy shift during the oil crisis, providing the unusual spectacle of an alliance between business and the left.

Another interesting example of pro-Arab sympathies was the formation of a Diet group of representatives from different parties promoting ties with the Palestinians in 1979 (Caldwell 1981a, 209, 219, 261). The Parliamentarians' League for Japan/Palestine Friendship was founded by former foreign minister Toshio Kimura (W. Campbell 1984, 149). Indeed, this was the organization that formally invited Yasser Arafat to visit Japan, although this invitation must have reflected a decision within the government, because the league has consistently been dominated by Liberal Democratic party members. About 100 of the 763 Diet members belonged to the organization (Kuroda 1983, 29; Kuroda 1985, 32). There is also a smaller group, the Parliamentarians' League for Japan-Israel Friendship, with about fifty members (Shillony 1985–1986, 19).

Conflicts within the LDP were considerably more important than differences among the parties.

> The major components of the LDP are its factions—well established, formally organized groups of Diet members, each headed by a senior LDP leader. Factions are sometimes identified with substantive policy orientations. . . . But the main cement holding factions together is reciprocal personal interests. . . . Sometimes . . . LDP leaders take publicly conflicting positions in their maneuvering for the succession.
>
> Similar maneuvering—and similar disarray—was evident in the Japanese response to the Arab oil export cutbacks in the fall of 1973, as several ministers took conflicting stances in their search for initiatives that could enhance their prestige and party standing. (Destler et al. 1976, 61, 63)

Indeed, all the major decision makers—Tanaka, Ohira, and Nakasone—were also faction leaders. Although the factions as such did not take different positions on the Arab-Israeli conflict, the personal disputes among top Japanese political leaders shaped the response of the government in important ways.

Perhaps the clearest example of the effects of this system came after the November 22 statement, which said that Japan would send an envoy to the Middle East. But no leader wanted to risk the journey—the mission was likely to fail and would injure the political reputation of whoever was involved. So Miki was chosen because he was expected to retire soon and had nothing to lose politically. Ironically, the mission was a great success, and although practically no one went to the airport to see him off, most of

Japan's political elite was there to welcome him home. More important, other LDP leaders, initially Nakasone, but others as well, made similar trips to the Middle East, each promising sizable Japanese economic assistance. These offers were not well-coordinated and seem to have been motivated as much by personal ambition as by national interest. In fact, when the crisis subsided, many pledges were not fulfilled, as Juster explains: "Bureaucrats within the Finance Ministry responsible for dispensing Japan's funds and greatly concerned with her dwindling foreign exchange reserves, as well as officials of the Foreign Ministry, immediately began undoing some of the many agreements, much to the anger and dismay of the Arabs" (1976, 81; see also Juster 1977, 305, 308–309; Akao 1983, 266; and Caldwell 1981a, 228–237).

The LDP's dominance, combined with its recruitment process, means that the same individuals remain in high government positions for a long time; thus, it is not surprising that Ohira and Nakasone, the two major antagonists, later both became prime minister. The process adds a certain stability to Japanese politics and policies—especially as contrasted to the United States—so once the government decided to alter its policy during the embargo, the decision was unlikely to be casually reversed.

Bureaucratic Organizations

Within the government, the oil crisis triggered a series of intense bureaucratic battles over both control and content of policy. The problem was complicated because the Japanese government did not seem to have a clear-cut idea of how to make such crisis decisions. In addition, because leaders of several LDP factions were also ministers of competing departments, it is sometimes difficult to separate personal from organizational political conflicts.

Initially, the Foreign Ministry retained control of responses to the crisis. Baker argues that the ministry did not really become concerned about the issue until the end of October, perhaps because it expected a short war and, with seventy-nine days of oil supply on hand, was not worried (Baker 1978, 40; Juster 1976, 49). Caldwell, however, cites a report by the ministry's Middle East bureau earlier in the month warning that a simple reiteration of support for Resolution 242 would not exempt Japan from oil cutbacks (1981a, 192–193). This prediction was confirmed by the October

24 production cuts for the Arabian Oil Company. The Foreign Ministry estimated a 40 percent shortfall in November and December—a gross overestimate (Yoshitsu 1984, 2).

Following the ministry's usual procedure, an advisory group was formed, consisting of the heads of "most of the bureaus" and led by Vice-Minister Hogen (Fukui 1977, 10; Juster 1976, 53). The Foreign Ministry in general, and Hogen in particular, had a reputation of being concerned with national dignity, and Hogen reportedly cautioned his group not to overreact to the crisis (Caldwell 1981a, 192). Yoshitsu contends that, in practice, the decisions were made by Foreign Minister Ohira, Vice-Minister Hogen, and Deputy Minister Togo and that in late October they decided to reject pressure from the business community to break ties with Israel in favor of a simple restatement of Japanese policy to the Arabs (1984, 3). A more complex version of events holds that Hogen's group embarked on a systematic search of past Japanese actions, looking for precedents that would increase its freedom of choice (Caldwell 1981a, 192–194).

> Except for its small Middle East office, which had been upgraded in mid-1973, the ministry saw the problem largely in terms of relations with the United States and other industrialized countries rather than with Middle East countries. Maintaining close ties with the United States, therefore, was its first objective. Until the US response became clear, it was deemed advisable to stick to previous positions. (Nau 1980b, 8)

Thus Japan's initial response of restating its position to the Arab ambassadors on October 19, Hogen's *note verbale* of October 26, and the Nikaido statement of November 6 presumably reflects the influence of the Foreign Ministry. Baker notes, for example, that the October 26 note ignored two major Arab demands—Israeli withdrawal from all the 1967 territories and Palestinian rights—but that even so there was opposition within the Foreign Ministry to the phrase expressing sympathy with the Palestinians (*Mainichi Shimbun*, November 16, 1973; cited in Baker 1978, 40; note Yoshitsu's contention that these issues were in fact included in the note [1984, 3–4]).

The Japanese Foreign Ministry is thus unique among the foreign ministries of the five countries studied here, because it opposed a pro-Arab policy. We have previously noted Spiegel's argument that organizations defining the Arab-Israeli conflict as a Middle Eastern dispute tend to support the Arabs because this makes their relations with other countries in the region easier (1985, 5–6). Why is the Japanese Foreign Ministry an exception? It was never pro-Israel; it was pro-American, which is quite different. If the Middle East office had been influential within the ministry, it would probably have argued positions similar to those of other foreign ministries. In the Japanese case, however, the dominant regional bureaus within the ministry were the North American divisions. Indeed, a Middle East–Africa section was not created until 1961 and became a full-fledged bureau only in 1965 (Kuroda 1983, 4).

Because of Japan's relative isolation from Western countries such as the European states, moreover, there was not much concern within the Foreign Ministry about their views; this also strengthened the hand of the Americanists. Indeed, one American observer commented that the Foreign Ministry was responsible for maintaining good relations with the United States, while MITI was responsible for accommodating the Arabs (personal interview, June 1984). Thus, as long as the Foreign Ministry retained control of the situation, any pro-Arab shift was unlikely.

The Japanese Foreign Ministry differs in still other ways from some of the other ministries we have examined. It was unable to retain control of its government's responses for long (Yoshitsu's account disputes this [1984]). The Foreign Ministry had some fundamental political weaknesses. Historically, its power and independence have varied widely. It was probably strongest just after World War I but was greatly weakened by a struggle with the armed forces just before World War II. It then staged a comeback during the Occupation, only to suffer somewhat in the reaction against American dominance and the heavy stress on economics in Japanese postwar foreign policy (Nish 1983, 332–341; Tsurumi 1976, 124). It is also by reputation weak in economics; unlike other ministries, it does not routinely have its personnel take midcareer courses in areas such as econometrics or statistics, for example. Moreover, a 1969 reorganization integrated political and economic decision making within each regional bureau, further diluting whatever economic expertise remained (Fukui 1978, 86–87). The ministry is one of the weakest government bureaucracies in terms of domestic political power, and its central policy, to stay close to the United States, is not particularly popular in Japan (Fukui 1977, 4); as a result it tries to stay aloof from public opinion (Juster 1976, 16–17).

A more specific difficulty in the case of the embargo was a lack of competence on the Middle East. The Japanese foreign service had in all only about thirty Arabists. There were few embassies in the Middle East, and their staffs were small and particularly inexperienced in oil questions. A significant proportion of their staff members were from Japanese trading companies and were not professional foreign service officers at all (Henry Nau, personal interview, June 1984; Nau 1980b, 8–9). Moreover, the best Japanese foreign service officers usually specialized in areas other than the Middle East.

Several factors seem to have contributed to this lack of expertise in the industrial country most dependent on Middle Eastern oil. In view of Japan's lack of historical connection with the Middle East, any expertise would have been developed only after World War II. The Japanese foreign service culture, perhaps reinforced by national snobbery, encouraged talented people to be concerned with industrial countries rather than those in the Third World, referred to within the service as "west of Burma" (Fukui 1977, 27; Sasagawa 1982, 34–35); this is not surprising in a country very concerned with establishing itself as a modern state. The foreign service also tends to produce few specialists such as Arabists (Fukui 1977, 28–29). Japan's general dependence on the United States and avoidance of political issues also contributed to this tendency.

A number of Japanese specialists agree that this situation has radically changed since 1973. The Foreign Ministry has hired more Middle East experts and has attracted high-quality officials by moving personnel in the area up the career ladder; the last three heads of the African and Middle East division of the Foreign Ministry had all previously been economic counselors in Washington (personal interviews, Henry Nau and others, June 1984). Nevertheless, Kuroda quotes Ryohei Murata, a former head of the division, who wrote in 1981 that Japan still had only fifty-six Arabic-speaking specialists in the ministry and only a third as many diplomatic personnel stationed in all the Arab countries as in the United States (Kuroda 1983, 5–6; Kuroda 1985, 30). Murata added that every year another four Arabic-language specialists are added to the Foreign Ministry (compared to twenty for the United States and ten for Britain) and that the ministry hopes to have a hundred Arab specialists by 1990.

As mentioned earlier, the Foreign Ministry was not totally united on its position on the Arab-Israeli conflict. The Middle East division of the Middle East and African Affairs Bureau proposed a pro-Arab shift and, despite its lack of weight within the organization, argued its case strongly. The two North American divisions of the American Affairs Bureau, which usually had a good deal more clout, were hampered by a lack of both information and concrete reassurances by the United States. Nonetheless, they argued that Japan's links with the United States must not be weakened and that Japan must be seen as "a stable and dependable ally, rather than a malleable and irresponsible one." They also cited the possible effects of a Jewish boycott of Japanese goods (Miyoshi 1974, 150).

The advisory group of bureau chiefs chaired by Vice-Minister Hogen was divided about equally on the issue (Juster 1977, 299–300, 302). As the crisis wore on, however, the Hogen group moved toward opposing a strongly pro-Arab position, largely because of Hogen and the director-general of the Treaties Bureau, normally the most influential bureau in the Foreign Ministry (Juster 1976, 64–65; Fukui 1977, 17–18). At the same time, sentiment within the ministry as a whole began to shift toward a pro-Arab position (Juster 1976, 56).

The other reason the Foreign Ministry lost control of the situation was the power of its chief adversary, the Ministry of International Trade and Industry (MITI). This unique organization covers a remarkable range of functions. As a former vice-minister noted:

> Among the problems we handle are imports and exports, economic cooperation, the exchange of capital, equipment investments, factory location, and environmental problems. We cover businesses from the largest down to the small- and medium-size enterprises. We cover industries from manufacturing to distribution, including the machinery industry, chemical industry, heavy industry, and textiles. Recently leisure industries, even golf, have been included. The advertising and newspaper industries are also included. . . . We also deal with consumer complaints and problems concerning energy and power resources. We have sixteen laboratories working [on] manufacturing technology, and patents are also a big field for us. (Ojimi 1975, 101)

MITI's close alliance with Japanese business gives it formidable political clout, and it is continually interested in expanding into new areas of policy (Ojimi 1975, 104). Moreover, especially since the Oil Law of 1962, MITI has been the central agency in determining Japanese oil policy (Tsurumi 1976, 114–118; Caldwell 1981a, 15–16).

Despite all this, MITI's bureaucratic and political influence had been severely diminished in the decade preceding the oil crisis.

One issue after another plagued the ministry in this era—industrial pollution, revolts against its administrative guidance, charges of corrupt collusion with big business, inflation, public dismay at some of the consequences of its industrial location policy, . . . and serious damage to relations with Japan's main economic partner, the United States. . . . By the mid-1970s the ministry began to show renewed strength. . . . The oil crisis and all of its ramifications gave the ministry a new lease on life. . . . One official characterized 1968–69 as the worst year in MITI's history, and Vice-Minister [Yoshihiko Morozumi] (1971–1973) referred to the years leading up to the basic reform of the ministry in July 1973 as a "long, dark tunnel." (Johnson 1982, 275)

MITI had been involved in the efforts by the "resource faction" of Japanese business leaders to alter policy before the oil crisis. In July 1973 it established a new Resources and Energy Agency, linking a number of separate bureaus together for the first time. Its first major act was to issue a white paper calling for promotion of closer relations with oil-producer states, conservation, and diversification of energy sources (Caldwell 1981a, 182–185). But the oil crisis occurred before it had a chance to do much with this ambitious agenda.

As noted earlier, the Japanese government became concerned when Japanese oil imports were reduced. MITI Vice-Minister Yamashita set up an ad hoc group of staff members from the Energy Agency and the Middle East and American divisions, much like that in the Foreign Ministry but apparently quite independent. Evidently, members of this group knew even then that Japan was unlikely to be classified as a "friendly" nation by the Arabs. Initially, they wanted information about the impact of oil-shipment cutbacks on the Japanese economy, but MITI bureaucrats proved unable to provide it, partly because of the numerous uncertainties and partly because of the oil scandal that was wracking the Energy Agency. The result was a series of "contradictory and pessimistic statements"; in particular, Vice-Minister Yamashita and Energy Agency head Yamagata differed in public on the likely impact of the oil weapon on the Japanese economy (Caldwell 1981a, 191–192, 197–198).

After some initially optimistic assessments, MITI oil-supply projections were consistently too pessimistic. These estimates, in turn, supported the case of those within Japan who wished to adopt a more pro-Arab position.

MITI's miscalculations aroused a good deal of discussion in Japan after the crisis. Some observers attributed them to inadequate information; for example, some Foreign Ministry officials apparently knew that Arab oil cutbacks were less severe than they acknowledged in their public announcements, but this information never reached MITI. Less charitable interpretations suggest that MITI deliberately created an artificial crisis to justify price raises for the domestic oil companies and to encourage passage of the Petroleum Supply and Demand Normalization Law, which increased MITI's bureaucratic power and was passed in December, with Yamashita referring to the Japanese oil industry as "the root of all evil" (Tsurumi 1976, 118–121; Caldwell, 1981a, 198–202, 267, 321–332; Baker 1978, 34–38, 49; Johnson 1978, 297–298).

In the midst of all this, MITI found itself pressured on another front. MITI had been involved in a bitter competition with the Fair Trade Commission for control of the oil and broader manufacturing sectors (Nau 1980b, 8). At issue was whether or not the antimonopoly law should be enforced. Tanaka had appointed a very tough head of the commission in October. On November 27, 1973, officials of the commission raided the offices of the Petroleum Association of Japan and of twelve oil companies, seizing documents they claimed proved illegal collusion among the companies. The companies replied that they had acted together only at MITI's directives. MITI denied it. On May 28, 1974, the companies were all indicted in the "black cabal" case, which dragged on until 1980. Despite all this, MITI's domestic political power was considerably increased by the oil crisis. In Johnson's view: "The significance of the oil shock . . . lies in the fact that it once again reminded the Japanese people that they need their official bureaucracy" (1982, 297, see also 297–301; also Tsurumi 1976, 121–123).

Juster argues that MITI did not simply adopt a pro-Arab position. MITI personnel in the West Europe–Africa–Middle East Division, in addition to staff from the American-Oceanic Division of the International Trade Policy Bureau and the Petroleum Department of the Natural Resources and Energy Agency, all worked on the problem. They were told by the major oil companies that Japan would be treated fairly (although the press was portraying the majors as the villains). MITI officials were also very conscious of Japanese economic dependence on the United States, and most of the oil companies were American.

MITI officials seem to have been divided on the issue (Caldwell 1981a,

210, 268). Juster argues that, on balance, they favored improving relations with the Arabs but agreed with their colleagues in the Foreign Ministry in opposing a pro-Arab position. Vice-Minister Yamashita also took this position. The key decision was made by Nakasone himself, who, "inspired by the sense of domestic unrest, made a move uncharacteristic of a minister by overruling his ministry and reversing their position to one of support for a pro-Arab shift" (Juster 1977, 299–300).

The internal power balance was shaped by a desperate need for information. "The biggest constraint for groups from both ministries was the widespread absence of information about the exact nature and extent of cutbacks in oil supply—a direct result of unfamiliarity and lack of contact with the Arab nations" (Juster 1977, 298). One author writes that "several officials commented that they experienced more of an information crisis than an oil crisis" (Juster 1976, 85; see also Caldwell, 1981a, 219).

Because the Foreign Ministry seemed incapable of obtaining this information through conventional channels, three separate high-level envoys were sent to the Middle East. Prime Minister Tanaka and a few Foreign Ministry officials had suggested Keiichi Morimoto, a former Arabist foreign service officer and later a businessman in the Middle East. Vice-Minister Hogen agreed but also suggested sending Hideji Tamura, a former ambassador to Saudi Arabia who was still in the Foreign Ministry. They both left on November 8. Sohei Mizuno, president of the Arabian Oil Company, was going to the Middle East anyway; "MITI officials may have suggested that he, too, privately contact Arab producers." All three were asked in briefings by the Foreign Ministry before departure to try to determine the "minimum conditions" Japan would have to meet in order to be classified as "friendly" (Caldwell 1981a, 202–203). The three deliberately kept apart from one another. "Each of these emissaries had long ties to and private sources of information in the region, and there was good reason to send more than one. But these individuals also may have played a role in the intensifying bureaucratic struggle in Japan" (Nau 1980b, 9).

William Campbell quotes Saudi officials as saying that this series of visits was particularly important for the Japanese because they were able to talk with the top level of the Saudi elite:

> In addition to King Faisal, the members of the mission met Kamal Adham, Faisal's brother-in-law and close advisor; two of Faisal's sons—Saud, who dominates the Supreme Petroleum Council and who became foreign minister in

1975, and Turki, who was intensely tutored by Adham and was to become head of Saudi intelligence in 1978; and Bandar, the son of Prince Sultan, the minister of defense who is second in line to succeed King Fahd to the monarchy. (1984, 143)

At the same time, American pressure increased. Sensing the disarray in the Japanese government, the United States asked Japan to delay a decision until Kissinger's visit on November 14. Nakasone argued for a more pro-Arab position. Ohira, the foreign minister, was able to secure a delay, but time was running out.

Kissinger's visit gave his allies in the Japanese government no new ammunition; indeed, it seems rather to have strengthened the hand of those who felt it was imperative to distance themselves from the Americans (Caldwell 1981a, 208). At a cabinet meeting on November 16 Tanaka and Ohira were able to get agreement to fulfill the former's promise to Kissinger not to alter policy for a month (Baker 1978, 43). But this proved to be the last clear-cut victory for the Foreign Ministry. On November 18 OAPEC exempted the European Community from the December 5 percent cut but took no action toward Japan. "This last OAPEC move seems to have been unexpected and proved devastating to the authority of the Ministry" (Baker 1978, 48; see also Caldwell 1981a, 211).

The Foreign Ministry's control of the situation quickly began to unravel. Two of the independent Japanese envoys, Tamura and Morimoto, returned on November 15 and 18 respectively; Morimoto had actually reported by telephone to the Foreign Ministry before Kissinger's arrival. "[Tamura and Morimoto] brought similar results: the Saudis wanted a Japanese statement that, if Israel continued to refuse to withdraw from occupied territory, Japan would have to reconsider its relations with Israel. Tanaka ordered the cabinet to begin drafting such a statement" (Nau 1980b, 12; see also Juster 1977, 301; Caldwell 1981a, 203–204, 208).

Mizuno, president of Arabian Oil, returned on November 19. He spoke directly to Tanaka, Ohira, and Nakasone, as well as to the leaders of the Keidanren rather than to bureaucrats (Juster 1977, 303). We have noted earlier that his position as head of a national policy company gave him unusual weight. He brought back even harsher reports than did Tamura and Morimoto; the Saudis had demanded a "clear indication of policy change toward Israel" before November 24 (when an Arab summit was scheduled) and threatened an OAPEC embargo against Japan (Yoshitsu 1984,

5; see also Caldwell 1981a, 211–212). More significant, he had also discovered that the Arab leaders had never received the Japanese November 6 statement (Nau 1980b, 12), which, in turn, may have explained why Japan had not been exempted from the December cuts as the Europeans had. "The discovery that the Foreign Ministry had not even ensured the communication of the November 6 policy statement to the most influential of the OAPEC members, Saudi Arabia, could well have been the decisive blow to the Ministry's credibility" (Baker 1978, 48). The shift of influence was symbolized by the fact that MITI, not the Foreign Ministry, proposed the issuance of a policy statement before the November 24 Arab summit. Although it was drafted by the Foreign Ministry, the wording was strengthened at the last moment owing to pressure from business leaders and Nakasone (Caldwell 1981a, 213–214).

Mizuno and the leaders of the Keidanren agreed that a pro-Arab policy statement was essential, and their position was leaked to the press. The public was so upset at the possibility of oil shortages and economic hardship that the Tokyo police reportedly pushed for a pro-Arab position to avoid domestic disturbances (Juster 1977, 303). The result was the November 22 statement, which essentially complied with Saudi demands; the Foreign Ministry was able to change the wording only from reconsidering "relations" with Israel to reconsidering "policy" (the former implies breaking diplomatic relations, the latter could involve simply stating another policy). The Foreign Ministry's prestige was so low that the cabinet did not want to send Ohira to the Middle East, and at the end of December Tanaka decided to give the Overseas Economic Cooperation unit to a new cabinet ministry-without-portfolio (Baker 1978, 48).

The November 22 statement did not settle all the policy conflicts, however. There was considerable discussion about whether Japan should buy oil directly from the Arabs or continue to rely on the major international oil companies. Both the Economic Affairs Bureau of the Foreign Ministry and the Natural Resources and Energy Agency of MITI supported staying with the majors, but no one was prepared to defend the majors publicly or to oppose sending an envoy who would make agreements for direct purchases (Juster 1977, 305).

But when the Arabs classified Japan as a "friendly" country on December 25, public anxiety over the issue declined considerably. As a result, the Foreign Ministry was able to regain control of foreign policy in this area (Nau 1980b, 20). Ironically, the success of the very policy the ministry had opposed had allowed it to regain its usual dominant position. The general decline of public concern reestablished the usual pattern of decision making, one that is dominated by bureaucratic organizations (Juster 1977, 309).

Researchers differ radically about the relative strength of the Foreign Ministry during the oil crisis. The key issue is what the Foreign Ministry advocated. Juster (1976, 1977), Nau (1980b), and Baker (1978), among others, argue that the Foreign Ministry, and particularly Ohira, opposed the terms of the November 22 statement as too pro-Arab. They see this period as characterized by bureaucratic competition under circumstances of incomplete information and strong external pressures, resulting in the decline in the ministry's influence as symbolized by the November 22 statement. They conclude that when a crisis reaches such proportions, conventional bureaucratic organizations are likely to lose power, particularly those like the Foreign Ministry, which have relatively weak domestic power bases.

Yoshitsu (1984) contends that the Foreign Ministry had decided late in October that the terms of the November 22 note were appropriate; therefore, its adoption represents a victory rather than a defeat for the ministry. If indeed the October 26 note to the Arab ambassadors was identical to the more famous November 22 statement, then the power of the Foreign Ministry certainly remained strong throughout this period. If Yoshitsu is correct, the debate during the crisis was about concessions. Were they sufficient or should more be added? Because no changes were made, this was evidently another Foreign Ministry victory.

Yoshitsu's interpretation is intriguing, but there is still insufficient public evidence to warrant its adoption. He makes another important point by implication, however. The central issue in the Japanese response to the oil weapon is the relative lack of policy change. Presumably, the major reason for this was U.S. pressure, but the efforts of the Foreign Ministry were probably also important, particularly once the crisis was over and the decision-making process had returned to normal.

The Influence of Individuals on Japanese Policy

The crisis atmosphere and the corresponding change in decision-making style made the role of individual preferences much more important than usual in Japan, as Juster writes: "The actual degree of Japan's 'innovative' action may depend greatly on the personalities and idiosyncrasies of the few policymakers involved in a particular situation" (1976, 44). Most of the influential people were in the government, but there were exceptions. For example, Sohei Mizuno, head of the Arabian Oil Company, played an important role in his private diplomacy to the Middle East and perhaps an even more important one by his direct pressure on Tanaka after his return (Baker 1978, 44). The two main protagonists were, nevertheless, in government: the foreign minister and the minister of industry and international trade. Although each was required by his job to participate in the debate, they both seem to have been impelled by personal considerations as well.

Minister Ohira was strongly pro-American. The Foreign Ministry itself was increasingly divided on how to resolve the crisis, but these differences do not seem to have been reflected in Ohira's position, suggesting that its source was personal to him rather than to his office. Minister Nakasone had been appointed to head MITI by Tanaka in July 1972 as part of a last-minute political deal to seal Tanaka's own victory within the LDP. Tanaka had won, in part, because he supported Japanese nationalism in the face of the "Nixon shocks" of 1971 (Johnson 1982, 292–294). It was therefore hardly surprising that Nakasone identified with Japanese nationalism and was less concerned with the United States than other Japanese politicians were (Baker 1978, 41). Also, as Juster writes, "beyond this political rationale, Nakasone . . . had personal ambitions to fulfill. He was the leader of a medium-sized, anti-mainstream faction and hoped to increase his popularity as well as his chances for the premiership by differentiating his policy from Ohira's and, thus, appealing to the populace as a strong, forthright leader during the present turmoil" (1976, 64).

But Nakasone's Middle East policy position was not simply opportunistic; it was consistent with his earlier behavior. Nakasone had been one of the first senior Japanese politicians to support the efforts of the "resource faction" of Japanese business leaders before the oil crisis. He had visited the Middle East in 1973 before the embargo and promised that Japan would never join a "consumer alliance." Later he also advocated a policy "standing on the side of the oil producers," saying that "securing oil is Japan's legitimate right" (Caldwell 1981b, 69–70; Caldwell 1981a, 180–183; Carvely 1985, 48–50). Nakasone worked hard during the crisis, building support for a policy shift on the Middle East. At the critical point, when the November 22 statement was being drafted, he personally called Tanaka to argue for strengthening the Foreign Ministry version (Caldwell 1981a, 210, 214). Nakasone's stance during the embargo was clearly important in boosting his political career (personal interview, Henry Nau, June 1984). Nor was Nakasone simply pro-Arab. As prime minister, he publicly called for the Arab states to recognize Israel, and he brought the United States and Japan closer together in foreign policy than they had been for some time (Shillony 1985, 78; Carvely 1985, 145–147).

Vice-Premier Miki seems to have been important as well. He was older than the other major politicians involved; in fact, as noted earlier, he was thought to be at the end of his career. He was also the most progressive of the LDP leaders (Yoshitsu 1984, 6). He apparently supported Nakasone's argument for a more pro-Arab policy and may well have been the decisive factor in effecting this policy change—he publicly supported it on November 16 (Caldwell 1981a, 209). Miki also was persuaded to undertake the visit to the Middle East after the November 22 statement when no other politicians wanted to risk failure and thus injure their careers. Perhaps Miki accepted the challenge because his own career was practically over, or perhaps because he thought his progressive views would appeal to the Arabs (Juster 1977, 305; Yoshitsu 1984, 6). Juster argues that Miki's discussions with Faisal were critical, that Miki stressed how oil cutbacks would damage the economy of Southeast Asia and make the area more vulnerable to communism, and that Faisal, as a result of these arguments, sent his brother-in-law to persuade President Sadat to agree to alter OAPEC strategy (1976, 75–76). At any rate, Miki's visit was clearly a personal triumph; by the time he returned, Japan had been classified "friendly."

Miki, in fact, became Japan's central negotiator during the crisis. After his return from the Middle East, he went to Washington and apparently laid the basis for the close cooperation between the United States and Japan during the Washington Energy Conference (Nau 1980b, 20). While in Washington Miki also called for Israel's complete withdrawal from all occupied territories (Baker 1978, 59–60), personifying the two central thrusts of Japanese diplomacy during this period: establishing close ties with the Arabs while retaining good relations with the United States.

In 1974 Miki became prime minister, rather to everyone's surprise, as a compromise candidate to replace Tanaka. His views on the Middle East were one important reason for Japan's willingness to allow the PLO to establish a Tokyo office, which opened in 1976 (Yoshitsu 1984, 14–15).

Logically, Prime Minister Kakeui Tanaka should have been a leader during this period, but he appears not to have been. The dominant figure in Japanese politics of the decade, Tanaka had been MITI minister before becoming prime minister in 1972, and his intelligence, drive, and ambition were well known. On the one hand, he was particularly sensitive to the growing political clamor over the oil crisis and felt it imperative to do something about it (Nau 1980b, 10). Tanaka thus helped provide impetus for action. On the other hand, he seems not to have been involved in the policy discussions, perhaps because of his lack of experience in foreign affairs, perhaps because of possible political consequences of close involvement. Some have argued that his inexperience in foreign affairs made him particularly susceptible to the influence of business's "resource faction," perhaps shown by his willingness to receive a delegation on short notice during the Kissinger visit (Caldwell 1981a, 207, 209, 268). At any rate, Tanaka seems to have been more concerned about creating a consensus than about what policy was adopted. He was certainly central to the basic policy decision to issue the November 22 statement, decisively influencing it toward the stronger position advocated by Japanese business groups (Juster 1977, 301; Caldwell 1981a, 220; personal interview, Henry Nau, June 1984).

Impact of the Oil Weapon on Japanese Middle East Policy

There is not much literature available in English on the making of Japanese Middle East policy. In addition, this rather limited literature has produced radically different interpretations concerning both substance and process. Some argue that the policy changes were quite significant, that Japan fundamentally changed its policy (or perhaps developed a policy for the first time) and broke with the United States to adopt a pro-Arab position, as the supply theory of sanctions would predict. Others argue that there was little change at all, that the basis had been laid before 1973, and that the only change was that policies were enunciated more clearly as the Middle East became a more important issue for the Japanese. In terms of process, some see the issue as involving a wide range of Japanese organizations and individuals, with conflicting decisions being made because of insufficient information and high stakes. Others argue that the government engaged in a massive information search and that, although this was a difficult and exhausting process, the government was eventually able to formulate policy in a relatively consensual manner (personal interview, Martha Caldwell Harris, June 1984). A third position is that the Foreign Ministry essentially made the decisions itself (Yoshitsu, 3–4).

It seems to me there are two separate issues to explain: First, why did Japan adopt a public position that was more pro-Arab than other industrial states? Second, why did Japan not meet further Arab demands? Most discussions of the case have focused on the first issue, but in some ways the second is more interesting. Sherlock Holmes once claimed that it was the fact that the dog did *not* bark which was important. Similarly, the supply theory of economic sanctions leaves us to wonder why the Japanese did *not* accede to most of the Arab demands.

Japan seemed to be the perfect target for the oil embargo. It was the OECD country that depended most on Middle Eastern oil, it had no historic interest or involvement with the Middle East, it had no internal Jewish pressure group working for Israel, and it had a reputation of maintaining a low-profile foreign policy that yielded readily to external pressure. This combination of factors has made Japan the most pro-Arab of the industrialized countries, with the possible exception of postrevolutionary Greece. Indeed, Japan is generally seen as the major success of the embargo.

But, in retrospect, the oil weapon produced remarkably little change in Japanese policy toward the Arab-Israeli conflict. Japan was the industrial country most supportive of the Palestinian cause both before and after 1973. Japan also threatened to reconsider its policies if Israel did not withdraw from the occupied territories, but it never carried out this threat. Economic relations with Israel increased rather than decreased over time. Japan never sent military assistance to the Arabs, and it broke many of its pledges of economic assistance proffered during the crisis. Japan's pro-Arab policies are no doubt partly a result of its apprehensions over oil supplies, but the use of the oil weapon reinforced rather than created this concern and did not result in significant policy changes.

The central question in the Japanese case, then, is why there was so little policy change when the embargo seemed to have such enormous power. One major reason undoubtedly was the offsetting external pressure of the United States, the major ally of both Japan and Israel. Kissinger made a serious effort to prevent the Japanese government from adopting a pro-Arab stance. Ironically, he feels he failed, although according to my analysis he succeeded. Japan remains within the general Western consensus of supporting the existence of Israel, not recognizing the PLO, and relying on Resolution 242 (albeit a different interpretation) as the cornerstone of a negotiated settlement. Arab oil leverage has not succeeded in moving it away from these positions.

The United States was in an unusually strong position to influence Japan because pro-American positions had become practically second nature to most Japanese bureaucrats, in MITI as well as in the Foreign Ministry. Thus, even the relatively innocuous pro-Arab rhetoric of the November 22 statement would probably never have been approved except for the involvement of leading politicians, particularly Nakasone. (Note that Yoshitsu [1984] attributes this change to decisions within the Foreign Ministry.) Thus, the embargo does seem to have produced this rhetorical change of policy. As soon as the crisis dissipated, however, the bureaucrats took over again, and their attitudes do not seem to have been changed much by the experience.

It can also be argued that, although the direction of Japan's policy has not changed much, the embargo gave Middle East policy a much more important position in Japanese thinking (Henry Nau, personal interview, June 1984). Moreover, given the general inertia of Japanese policy, the Japanese justifiably see this shift as significant, although to outsiders the change seems fairly trivial.

Finally, we are left with the question of how the Japanese would have reacted if the Arabs had continued to use the oil weapon and had succeeded in severely cutting back oil supplies to Japan. To put it differently, if the Arabs had "really" wanted to alter Japanese foreign policy significantly, could they have done so? Was their failure the result of limited motives, Japanese resistance, or the inherent difficulties of economic coercion? Respondents felt that Japan, more than any other country in this study, would have been willing to make further concessions to the Arabs under such circumstances (personal interviews, June 1984). But it is hard to escape the sense that Japan's political vulnerability has been overestimated, given its lack of real change in the face of very impressive threats.

In some other countries the long-term shift in Middle East policies seems best explained by the new wealth of the Arabs and the consequent increases in Arab export markets and capital investments. This explanation seems less applicable to the case of Japan, both because of its greater dependence on Middle Eastern oil and because of its more impressive general economic performance. Economic activities with the Middle East seem to have remained a means to an end for the Japanese, a way to increase access to oil supplies in the future, whereas in the European countries such activities became a key part of their economies. There is no doubt that new Arab wealth facilitated the shift in Japanese policy (personal interview, Henry Nau, June 1984), but for Japan, unlike the other countries studied, the issue of oil supplies remained paramount.

The Japanese diplomatic response, however, was not to break ranks with Western countries, desperately currying Arab favor with irresponsible gestures. Instead, Japan has continued its earlier policy of staying just a little more pro-Arab than its Western allies. The oil weapon didn't really "teach" the Japanese anything new. They already knew they were dangerously dependent on Middle Eastern oil, and they had already taken steps, albeit quietly, to rectify this situation by diplomatic gestures. The oil crisis had major short-term effects on Japan, but in the longer run it seems to have merely confirmed to the Japanese the wisdom of their earlier strategy.

6

The United States: Direct and Indirect Target

The United States was both the direct and the indirect target of the oil weapon. As the first country embargoed by the Arab oil producers, it was the direct target. As Sheik Ahmed Zaki Yamani and Muammar Quaddafi stated separately, however, Europe and Japan were pressured not only to alter their policies but also to influence the United States (Yamani interview with Hans Hielscher in *Der Spiegel*, December 3, 1973, cited in Szyliowicz and O'Neill 1975, 204; Qaddafi interview with Eric Rouleau in *Le Monde*, October 29, 1973, 1–5, cited in Bouchuiguir 1979, 163–164). Finding the United States fairly impervious to direct pressure, the Arabs used the United States' allies as surrogate hostages.

The United States was also under severe cross-pressure both because it was the country most easily influenced by Zionist pressures and because, as a superpower, it was particularly reluctant to change its policy as a result of coercion (Scheinman 1976, 11). As the superpower supporting Israel, the United States was the only country capable of persuading Israel (the real indirect target of the oil weapon in all its forms) to alter its policy and thus was the only country outside the region that could really make a difference in the Middle East (Sankari 1976, 265, 273).

Historical Background

As a society, the United States has long been involved in the Middle East through activities such as missionary work. The American government did not really become interested in the region until after World War I, however, when the issues of oil and colonialism became important. American concern increased with the founding of Israel, and the issue has been a staple of foreign policy ever since. William Quandt notes that "few issues have been considered more critical to American foreign–policy makers than the Arab-Israeli dispute" and attributes American involvement to three issues: U.S.-Soviet relations, Israel, and oil (1977, 1, 14).

The United States has a unique role in the Arab-Israeli conflict as Israel's closest (and increasingly its only) ally. In particular, it has given an enormous amount of aid to Israel over the years, both economic and military—much more than to any other country. Its governmental assistance has been massive and unprecedented, even for the United States. Baldwin notes that Marshall Plan participants received an annual average of slightly more than twenty dollars per person from 1948 to 1952, and Third World recipients have rarely gotten as much as one dollar per person per year, but "Israel, with a thirty-year annual average of nearly two hundred dollars per capita aid, is clearly in a class by itself" (1985, 638–640, 675). Even in absolute terms, Novik shows that between 1946 and 1979 Israel received more aid than any other single country except for Vietnam (1986, 98, 143–145). The "special relationship" between the United States and Israel has varied over time, however.

Lyndon Johnson was not particularly interested in the Middle East; it was relegated to a secondary role in his foreign policy until the 1967 war. His administration was preoccupied with the growing war in Vietnam and was unwilling to get involved in the Middle East. Indeed, when the June war broke out, the United States had no ambassador in Cairo; Ambassador Lucius D. Battle had gone to Washington to take over the Near East Bureau of the State Department, which in turn had had no director for several months (S. Spiegel 1985, 119–120, 136).

The major policy innovation during this period was that the United States became Israel's principal arms supplier. Eisenhower had refused to sell arms to Israel after Egypt's arms deal with the Soviets in 1956. Kennedy had agreed to an initial sale of antiaircraft missiles, and the Johnson administration supplied Patton tanks in 1966. Military aid to Israel went from practically nothing in 1962 to $136.4 million in 1967, although Egypt and Jordan each received more economic assistance than Israel during that same period (Roehm 1980, 33–34, 44). The really large arms transfers began after the 1967 war, when France decided to stop selling weapons to Israel.

Although the U.S. government quickly recognized the seriousness of the 1967 crisis, it did not act either to maintain the United Nations Emergency Force in Sinai or to keep the Straits of Tiran open until after Nasser had taken strong public positions on both issues. The United States tried to restrain Israel but was unable to furnish any reason to justify restraint. The one possibility, a multinational fleet to open the Straits of Tiran, was supported by some advisors but opposed by both the Near East Bureau of the State Department (because of Arab reaction) and the Defense Department (because of commitments in Vietnam); only the British, Dutch, and Australians were willing to participate (S. Spiegel 1985, 139–150). Thus, in retrospect, it is hardly suprising that the Israelis decided to attack. (Mearsheimer [1983, 143–155] provides a useful summary of the literature on the Israeli decision.)

The United States had opposed an Israeli attack because it feared being drawn into the war. Once the Israelis were successful, however, the United States decided to try to translate the new status quo into a Middle East settlement and adopted a new policy whereby the 1967 Israeli territorial gains would be traded for a comprehensive peace agreement, although Israel never agreed to the terms of such a settlement (Safran 1978, 418, 424–425).

The first result of this process was a draft U.N. resolution sponsored by the Soviet Union and the United States. It called for an immediate Israeli withdrawal in return for "a vague declaration of non-belligerency." The Arabs immediately rejected it, although, ironically, it was the most pro-Arab resolution offered by the superpowers before or since (S. Spiegel 1985, 155). Negotiations resumed in the next U.N. session, leading eventually to U.N. Security Council Resolution 242 (Quandt 1977, 37–71; S. Spiegel 1985, 155–157).

As part of this diplomacy, U.S. arms transfers to Israel increased drastically. Skyhawk fighter-bombers were sold in 1968. After the 1967 war Johnson restricted arms sales for more than a year in an effort to encourage the Soviet Union to join in a Middle East arms control agreement, but the Soviets never reciprocated. Just before leaving office—in response to the Soviet rearmament of Egypt and Syria—Johnson approved the sale of fifty Phantom jets to Israel, effectively doubling the capability of its air force (Roehm 1980, 34–35; Pollock 1982, 36–38; Safran 1978, 584; Reich 1977, 79–99; Steinberg 1983, 124–125).

The Nixon administration was more concerned about the Arab-Israeli dispute than its predecessor had been, feeling that it was a powderkeg and that the United States should not simply let the Israelis sit on their 1967 gains and hope the Arabs would negotiate (Safran 1978, 432; Steinberg 1983, 119–120). Moreover, the Middle East was viewed in terms of Soviet-American competition rather than as a regional problem (Pollock 1982, 57–58). The Nixon administration therefore chose the Middle East as its primary foreign policy issue, and the Rogers plans (named after U.S. Secretary of State William Rogers) became the formal U.S. position (Rustow 1982, 161). The first Rogers plan was actually the result of informal talks between the Soviet Union and the United States. When the Soviets and Egyptians rejected the American proposal, the United States went public in an effort to encourage further negotiations (Whetten 1976–1977, 14–16; S. Spiegel 1985, 181–189), to establish a position separate from Israel in the eyes of the Arabs (Safran 1978, 585), and to "put the Israelis on notice that their increasing appetite for territory would not receive United States support" (Quandt 1973, 273). In order to get bargaining leverage with Israel, the United States took no action on Israeli requests for arms, particularly for aircraft; arms promised in 1968 were not delivered until after 1970 (Quandt 1978, 124; Steinberg 1983, 124–126).

The Rogers plan called for indirect negotiations between the Israelis and the Arabs leading to an agreement based on the American reading of U.N. Resolution 242, a package settlement in which Israel would withdraw from most of the 1967 territories in exchange for agreements on other disputed areas. The Arabs and Soviets argued that the Israelis should withdraw before negotiations could begin; the Israelis wanted face-to-face negotiations with the Arabs and guarantees for the agreements. The Rogers plan eventually was rejected by all participants except Jordan (Reich 1977, 99–114).

In 1969 Egypt began a war of attrition against Israel in the Suez area. The Israelis successfully retaliated with their air force and then escalated to deep-penetration bombing of Egypt, both to enforce a cease-fire and to try to topple the Nasser regime. Israel's decision was probably influenced by the delivery of Phantom aircraft from the United States, which made such attacks possible (Shlaim and Tanter 1978). In response, the Soviets sent large numbers of troops and surface-to-air missiles to Egypt, and Soviet pilots began flying combat missions. The Israelis asked for more U.S. support, and the United States agreed, in return for Israeli acceptance of Resolution 242 and a cease-fire in 1970, a plan that Egypt also accepted. Then the United States found that the Egyptians and the Soviets were violating the cease-fire, perhaps in part because, just before it had been announced, the Israelis had pretended to attack the Nile Valley to lure Soviet fighters into an ambush; Israeli pilots shot down four of them (Whetten 1976–1977, 19; Mangold 1978, 123–124). For the first time the Soviet Union had successfully waged a war, albeit a limited one, in the Middle East (Reich 1977, 114–171): It had forced the Israelis to end their bombing and had moved Egyptian defense lines up to the Suez Canal.

In 1970 militant Palestinians staged a series of spectacular airline hijackings. This in turn led to a civil war in Jordan between Hussein and the Palestinians, which the Jordanian monarch won decisively. When Syrian tanks entered Jordan, presumably with Soviet knowledge, the United States and Israel prepared to intervene with air and land forces. Syria abandoned its invasion (Mangold 1978, 107–111). The Jordan crisis strengthened the image of Israel as a faithful and useful American ally against the Arab clients of the Soviet Union. Moreover, the United States took an increasingly hard line on Palestinian terrorism. "The end-result was a virtual metamorphosis in US-Israeli relations" (Garratt 1982, 88; see also Safran 1978, 455–456; S. Spiegel 1985, 196–203; Pollock 1982, 114–115).

From the United States' point of view, the Middle East stabilized after the Jordan crisis. Nasser was dead, and the Egyptian-Israeli truce held. Sadat expelled ten thousand Soviet advisors in 1972, and King Hussein controlled Jordan and expelled the Palestinians. A new government in Syria seemed more moderate. The United States became complacent, committed to the notions that time was on the side of Israel and that the military balance was more important than Arab political discontent. In addition, Washington increasingly began to connect Soviet activity in the Middle East to other parts of the world through "linkage" (Sicherman 1978, 16–20), which complicated the settlement of any particular dispute. Thus, the United States supported a new version of the Rogers plan and moved toward verbal acknowledgment of the Palestinians (referring to the "legitimate interests of the Palestinian people"), but it did not pressure Israel to make concessions, even after the Vietnam war was "settled" in January 1973 and Kissinger turned to the Middle East (Shadid 1981, 95–98; Pollock 1982, 115–116). The U.S. policy of "standstill diplomacy" did not change until the 1973 Arab-Israeli war (Quandt 1977, 72–164).

U.S. arms aid to Israel mushroomed. From 1962 to 1968 the United States gave an average of less than $23 million per year for weapons. The comparable figure for 1969–1973 (not including the emergency $2.2 billion to pay for the October war) was $310 million (Roehm 1980, 44; Steinberg 1983, 126).

In the meantime, concern was growing about an energy crisis—the United States was becoming more dependent on imported oil, and prices were rising. In a speech to the Middle East Institute on September 30, 1972, almost a year before the oil crisis, Yamani, the Saudi oil minister, offered to expand Saudi output to 20 million barrels per day in return for "a privileged status for Saudi investments in the United States and a basic modification of American policy toward Israel" ("World Energy Demands" 1972, pt. 1, 95–105; see also Rustow 1977, 507; Stookey 1975, 240; Daoudi and Dajani 1985, 129). Quandt says the proposal was not accepted because it would have encouraged Europe to make similar agreements, not because of pressure from Israel (1976b, 282–283). James Akins, American ambassador to Saudi Arabia, reportedly began a program in 1973 to link the two countries economically under Kissinger's guidance before the embargo. The program apparently selected American corporations that would be helpful to Saudi Arabia and gave them special assistance from the embassy ("U.S. Steering Toward Special Relationship with Saudi Arabia" 1973, 1, 3).

The Oil Crisis

The U.S. intelligence community did not expect a Middle East war in the spring of 1973. The U.S. government had only two hours' warning, and one hour after the attack the Watch Committee of the National Security Council could "find no hard evidence of a major, coordinated Egyptian-

Syrian offensive." This surprise can be attributed to a belief that it would not be rational for the Arabs to start a war, to overconfidence in Israeli intelligence, and to very good Egyptian deception. The United States deliberately did not accuse the Egyptians of aggression (note the contrast with Dutch policy). Instead, for the first week, it tried to stop the war, delaying military supplies to Israel to encourage a cease-fire, in part because it feared an oil embargo if it openly resupplied Israel (Sorley [1983, 90] says this factor was "probably of greatest significance by far"). Israel actually agreed to a cease-fire, but Egypt rejected it. President Nixon then decided on October 13 to institute an airlift to re-supply Israel with arms (S. Spiegel 1985, 250–256; Quandt 1977, 168–183; Quandt 1978, 126).

There has been considerable controversy about how much warning, if any, the U.S. government had of the impending oil embargo. The possibility had certainly been discussed extensively. In April 1973 James Akins, on loan to the White House from his position as director of the State Department's Office of Fuels and Energy Policy, published an article entitled "The Oil Crisis: This Time the Wolf Is Here" in *Foreign Affairs,* the premier journal of the U.S. foreign policy establishment. He asserted that "in 1972 other Arabs in responsible or influential positions made no less than 15 different threats to use oil as a weapon against their enemies" (1973, 467). In May Senator J. William Fulbright (D-Ark.) suggested that military action might be necessary to secure oil supplies in the Middle East (Mangold 1978, 72). Robert Hunter wrote an article for the *New York Times Magazine* in September called "Can the Arabs Really Blackmail Us?" (Rustow 1982, 154–155). Charles Yost, former U.S. ambassador to the United Nations, also discussed the issue (Nau 1977, 332n). A respected English source said in August 1973: "It would now seem that the main question is not so much *whether* the [Arab oil-producing states] will take action to limit their oil output but *when* they will do it and by *how much.* Some observers expect the decision to be taken before the end of the year" (Economist Intelligence Unit, August 17, 1973, 3, emphasis in original).

In addition, conservative Arab leaders, in particular the Saudis, transmitted warnings that were noted but not heeded in Washington. Quandt reports that early in 1973 "visitors to Saudi Arabia began to report that King Faisal, for the first time, was speaking of using the oil weapon to bring pressure on the United States unless Israel were forced to withdraw from Arab territory" (1977, 156). In March the emir of Kuwait asserted that his country would use its oil wealth in any struggle against Israel (*New York Times*, March 14, 1973, cited in Daoudi 1981, 334). In April Faisal sent a mission led by Oil Minister Yamani and Deputy Prime Minister Saud bin Faisal to Washington, where Israel and oil were publicly linked for the first time (al-Sowayegh 1980, 208); Yamani spoke to Kissinger, Rogers, and Secretary of the Treasury George P. Shultz and then went public with the issue in the *Washington Post* (Sheehan 1976, 66). The warning was repeated twice by Saudi Foreign Minister Omar Saqqaf during a visit to Los Angeles in May (Economist Intelligence Unit, August 17, 1973, 2–3).

The issue at this time was whether the Saudis would increase production as the Americans hoped; no one was threatening to cut off oil to the United States altogether (Economist Intelligence Unit, May 2, 1973, 2; Bouchuiguir 1979, 121). Both the threats and demands were severely limited; the Saudis were asking the United States to pressure Israel to adopt their interpretation of Resolution 242, not asking the U.S. to stop arms shipments (Korany 1986, 101). On May 3 and 23 Faisal linked possible use of the oil weapon to the Arab-Israeli situation (Stobaugh 1976, 182–183; Sampson 1979, 291–292). This position was reinforced by a brief, symbolic halt in oil production by Algeria, Kuwait, Iraq, and Libya on May 15, the anniversary of the creation of Israel (Ali 1976, 108).

After the outbreak of the 1973 war, Saqqaf reportedly warned Kissinger on October 12 that oil production would be cut if the United States supplied arms to Israel, although Kissinger does not mention the meeting in his memoirs (Dowty 1984, 249). On the same day a letter from the Arab-American Oil Company (ARAMCO), hand-delivered by attorney John J. McCloy to the White House, asserted that the Saudi government had told ARAMCO that it would reduce oil production and spoke of a "more substantial move" if the United States continued to support Israel. One observer has called this "the Saudi Government's last warning to the U.S. government" (Bouchuiguir 1979, 223–224).

But the Saudis were not consistent. In February 1973 Saqqaf said that Saudi Arabia would not use oil as a political weapon. As late as August 1973 King Faisal and his son, Prince Saud, publicly argued that such a use would be unwise (Daoudi and Dajani 1985, 128–129; Garratt 1982, 89). Also in August, Faisal's brother, commander of the Saudi National Guard, said that "World Zionism" wanted to weaken the oil weapon by causing it to be applied prematurely (Bouchuiguir 1979, 130–131). As a result, the threat of an oil embargo became less useful. One observer concludes that Faisal was trying to use the threat of the oil weapon not to end the U.S. connection with Israel, but to get some tangible result to offset the rise of radicalism in the Arab world (Daoudi 1981, 337); another says Faisal did not want to alienate U.S. and Saudi moderates (Korany 1986, 100).

This was apparently too subtle for the U.S. government, some of whose members saw indecision in such contradictory actions: "Every time that King Faisal would issue a threat, some well-meaning member of his entourage would rush to reassure the Americans that this was merely being said for Arab consumption and that Saudi Arabia would never act against its good friend the United States" (Quandt 1976b, 284; see also Quandt 1977, 160n). The best study of the topic concludes that Faisal's position did not really harden until October 1973, only days before the oil weapon was used (Daoudi and Dajani 1985, 132); if this is true, it is hard to fault American intelligence too much.

Aside from contradictory signals, there were other reasons the Arabs' warnings were not heeded. Malcolm Peck points out that the United States has historically not understood Saudi Arabia well (1980, 232–239). Although the U.S.-Saudi relationship has been important for a long time, it has involved relatively little direct contact, partly because Saudi Arabia preferred to keep foreigners out to avoid disruption of its society. In addition, the scholarly study of Saudi Arabia in the United States has been rather weak.

Within the Nixon administration, secretaries Rogers and Shultz thought the Saudis were bluffing. Others took the warnings more seriously but did not see what they could do without alienating Israel (Peck 1980, 232; Kassis 1981, 70–71). Intelligence field reports on Saudi willingness to use oil as a political weapon were not reflected in final intelligence reports that relied heavily on embassy reporting, which stressed general Saudi cooperativeness with the United States (Kassis 1981, 66–67; Dowty 1984, 206).

Ironically, one reason the embargo caught Washington by surprise was its tardiness. The war began on October 6, but (despite considerable Arab rhetoric before the war about the oil weapon) even the radical states did not interfere with petroleum shipments; Libya, for example, declared it needed oil revenues to finance the Egyptian and Syrian war effort (Bouchuiguir 1979, 146–150). Quandt recounts that "day by day, Washington

officials watched in amazement while the Saudis allowed ARAMCO to pro-
ceed with business as usual. By October 13 the United States was overtly
helping Israel by means of a highly publicized airlift of arms. Still the Saudis did nothing" (1982a, 10; see also Kissinger 1982, 873; Kassis 1981,
88–89). Reportedly, U.S. intelligence had originally predicted an embargo in the first week of the war, then believed a military airlift would
cause one (Dowty 1984, 249); both predictions were incorrect. In any
event, the United States was able to resupply weapons to Israel with impunity, but the American decision to pass a $2.2 billion aid bill to pay for
them triggered the embargo. The U.S. government may perhaps be for-
given for not expecting this fine distinction.

     Quandt asserts that Nixon and Kissinger understood that the arms re-
supply might well bring about the full-scale confrontation with the Arab
world that, so far, had been avoided (1977, 184–185), but "the fear was
not prominent" (Dowty 1984, 240). They were prepared to take the risk
because the arms resupply was aimed at what they viewed as the Soviet
Union's attempt to alter the power balance in the Middle East, an important U.S. national interest. From their point of view, the Arabs were not
the main target. Indeed, as soon as the airlift started, the U.S. government
began to plan for an oil embargo, a prescient move, because on October 14,
the day the airlift was publicly announced, Kuwait called for an OAPEC
meeting, which took place on October 16 (Stookey 1975, 255).

     When the question of an aid bill to pay for the arms was raised in the
United States on October 16, Kissinger and Nixon argued for a large
amount. In doing so, they sought to gain for the United States as much
credit with Israel as possible, because Israeli and U.S. aims would diverge
during peace negotiations. They also felt that the Arabs were already upset
and would not be affected by an aid bill (Sheehan 1976, 69). At noon the
same day Kissinger received a letter from King Faisal saying that the
United States should stop sending arms to Israel and should force it to
withdraw, or U.S.-Saudi relations would become "lukewarm" (Quandt
1977, 184–188). A U.S. reply was never sent, with officials thinking that
anything the United States said would simply anger the king more (Peck
1980, 231–232). On October 17, the day production restraints were announced, Arab diplomats met with President Nixon in an unsuccessful
effort to frame an agreement on ending the war (Daoudi and Dajani 1985,
137–138).

In retrospect, the United States' cavalier behavior is surprising. But it must be noted that although U.S. intelligence expected an embargo, it was not expected to have much impact. In fact, the embargo was initially seen as symbolic rather than significant. Perhaps this is why James Akins describes the government's reaction as "soporific" (Dowty 1984, 250).

The Arabs' actual intentions in using the oil weapon have been seriously debated. This study assumes the Arabs wanted to influence Israeli policy. Nonetheless, there is another argument, that the major purpose of the production cutbacks was to make the September oil-price increases effective (Rustow 1982, 156–157, 160; Feith 1981). (For contrary arguments, see Nau 1977, 335, and Bouchuiguir 1979, 283–325; Daoudi and Dajani 1983, 106, and Daoudi and Dajani 1985, 148, contend that the Arab goals became economic as the embargo went on.) This latter argument implies either that the price hikes were timed in connection with the Yom Kippur war or that the production cutbacks and embargoes would have been applied even if the war had not broken out; J. B. Kelly notes that the Arab oil producers had threatened an oil embargo in 1970 and again in 1971 to raise oil prices (1976, 457). Neither contention seems persuasive to me, but we simply do not have sufficient evidence of what the Arab governments really wanted. At any rate, this study is concerned with the reactions of the target states, so, strictly speaking, Arab motives are a secondary issue. The target governments, including the United States, clearly *believed* the embargo was linked to the Arab-Israeli conflict and acted accordingly (personal interview with a former senior official, February 1985).

While the Arab ministers debated the embargo on October 17, Saqqaf delivered King Faisal's letter to Nixon and Kissinger, threatening an oil embargo against the United States if it sent more aid to Israel. Saqqaf was told that the United States "was committed to helping Israel" (Kelly 1980a, 397). The die had been cast.

Saudi supporters contend that the embargo was a moderate policy, at least in contrast to the alternative of nationalizing all foreign oil interests in the Arab world (al-Sowayegh 1980, 209). In the initial OAPEC declaration of October 17, Saudi Arabia wanted the embargo against the United States to be optional rather than required, still hoping to persuade America to reduce its support of Israel. Thus, on October 18, the Saudis cut their exports only 10 percent and did not embargo the United States. On October 19, however, Nixon asked Congress for $2.2 billion in military aid to Israel, reportedly the average of U.S. and Israeli estimates of the cost of the war (Pollock 1982, 175; Dowty 1984, 268–269). The U.S. action seems to have been the final straw. On October 20 Saudi Arabia announced a ban on exports to the United States (Lenczowski 1976a, 13–14; Bouchuiguir 1979, 158–162).

Kissinger learned of the embargo en route to Moscow to negotiate the terms of a U.N. resolution for a Middle East cease-fire. His mission was successful, and a cease-fire was declared on October 22. There were no provisions for observation or enforcement, however, and the Israelis continued to encircle the Egyptian Third Army in the Sinai. The Soviet Union on October 24 threatened military intervention if Israeli encirclement continued. The United States responded by raising the alert status of its forces around the world in an effort to deter Soviet involvement and pressured Israel, apparently threatening to cut off arms supplies and to supply the trapped Egyptian Third Army itself from Egyptian bases. The result was a new cease-fire that held (Quandt 1977, 191–200; Quandt 1978, 126; Whetten 1976–1977, 32; S. Spiegel 1985, 258–267; Sicherman 1976, 56–57; Mangold 1978, 127–131; Pollock 1982, 177–178).

Once the cease-fire was in place, the United States had to decide on a policy for the ensuing negotiations. The result involved several principles. First, the United States should be actively involved rather than waiting for the parties to solve the problem. Second, the American goal would be a process of negotiations among the parties rather than an American statement of what the final peace result should be. Third, in order to function as an independent negotiator, the United States had to resist pressures from both its allies and the oil embargo to adopt a pro-Arab position. "Much of Kissinger's tactical maneuvering of succeeding months was aimed at ensuring that the United States could act free of the multiple pressures, domestic and international, generated by the October war. On the whole, he was remarkably successful" (Quandt 1977, 210). A fourth principle concerned the Arabs, who would participate in negotiations only insofar as the talks yielded more results than another war would; therefore, concessions would have to be made and aid supplied. Kissinger and Nixon preferred to work with Egypt, Jordan, and Saudi Arabia; they were slower to get involved with Syria and the Palestinians. The fifth principle centered on the U.S.-Israeli relationship as the fulcrum of U.S. leverage; it could not be abandoned. Therefore, pressure on Israel for concessions would have to be balanced by increased aid. Sixth, the Soviet Union was to be left out of the process, and, finally, the negotiations would use "step-by-step" diplomacy, whereby a series of limited agreements—reached through bilateral channels—would build both confidence and a base for further agreements (Quandt 1977, 208–213).

The relationship between all this and the embargo was uncertain, however, in part because Saudi Arabia had trouble communicating its desires clearly to Washington. On November 11 a "not official" message from Yamani asked Kissinger to force Israeli troops to withdraw from the Sinai. On November 17 King Faisal reportedly required that Israel deny sovereignty over Jerusalem and that the termination of the embargo be approved by Syria, Kuwait, Egypt, and Algeria. The next day "an aide to one of the senior princes" said that the 1967 borders were the ultimate aim but that the embargo might be modified if there were "effective and tangible" moves such as the beginning of an Israeli troop withdrawal. On November 19 Yamani said a "timetable" for Israeli withdrawal was required. Then the Kuwaiti communiqué of December 8 required a timetable for withdrawal from all occupied territories, including Jerusalem; Kissinger believed the Saudis had changed their minds under pressure from the radical Arab states (1982, 878–879, 833). These contradictory signals probably weakened the Arab position.

Kissinger's first concern was to bring the parties together in Geneva for negotiations. The conference was held in December, but Syria refused to attend, and it ended with nothing more than propaganda speeches on all sides. Israeli Defense Minister Moshe Dayan asked Kissinger to come to the Middle East in January, and President Sadat asked him to help Egypt and Israel reach an agreement without returning to Geneva (S. Spiegel 1985, 273). Intense negotiations produced a Sinai disengagement agreement on January 17, actually a complex set of agreements in which Egypt and Israel agreed on the locations of their troops and the Egyptians symbolically retained control of both sides of the Suez Canal. As part of this agreement, Sadat flew to Saudi Arabia to ask Faisal to end the oil embargo. Kissinger tried to get a similar agreement with Syria but found it more difficult; the Israelis feared more U.S. pressure, and Syria's President Hafiz Assad was more intransigent than Sadat. Nonetheless, by February Israel and Syria were negotiating (Quandt 1977, 214–230).

This did not end the embargo, however. The leaders of Algeria, Saudi

Arabia, Syria, and Egypt met in Algeria, decided that further progress on Syrian-Israeli relations was necessary, and sent the Egyptian and Saudi foreign ministers to Washington to apply pressure on this issue (Bouchuiguir 1979, 207–208). Different authors have analyzed this decision differently. Quandt (1977, 231–232) says that Nixon and Kissinger were angry but decided to go ahead with their diplomacy anyway and stopped asking that the embargo be ended (see also Kissinger 1982, 893–894). Sheehan sees the Arabs' statement as a major retreat from their original terms of a total Israeli withdrawal from the 1967 territories, attributing this to their fear that the embargo was causing economic harm to Europe and not really influencing the United States (1976, 116).

On March 6 Nixon once again stressed that the embargo would have to be ended in order to allow U.S. diplomacy to work. After some hesitation most Arab countries ended their embargoes on March 18. Nixon and Kissinger had used their lack of success with the Syrians and Israelis to their own advantage, although the Saudis warned the United States that if no progress was made in two months, the embargo might be reimposed.

Kissinger's problem was how to bring the Syrians and Israelis together. He tried to bring Arab pressure on Syria and spent a lot of time in Algeria and Saudi Arabia as well as in Egypt. On Israel he used the carrot of aid and the stick of some very tough language from Nixon, which included threats to reexamine the U.S.-Israeli relationship. The negotiations were extraordinarily difficult. At least twice Kissinger despaired of an agreement, drafted departure statements, and packed to go home. But on May 31 the agreement was signed (Quandt 1977, 229–244).

> Kissinger subsequently denied the "linkage," but in fact his Syrian shuttle was the price he paid to end the embargo—and (in terms of substance) not a high price at that. (He would have undertaken the shuttle anyway—to strengthen Sadat and to end the war of attrition in the Golan Heights.) The Saudis were rewarded with a grandiose American commitment to industrialize their kingdom and to sell them large quantities of modern tanks, naval vessels, and fighter aircraft. (Sheehan 1976, 116)

The Arabs ended the embargo before seriously damaging the American economy, apparently deliberately. One Saudi minister is quoted as saying: "If the embargo were to remain, we should see a major recession in America. That, in turn, would affect all of us adversely. Our economies, regimes, our very survival depend on a healthy U.S. economy" (*Newsweek,* March 25, 1974, cited in Ali 1976, 118). Another authority argues that the Arabs ended the embargo because it was improving America's international competitive position since the United States was less dependent on imported oil than the Europeans and Japanese were (Rustow 1974, 150).

U.S. Policy Toward the Middle East Since 1974

Part of Kissinger's strategy to prevent a new oil embargo was to foster greater economic interdependence between the United States and the Arab oil-producing states, especially Saudi Arabia. There was also some political movement. In April 1974 the United States signed an agreement on economic, technical, and military cooperation with Saudi Arabia, an agreement it had declined to sign the previous year, before the oil embargo. In June a joint economic commission was created with Saudi Arabia (S. Spiegel 1985, 226). Soon afterward the United States voted for a Security Council resolution condemning Israel's raids into Lebanon without mention of the raid that had provoked them. It had abstained on a similar resolution a year earlier (Szyliowicz and O'Neill 1975, 203, 207–208). Arms sales were encouraged, and massive building projects, sometimes for entire cities, were carried out under the direction of the Army Corps of Engineers, among others (Sheehan 1976, 76–77).

Gerald Ford became president in August 1974, but there was little policy change. Kissinger tried to negotiate a settlement between Israel and Jordan to undercut the PLO. But this idea was not pushed strongly, both because of strong domestic resistance within Israel to relinquishing the West Bank and because the basis for a settlement was unclear. Two inconclusive meetings were held with PLO representatives (personal interview with a former senior official, February 1985), but in October at the Arab summit in Rabat members voted unanimously that the PLO was the sole legitimate representative of the Palestinians, effectively torpedoing this effort. Alfred L. Atherton, Jr., who participated in these negotiations, feels that a historic opportunity was missed (1984, 1202).

Kissinger then worked on a second-stage agreement between Israel and Egypt, involving Israeli withdrawal from the Mitla and Gidi passes and the Sinai oil fields. In March 1975 relations between the United States and

Israel hit a new low as the negotiations floundered; Ford spoke publicly of a "reassessment" of U.S. relations with Israel and suspended arms sales. Kissinger seemed to be seriously considering abandoning "step-by-step" diplomacy and adopting the "Geneva option," trying to negotiate a total political settlement for the Arab-Israeli dispute at a conference of all concerned nations. In retrospect this was probably just a tactic to influence the Israelis, because it would have meant deliberately introducing the Soviets into the issue, which Kissinger adamantly opposed.

In the short run, Israel was able to resist U.S. pressure because it had been rearmed and did not need resupply immediately. In addition, Israel had strong congressional support within the United States, particularly shown in a letter that explicitly opposed President Ford's "reassessment" policy and was signed by seventy-six of the one hundred senators. Israel still needed American credits to pay its staggering arms bill, however, and the threatened end of step-by-step diplomacy raised the specter of another Geneva conference, which the Israelis detested. Thus, when Kissinger added incentives to his threats, the Israeli government agreed to some concessions (Pollock 1982, 185–196; Sheehan 1976, 164–178; Reich 1977, 312–325; Wheelock 1978, 129–134; Ben-Zvi 1984, 12–21).

Sadat also made some concessions, and in September 1975 the complex Sinai II agreements resulted in further Israeli withdrawals from the Sinai. Egypt agreed to the presence of U.N. observer forces and to allow Israeli cargoes through the Suez Canal. More important, the United States agreed to increase arms shipments to Israel, guarantee oil supplies until 1980, and even to commit civilian observers to Sinai, a controversial move so soon after Vietnam. Israel was given a veto over participants in any future Geneva conference, and the United States agreed to seek to "insure" that any "substantive negotiations" be directly between Israel and the Arabs. The United States also agreed not to negotiate with the PLO until the organization agreed to U.N. resolutions 242 and 338 and recognized Israel's right to exist; this effectively foreclosed an independent U.S. diplomacy with the PLO and was perhaps the United States' major long-term concession. (Some of these provisions were in a series of secret agreements among the United States, Israel, and Egypt, but they became public a few days after the other agreements were announced [Quandt 1977, 253–276; S. Spiegel 1985, 283–305; Safran 1978, 554–560; Bahbah 1982, 114–120].)

Sinai II was qualitatively different from the 1974 disengagement agreements because it provided not only for a military disengagement separating forces and describing a new status quo, but it also provided the first steps toward increased accommodation between the parties and it moved in the direction of an overall political settlement. It also clearly engaged U.S. prestige, participation, and expenditure in the continuing search for peace in the Arab-Israeli zone of the Middle East. It thus formalized the increased U.S. role and involvement, which had characterized the post–October War period. (Reich 1977, 351)

American policy on the Palestinian issue ran on a somewhat different track. On November 12, 1974, a U.N. resolution was passed "recognizing the rights of the Palestinians to independence and sovereignty in Palestine. . . . Of all the developments in the Middle East since the creation of Israel, the passage of the Palestine resolution is the most significant" (Ali 1976, 120). In November 1975 the United Nations passed a resolution stating that Zionism was racism. Governmental and public reaction in the United States was strongly negative.

But that same month, Harold Saunders, deputy assistant secretary of state for Near Eastern Affairs, told a congressional committee that the Palestinian issue was central to the Arab-Israeli conflict and that the "legitimate interests of the Palestinian Arabs must be taken into account in the negotiating of an Arab-Israeli peace." Although Kissinger had approved the statement, the reaction from Israel and its American supporters was so strong that he disowned it as an "academic exercise." In March the United States did not veto PLO participation in a U.N. Security Council debate on the Middle East, although Ambassador Daniel Patrick Moynihan wanted to do so. Ironically, the United States then had to veto the resulting resolution. William Scranton, in replacing Moynihan, delivered a speech criticizing Israeli policies in East Jerusalem and on the West Bank. In the end, however, despite these tentative feelers, the United States was unwilling to follow up on this apparent attempt to move toward the PLO (Quandt 1977, 278–279; S. Spiegel 1985, 305–307). Ford and Kissinger told Sadat that if Ford won the 1976 elections, they would abandon step-by-step diplomacy and seek a comprehensive solution (Burns 1985, 184, 187).

In early 1976 the situation in Lebanon began to fall apart as civil war and international conflicts merged into increasing chaos. For a time, with tacit U.S. and Israeli support, Syria attacked the Palestinians "to the amazement of all involved" (S. Spiegel 1985, 311). The United States reportedly accepted Israeli assistance to Lebanese Christian forces, which at that time were working with the Syrians against the PLO (Pollock 1982, 165). After the U.S. ambassador in Beirut was assassinated, presumably by dissident Palestinians, Americans were evacuated with the assistance of the PLO, apparently as the result of some quiet U.S.-PLO diplomacy, regardless of the agreements with Israel. Then Saudi Arabia intervened diplomatically, arranging for an agreement in which the fighting would stop and an Arab "peacekeeping force," mostly composed of Syrians, would guarantee order; the package included a reconciliation between Egypt and Syria (Quandt 1977, 281–284). This "Arab civil war" effectively ended Arab pressure on the United States, just in time for the 1976 elections. Israel's reputation was further strengthened by the Entebbe rescue mission in July 1976 (Safran 1978, 564).

As Arab wealth increased, the Arab economic boycott of Israel became a touchy issue. American corporations that wanted new Arab business were under pressure to certify their compliance with the boycott. The Anti-Defamation League of B'nai B'rith sued to force the government to reveal the names of corporations that had acceded to this pressure, and during the presidential election campaign Jimmy Carter attacked Ford on the issue. Ford suddenly declared he would have the Commerce Department make the names public; the next day Elliot Richardson, then secretary of commerce, said this would not be possible (S. Spiegel 1985, 310). Despite this byplay, the issue was not central to the campaign.

The Carter administration took major new initiatives on the Middle East, based on assumptions that differed from those of Kissinger:

> that a rapid resolution of the conflict was desirable given the persistent danger it posed to American interests; that direct external pressure from the United States was required to bring about a settlement; that such pressure should be aimed more pointedly at Israel than . . . in the past; that a comprehensive approach which dealt with the most intractable of the issues—Palestinian self-determination and the role of the PLO—was preferable to the step-by-step approach of Dr. Kissinger; and that the Soviet Union should be provided an incentive to cooperate in a settlement, since Moscow would very likely spoil any arrangement it did not approve. (Garfinkle 1983, 19)

Steven Spiegel actually argues that the oil weapon had more of an impact on the Carter administration than on the previous two. He also contends that members of the new administration believed that the United States faced an energy crisis and that, because they could not get their domestic energy program through Congress, they had to propitiate Arab oil producers. "As had happened before, the first new presidency to follow a crisis entered office prepared to develop new policies to deal with the changed conditions" (1985, 317; see also Feith 1980, 20).

Initially Carter tried to revive the comprehensive approach with a second Geneva conference. As part of that effort, he took a new initiative on the Palestinians, declaring in March 1977 in Clinton, Massachusetts, that the Palestinians must have their own "homeland" (Rustow 1982, 242; S. Spiegel 1985, 332; Safran 1978, 567–570), although he reportedly did not mean to send any signals to the PLO (Quandt 1986, 48–49). In August Carter approved a law making it easier for PLO members to visit the United States. He also used a number of prominent Americans in private life to communicate with the PLO (Shadid 1981, 133–134, 141–143). This U.S. policy shift encouraged the EC heads of government to issue a statement on June 29, 1977, calling for a Palestinian homeland (Ramazani 1978, 52; Sicherman 1980, 848–849).

In May 1977 Menachem Begin was elected prime minister of Israel, much to the surprise of the U.S. government. Unlike its predecessor, the Begin government interpreted the withdrawal provisions of Resolution 242 to apply only to the Sinai, not to the West Bank and Gaza (Atherton 1984, 1205). American-Israeli relations deteriorated. Carter opposed arms sales to Israel primarily because of his attitude toward arms sales in general, although he was also unhappy about Israeli settlement policies in the occupied territories.

In October 1977 the Soviets and Americans issued a joint communiqué specifying the terms under which they would cochair a new Geneva conference including "representatives of the Palestinian people." Because most states recognized the PLO as the representative of the Palestinian people, this was interpreted by some as a move toward recognition of the PLO by the United States (Shadid 1981, 136–137). Both Israel and the Israeli lobby in the United States reacted strongly against the possibility of a U.S.-Soviet condominium in the Middle East. The resulting meeting with Moshe Dayan led to a compromise agreement whereby Israel would participate in a Geneva convention with a single Arab delegation that would include Palestinians. Syria refused to participate, although negotiations continued. Carter's Geneva diplomacy seemed stalemated (S. Spiegel 1985, 329–340; see also Vance 1983, 192–194; Quandt 1986, 96–134; Ben-Zvi 1984, 22–32).

Despite its agreement with the United States, Israel had always opposed a Geneva conference, preferring to negotiate directly with the Arabs. Begin found an unexpected ally in Anwar Sadat, with whom he had established good relations by informing him directly of a Libyan plot against Sadat's government. After private negotiations—some conducted through the U.S. ambassadors to Egypt and Israel (Quandt 1986, 146–147)—Sadat made a dramatic trip to Jerusalem in November 1977. This gesture forced the U.S. government to revert to the piecemeal strategy it had rejected, because an agreement between Egypt and Israel now seemed possible, even though the other Arab states refused to participate. American efforts to facilitate the negotiations led to the Camp David accords of 1978, which called for Israeli evacuation of the Sinai, full diplomatic recognition of Israel, and a series of timetables for further negotiations on other problems, although Begin refused to allow any formal connection between the Palestinian and Sinai issues (Quandt 1986; S. Spiegel 1985, 340–361).

The accords were undercut, however, by a fundamental misunderstanding between the United States and Israel: The Americans thought they had an Israeli commitment to freeze new settlement activity in the occupied territories during negotiations and said so to the Arabs, whereas Begin insisted he had agreed to a freeze for three months only (Vance 1983, 224–225, 228–229; Atherton 1984, 1206; S. Spiegel 1985, 362; Quandt 1986, 161–162). When the subsequent direct negotiations between Egypt and Israel broke down, President Carter flew to the Middle East himself and negotiated the Egypt-Israel peace treaty of 1979. One critical element here was the U.S. promise to guarantee Israel's oil supplies until 1990, extending by ten years the time specified in the Sinai II accords (Schneider 1983, 432–433; Bahbah 1982, 123–130).

Camp David accomplished several things. It effectively brought peace between Egypt and Israel. This may well have reduced pressure on the Israeli government to grant further concessions on other issues, particularly the Palestinians, although the Begin government would probably not have done so in any case. Camp David also established a piecemeal negotiating process that never produced much. Also, Arab reaction was negative, so

Egypt was isolated within the Arab world, although the break was never total. For example, the Saudis broke diplomatic relations with Egypt but did not evict the many Egyptian workers in Saudi Arabia. Nor was there any serious talk of a new oil embargo. Finally, Camp David survived the political demise of all its major participants.

Arab opposition to Camp David led the Carter administration to respond favorably in the summer of 1979 to an Arab proposal to alter U.N. resolutions 242 and 338 to call for "Palestinian rights" (Garfinkle 1983, 26–27). In the complex negotiations surrounding this initiative (opposed by both Israel and Egypt), U.N. Ambassador Andrew Young met with a PLO representative in New York. When asked about it, Young first said the meeting had not occurred, then admitted it had. Secretary of State Cyrus Vance demanded that Young resign, and Carter reluctantly allowed him to do so in August (Quandt 1980, 547). Shortly thereafter the United States voted for a U.N. resolution condemning Israeli settlements, including Jerusalem. Although this vote was in error owing to a communication problem between Washington and its U.N. delegation, the administration seemed willing to support the resolution if references to Jerusalem were deleted (S. Brown 1983, 501–503). In any event, the crises in Afghanistan and Iran effectively ended the Carter administration's efforts to resolve the Arab-Israeli dispute (Brzezinski 1983, 437–443). "It was ironic that as this administration achieved the greatest American diplomatic victory in the Middle East, other regional conflicts began overshadowing its success" (S. Spiegel 1985, 376).

When Ronald Reagan became president in 1981, his administration focused on possible Soviet aggression in the Gulf and essentially ignored everything else in the region. More than any other regional policies, Middle East policy was made by top-level political figures. Problems were thus defined in terms of worldwide issues rather than regional ones. Eventually, the sale of AWACS aircraft to Saudi Arabia, Sadat's assassination, and the Israeli use of American-made aircraft to destroy an Iraqi nuclear reactor demonstrated that this was not a viable strategy. Relations with Israel declined to the point that the Israelis believed their invasion of Lebanon had the tacit approval of Washington. In the ensuing debates, Secretary of State Alexander M. Haig, Jr., resigned at the height of the crisis. Relations between the United States and Israel continued to deteriorate as Israeli troops moved deeper into Lebanon, followed by the massacres in Sabra and Shatila and growing feeling that Israel intended to remain permanently in Lebanon. American troops were put into Beirut as a peacekeeping force, withdrawn, then brought back again after the massacres (S. Spiegel 1985, 412–418, 425; Ben-Zvi 1984, 41–57). Despite a call from Iran, the oil weapon was not used during this first, large-scale armed conflict between the Arabs and Israelis since 1974, presumably because of the world oil glut and the fact that the Arab combatants were limited largely to Palestinians (Schneider 1983, 497).

The Reagan administration had been much less sympathetic toward the European approach to the Arabs, particularly the PLO, than the Carter team had been (Garfinkle 1983, 67). But when the Reagan administration finally took the diplomatic initiative, the Europeans were pleasantly surprised. In 1981 the State Department established a secret contact with the PLO through John Edwin Mroz, an American researcher and consultant. Some groups within the administration supported the recognition of the PLO, perhaps as the price of its withdrawal from Beirut; Reagan promised Arab diplomats that he would address the Palestinian problem after the PLO withdrawal.

The 1982 Reagan plan attempted to deal with the Arab-Israeli issue directly. It looked much like earlier American plans but also included an explicit statement of the U.S. vision of a settlement, including a Jordanian-Palestinian state on the West Bank and in the Gaza Strip. Like most earlier U.S. plans, it was rejected by both sides (S. Spiegel 1985, 418–423; Rubin 1983, 367–389; Pollock 1982, 275–282; S. Brown 1983, 580–584, 612–617).

The Arab rejection of the Reagan plan, coupled with the resignation of Ariel Sharon over the Palestinian camp massacres, improved U.S.-Israeli relations immediately. The Reagan administration suddenly decided that Syria, not Israel, was refusing to leave Lebanon, and it reacted accordingly—even encouraging the Israelis to remain in Lebanon. But when a truck bomb killed 241 U.S. Marines in Beirut on October 23, 1983, the United States withdrew its troops.

External Influences on U.S. Policy

U.S. policy was influenced in different ways by a number of actors outside its own borders. Prominent among these external influences were America's European and Japanese allies, the Soviet Union, and the Arab states.

Pressure from Western Europe and Japan

Most observers assume that the Europeans encouraged the United States to reduce its support of Israel. Direct pressure, however, seems to have been vague and general: "With the possible exception of the Netherlands, our European allies were clear about what should follow the cease-fire: American pressure to induce Israel to return immediately to the 1967 borders. . . . None of them had any clear-cut idea of how to accomplish this" (Kissinger 1982, 717). Gil Carl AlRoy argues that European leaders were actually unhappy when the United States restrained the Israelis in the Sinai because this made the oil weapon credible (1975, 96). But there is no evidence that the U.S. government perceived this to be true (personal interviews with Henry Nau, June 1984, and a former senior official, February 1985).

We may assume that the Arabs applied pressure on the Europeans and the Japanese so that they in turn would exert their influence on the United States. In fact, however, the target governments felt they had little or no influence on U.S. policy toward Israel. Some regretted this lack of clout, and it was commonplace to attribute U.S. policy to the Zionist lobby rather than rational choice, but no one seemed to feel they could do very much about U.S. policy, even when it seemed to get America's allies into trouble not of their own making ("Using the Oil Crisis" 1973, 11).

The most significant action taken by the Europeans to pressure the United States was the series of decisions refusing the use of European bases for U.S. airlifts to resupply Israel. Kissinger in particular saw the airlift as important, not so much to preserve Israel (which he never felt was in much danger) but to keep pace with the Soviets' military shipments to the Middle East. Kissinger is reported to have wanted the United States to ship more daily tonnage to the area than the Soviets and was outraged if this did not happen. In any event, Europe's denial of bases was not a major problem for the United States. In fact, there was some quiet cooperation: The British allowed the use of their airfield for the SR-71 reconnaissance flights over the Middle East, the West Germans allowed weapons from U.S. forces to be sent indirectly to Israel, and Spain agreed to allow aerial tankers for the airlift to use its airbases. But when Portugal initially refused the use of the Azores, Kissinger was outraged, and Nixon sent the Portuguese premier "the toughest letter from one head of state to another that I have ever seen," according to a former senior official (personal interview, February 1985). Portugal reversed its position within twelve hours. The U.S. government therefore appears not to have felt much pressure at the time.

The Europeans may underestimate their impact on U.S. policy, however. In separate analyses, Henry Nau (1980a, 40) and William Quandt (1976b, 286) argue that the 1973 crisis was defined by the U.S. government as two political problems: (a) possible limits on U.S. freedom of action in the Middle East, and (b) concern about the weakening Atlantic alliance in terms of the East-West conflict. They contend that the threatened dissolution of the alliance caused the U.S. government to accommodate the Arab states by negotiating a cease-fire and a partial Israeli withdrawal from the Sinai. Nau adds that the United States then had to accept higher oil prices because asking for lower prices would only have given the Arabs more leverage. Of course, the higher prices were more of a burden for the other industrial countries, which were more dependent on imported oil.

Pressure from the Soviet Union

The U.S. government's approach to the oil crisis was dominated by concern for the Soviet Union, especially during the Kissinger era. Policy in this regard was complex and ambivalent, reflecting the variety of issues involved. On the one hand, both Nixon and Kissinger assumed that the Soviets had at least known of the Egyptian attack in advance and had not warned the Americans. Unlike the 1970 Jordan crisis, however, the 1973 war was not seen as a U.S.-Soviet confrontation. Concern about possible Soviet intervention encouraged the Americans to restrain the Israelis from scoring a major victory (although, because Kissinger's negotiating strategy called for a stalemate, this would probably have occurred anyway). Kissinger even reportedly tried to make a trade, increasing arms aid to Israel in return for dropping the Jackson amendment tying Soviet emigration policy to most-favored-nation trade status; the Conference of Presidents of Major Jewish Organizations agreed, but Jackson refused.

As the October war wore on, Kissinger tried to get the Soviets involved in a cease-fire proposal. Concern about Soviet arms shipments also caused him to push for a quick cease-fire. Once such shipments began, the issue of "sending a clear message to the Soviets" became an important justification for the U.S. arms shipments, which, of course, triggered the embargo (S. Spiegel 1985, 249–250, 257).

Pressure from the Arab States

The Arabs influenced U.S. policy by instruments other than the oil weapon. In particular, the 1973 war itself altered at least three assumptions that had governed U.S. policy toward the Arab-Israeli conflict. The war showed that (a) Israeli military power could not guarantee stability, (b) U.S.-Soviet détente could not prevent regional conflicts and possible superpower interventions, and (c) Arab capabilities had to be respected (Quandt 1977, 201).

Arab influence on U.S. policy was strengthened under the Carter administration, which took office concerned to avoid another embargo because of worry about the energy crisis as well as a more general desire to improve relations with the Third World. This administration was certainly more interested in obtaining a comprehensive settlement of the Arab-Israeli dispute, perhaps even at the expense of Israel, than those before or since (S. Spiegel 1985, 316–320; Feith 1980, 20). Such an analysis suggests that the real influence of external actors is as much in the heads of the targets as in concrete actions.

Societal Influences on U.S. Policy

Public Opinion

We have noted earlier the uncertain relationship between public opinion and foreign policy in a democracy. Nonetheless, public opinion does have an impact, at least by raising issues to prominence and perhaps by excluding certain options, if not necessarily by directly swaying government policy. As discussed in previous chapters, it is important to divide a discussion of public opinion regarding the Arab-Israeli dispute into two separate issues: Which side is favored? And what should U.S. policy be?

It is often assumed that American public opinion has always favored Israel over the Arabs. Fortunately, we have fairly good data on this issue from a series of Gallup public opinion polls, supplemented by a few others.

The data suggest several conclusions (table 5). At no time has sympathy for the Arabs approached that for Israel, but neutrality (sympathy for "neither side") has also often been high. When sympathy for Israel declines, the result is usually an increase in neutrality (perhaps better expressed as "a

Table 5.   U.S. Public Opinion: Sympathies in the Arab-Israeli Dispute

	Favor Israel (%)	Favor Arabs (%)	Favor Neither (%)	No Opinion (%)
October 1947	24	12	38	26
1967 War				
June 1967	56	4	25	15
January 1969	50	5	28	17
March 1970	44	3	32	21
October 1973	47	6	22	25
1973 War and Oil Weapon				
December 1973	50	7	25	18
December 1973[a]	54	8	24	14
October 1974	55	9	22	14
January 1975	44	8	22	26
April 1975	37	8	24	31
June 1977[b]	44	8	28	20
October 1977[b]	46	11	21	22
Sadat Trip to Jerusalem				
December 1977[b]	44	10	27	19
Breakdown in Israeli-Egyptian Talks: Sadat Visits U.S.				
February 1978[c]	33	14	28	25
March 1978	38	11	33	18
May 1978	44	10	33	13
Camp David Summit Begins				
September 1978 (1)	41	12	29	18
Camp David Agreement Announced				
September 1978 (2)	42	12	29	17
November 1978	39	13	30	18
January 1979	40	14	31	15
March 1979	38	12	33	18
April 1979[d]	47	11	19	22
December 1979[e]	49	6	16	29
October 1980	45	13	24	18
August 1981	44	11	34	11
January 1982	49	14	23	14
April 1982	51	12	26	11
Israeli Invasion of Lebanon				
June 1982	52	10	29	9
July 1982	41	12	31	16

Continued on next page

Table 5, *continued*

	Favor Israel (%)	Favor Arabs (%)	Favor Neither (%)	No Opinion (%)
Sabra and Shatila Massacres				
August 1982	52	18	30	f
September 1982	32	28	21	19
November 1982	39	23	38	f
January 1983	49	12	22	17
July 1983	48	12	26	14

SOURCES: Data from 1947 to 1970 from Gallup 1972, 687, 2068, 2181, and 2242, respectively. October 1973 and the first December 1973 data are from *Gallup Opinion Index* 1979a, 28; the second December 1973 data and January, April, and June 1975 are from Gallup 1978, 220, 408, 458, and 1121, respectively. October 1974 data from a Yankelovich poll found in Curtiss 1982, 190. October and December 1977 and all 1978 and 1979 data from *Gallup Report* 1982, 4, except for the Yankelovich polls of April 1979 and December 1979, which are taken from Gilboa 1985, 32–33, and the figures for March 1979, which are from *Gallup Opinion Index* 1979b, 8. Gallup data for October 1980 from Gilboa 1985, 32–33. Data for August 1981–July 1982 from *Gallup Report* 1982, 4. August 1982 and November 1982 figures from Novik 1986, 25; September 1982 and January and July 1983 from Gilboa 1985, 32–33.

NOTE: These questions are usually, but not always, asked only of people who state that they have heard of or read about the Middle East situation; in recent years that has been 90 percent of those approached.

[a]This seems to be a different set of results for the same survey; I have assumed that the first one, reported by the most sources (*Gallup Opinion Index* 1979a, 28; *Gallup Report* 1982, 4), is correct.

[b]*Gallup Opinion Index* 1978a, 4, gives different results for the surveys of June, October, and December 1977. These results are not repeated elsewhere, and I have assumed they are incorrect. Each would reduce reported strength of support for Israel from 4 to 7 percent, reduce support for the Arabs and for neither side 1 to 2 percent each, and increase the percentage who had no opinion from 4 to 11 percent. These differences seem too large to be simple errors, but they would not fundamentally alter the analysis here if they were correct.

[c]Telephone survey.

[d]Yankelovich poll.

[e]Yankelovich poll; the question was: "If war should break out between Israel and the Arabs, with whom would your sympathies lie?"

[f]"Favor Neither" and "No Opinion" responses not separated.

plague on both your houses"). Sympathy for the Arabs remains low but has slowly risen and is less subject to vagaries than is sympathy for Israel. The embargo seems to have helped Israel in American public opinion, at least in the short term (Gruen 1975–1976, 35; Gilboa 1986).

Public support for Israel declined irregularly through the 1970s. The reasons why the decline started so abruptly between October 1974 and January 1975 are unclear. Public support hit a somewhat suspect low (the survey was a telephone poll) after Sadat's trip to Jerusalem produced no peace agreement. Sympathy increased during the early 1980s, however, even rising above the 1967 level. This trend was interrupted only briefly by the 1982 Israeli invasion of Lebanon (Gilboa 1985, 41). Just after the Sabra and Shatila massacres, support for Israel dropped to an all-time low point, and support for the Arabs almost equaled it. Within a few months, however, the figures had returned to their usual pro-Israeli balance.

Public sympathy for Israel declined while the Carter administration, widely seen as rather unsympathetic to Israel, was in office. Similarly, public support increased under the Reagan administration, which was verbally more supportive of Israel. This is apparently an example of public opinion following foreign policy, a common phenomenon in foreign affairs. Because more than 30 percent of those polled continued to support Israel even under the Carter administration, they can be considered fairly hardcore supporters. Other observers put the figure at about 25 percent, call it the only veto group on the Middle East in American electoral politics, and note that it is mostly composed of non-Jews (Lipset and Schneider 1977, 29). A sympathetic administration can add another twenty to thirty percentage points to that level. The pollsters attributed the 1982 decline to increasing concern over the Israeli invasion of Lebanon; 49 percent disapproved and only 23 percent approved (*Gallup Report* 1982, 6). This decline in sympathy for Israel is roughly equivalent to that produced by an unfriendly American administration; again, we find a very significant core of support for Israel in American public opinion.

Although Americans remained fairly sympathetic to Israel, they were reluctant to get involved, not particularly sympathetic to U.S. military aid to Israel, and deeply opposed to possible use of U.S. troops. Between 1967 and 1975 the Gallup organization asked (phrasing the question somewhat differently at different times) what U.S. policy should be toward the Arab-Israeli conflict. The most popular reply was "stay out of conflict"; it drew from 41 to 61 percent of the response. "Negotiate" was

the response of from 7 to 14 percent; "work through the U.N." was suggested by from 2 to 11 percent. "Support the Arabs" was never supported
by more than 1 percent. The high point of support for Israel was in 1967
after the war, when 16 percent supported military aid and 5 percent supported sending troops. A year later, support for troops dropped to 1 percent; support for aid remained at about 10 percent (Gallup 1972, 2068,
2149, 2242; Gallup 1978, 408).

On more specific policy issues, several polling organizations have also
asked a more direct question, whether the United States should sell and/or
give arms to Israel. The results can be fairly described as chaotic (table 6).
After 1974 the Harris organization seems to be polling a different public

Table 6.   U.S. Public Opinion: Should the U.S. Sell and/or Give Arms
            to Israel?

	Yes (%)	No (%)	No Opinion (%)	Poll
November 1955	25	50	26	National Opinion Research Corporation
April 1956	19	63	16	National Opinion Research Corporation
June 1967	35	39	26	Harris
July 1968	24	59	17	Gallup
July 1971				
jet fighters	36	42	22	Harris
anti-aircraft				
missiles	39	40	21	Harris
October 1973	46	34	20	Harris
October 1973	37	49	14	Gallup
Winter 1974–				
1975	65	21	14	Harris
1976	65	23	12	Harris
November 1976	37	42	21	Roper
January 1977	59	19	22	Harris
March 1978	31	56	13	Gallup
1978	68	19	13	Harris
1980	75	15	10	Harris

SOURCES: Figures for November 1955–October 1973 and for March 1978 from Suleiman
1980, 32. Figures for 1976, 1978, and 1980 Harris polls found in Novik 1986, 23; January
1977 Harris poll figures from Trice 1978, 242, 242n. Data for Winter 1974–1975 and November 1976 from Lipset and Schneider 1977, 23–24.

NOTE: Some rows may not add up to 100 percent because of rounding errors.

from everyone else: All other polls, including earlier ones by Harris, show the public more or less evenly divided on the issue, whereas Harris after 1974 shows strong support for arms to Israel.

None of the explanations for these differing results is entirely satisfactory. In the first place, the wording of the question varies over the years, which is unfortunate given the public's apparent differentiation between selling and giving arms. A Harris poll in 1974–1975 found that 35 percent were in favor of selling military weapons to friendly nations, but only 22 percent favored giving military supplies to friendly nations (Lipset and Schneider 1977, 23); by July 1981 Harris's figures were 55 percent and 37 percent, respectively (Hastings and Hastings 1983, 304).

But these differences in wording do not account for the staggeringly divergent responses in the results of various polling organizations. Lipset and Schneider explain the differences by arguing that the public dislikes military aid, noting that it is more supportive of aid to Israel than to other countries and also that the 1974–1975 Harris questions were placed in the middle of a series of other questions on the Middle East, suggesting to respondents that the question was about Israel and not military aid (1977, 23–24). Whether this is true of all subsequent Harris polls is less certain, but these major differences confirm suggestions by specialists on public opinion polling that Harris polls on public issues cannot be regarded as valid unless confirmed by data from other organizations (personal interviews with public opinion specialists, July 1987 and August 1987).

I have chosen to accept the majority position on this issue and conclude that the polls do not confirm the monolithic public support for Israel so often pictured. The arms sales question was asked twice in October 1973, when the U.S. government was rearming Israel; the results differed, but both showed the public to be deeply divided. On the other side of the same issue, in July 1981 a Harris poll showed similar divisions on the sale of AWACS to Saudi Arabia, with 48 percent opposed and 42 percent in favor (Hastings and Hastings 1983, 305).

Conversely, there seems to be deep resistance among the American public to abandoning military assistance to Israel under direct Arab pressure (table 7). Richard H. Curtiss notes that university-educated professionals, and those with relatively high incomes, are somewhat more likely to support the Arab position. But they are also most opposed to concessions under pressure; the movement toward the Arabs came not after the oil em-

Table 7.   U.S. Public Opinion: Arab Oil and Arab-Israeli Policy

We need Arab oil for our gasoline shortage here, so we had better find ways to get along with the Arabs, even if that means supporting Israel less.

	Agree (%)	Disagree (%)	Not Sure (%)
October 1973	26	50	24
January 1974	23	61	16
December 1974	20	68	12
March 1975	26	61	13
January 1976	23	65	12
March 1979	39	55	6
October 1979	33	60	7

Now, if it came down to it and the only way we could get Arab oil in enough quantity would be to stop supporting Israel with military aid, would you favor or oppose such a move by the United States?

	Agree (%)	Disagree (%)	Not Sure (%)
December 1974	18	64	18
March 1975	23	55	22
January 1976	20	61	19
March 1979	31	57	12
October 1979	29	60	11

SOURCES: Hastings 1981, 94; Curtiss 1982, 201.

bargo, but following Sadat's journey to Jerusalem. As Curtiss explains: "It seems fair to deduce that American public opinion reacts negatively to any implication of bullying or blackmail on Middle Eastern issues. It will only change or be persuaded if given a relatively altruistic rationale for support-ing a step, even if that step is also in America's own self-interest" (1982, 201). Similarly, an Israeli observer suggests that American public support for Israeli policies might weaken significantly if the Arabs persuasively demonstrated a willingness to accept peaceful coexistence (Novik 1986, 35–36).

Although the American public has been divided over the question of military assistance for Israel, there has never been much support for send-ing U.S. troops (table 8), except in 1956—just after the Suez crisis—when Gallup found 50 percent of those polled supporting a promise to

Table 8.   U.S. Public Opinion: Sending U.S. Troops to Help Defend Israel

	In Favor (%)	Opposed (%)	Not Sure or No Opinion (%)	Poll
June 1967	24	56	20	Harris
October 1967	22	59	19	Harris
July 1968	9	77	14	Gallup
August 1970	38	38	24	Harris
October 1970	35	51	15	Harris
January 1971	39	44	17	Harris
July 1971	25	52	23	Harris
January–April 1971	11	44	45	Gallup
February 1973	31	52	17	Harris
1974	23			Harris
December 1974	27	50	23	Harris
December 1974 (leaders polled)	41	44	15	Harris
April 1975	12	42	46	Gallup
1976	23			Harris
July 1978	21			
November 1978	22			
January 1980	43			
February 1980	35			
August 1980	29			
February 1981	26			
July 1981	28			
October 1981	28			
December 1982	30			

SOURCES: Suleiman 1980, 30; Novik 1986, 22.

send U.S. troops if the Soviets sent troops first (Suleiman 1980, 29). In 1947 65 percent were in favor of sending a "U.N. army" if war broke out in Palestine, but only 3 percent recommended U.S. troops (Gallup 1972, 686). The Gallup question about alternative policies is probably not very useful here because it asks what the government should do when there is no particular need to send U.S. troops. But other public opinion surveys have asked about more extreme situations. For example, in March 1970 "only one-third of the American population was willing to send troops to prevent Israel from completely 'going under' to Russia and the Arabs, two-thirds were willing to send troops to fight in Berlin, in case Russia and East Germany closed approaches to West Berlin" (Raab 1974, 26). In

1975 the Chicago Council on Foreign Relations poll found only 24 percent of Americans were prepared to use force if Israel's existence were threatened, compared to 37 percent who would do so in defense of the Philippines and 33 percent who would support West Berlin (Howe and Trott 1977, 293; see also Suleiman 1980, 29–34).

In a sophisticated analysis of eleven such poll questions, a State Department public affairs analyst suggested that responses depend on how the question is asked. Only about 10 percent were willing to send U.S. troops to defend Israel if the question did not specify that Israel was in great danger and if simple military assistance was one alternative. Changing each of these conditions produces an extra 10 percent, so that about 30 percent of the respondents support the use of U.S. troops if no other military assistance option is available and if Israel is in serious danger of defeat (Curtiss 1982, 204–205; see also Lipset and Schneider 1977, 24). This support is not trivial, but it is also not overwhelming, as Robert H. Trice explains: "Frequent interest group reminders of the 'special relationship' between the United States and Israel notwithstanding, American . . . mass public opinion concerning the Arab-Israeli conflict does not qualify United States support for either Israel or the Arab states as a national tradition" (1978, 244).

This distinction between general sympathy and lack of support for government assistance shapes the impact of public opinion on policy. Thus, even in 1967, with popular support for Israel at its peak, the U.S. government felt free to impose an arms embargo on the warring powers, which obviously hurt Israel more than the Arabs. Subsequent U.S. involvement in mediation efforts "in effect ignored the preferences of the largest opinion cluster within the mass public," which preferred that the United States stay out altogether. Interestingly, as the negotiations became routine and then produced results under Kissinger in the 1970s, public opinion became more supportive of American involvement, despite opposition from the Israeli lobby. This suggests once again how public opinion is often led by government policy rather than vice versa (Trice 1978, 244).

At the same time, the general sympathy for Israel sometimes makes it hard for the U.S. government to aid friendly Arab governments. The 1967 arms embargo was designed in part to persuade the Soviets to halt arms sales to the radical Arab governments as well as to end the war. When the Soviets continued their arms sales after the war, Washington felt it necessary to rearm Jordan for fear it would turn to Soviet weapons. Government officials believed, however, that public opinion would not allow them to sell arms to Jordan without also selling them to Israel. The upshot was a package deal whereby Israel and five Arab countries received arms (S. Spiegel 1985, 158–164).

We have conflicting testimony about the relationship between the oil weapon and American public opinion. On the one hand, as one analyst describes it:

> Mr. Nixon, empowered by the Arab action, forced Israel to accept a cease-fire as it started to get the upper hand against its adversaries. This was pressure of a kind which American public opinion would not have let him apply had there been no Arab boycott or threat of one. Now, overnight, it had become a commonplace among Americans that one could no longer antagonize the Arabs for such peripheral considerations as the voting allegiance of a small, vocal Zionist minority. (Rand 1975, 318–319)

On the other hand, Quandt argues that Nixon and Kissinger were concerned that public opinion might turn against the Arabs and prevent a strengthening of ties with various Arab governments (1975, 45; 1976b, 289–290). In fact, the polls showed little if any change in public attitudes toward Israel (Garnham 1976, 295–309).

> In a survey of public attitudes carried out in December, 1973, almost immediately after the announcement of the Arab oil embargo, Americans tended to blame the energy crisis on the oil companies (25 percent), the federal government (23 percent), the Nixon Administration (19 percent) or US consumers (16 percent). While 7 percent blamed the "Arab nations," hardly anyone blamed the Israelis—the main focus of Arab concern and the motivating factor behind the embargo. Apparently, to most Americans energy is not their number one problem, and in any case the oil companies and "the government" are to blame. Somehow the message that the Arabs wanted to convey to Americans via the oil embargo has not been delivered adequately, if at all—at least not to the general public. (Suleiman 1980, 22)

Raab notes that American public opinion supported the aid to Israel, which was the catalyst for the oil embargo (1974, 28). But he concludes that American public opinion on Israel is not solidly structured and will therefore, as on most foreign policy issues, be responsive to government leadership:

> Consequently, if the national leadership pulls back, the American public will not demand support for Israel. If the leadership continues to insist on the crucial connection between Israel's survival and basic American interests, then the

American public will support, however reluctantly, the perceived imperatives of American foreign policy. The decisive consideration is and will remain the strength of Washington's will. (Raab 1974, 29)

Analysts often assume that American public opinion explains why U.S. policy has often supported Israel. In Quandt's view, "because public opinion generally favors Israel, Israel's security has become an important concern of the United States in a way that is not true for France or Great Britain" (1973, 282). In fact, however, as we have already noted, although American and British public opinion on the subject seem quite similar (table 9), this has not prevented the two governments from developing quite different policies toward the region.

Table 9.   U.S. Public Opinion: Sympathies in the Arab-Israeli Dispute, Compared to British and Canadian Attitudes

		Favor Israel (%)	Favor Arabs (%)	Favor Neither (%)	No Opinion (%)
United States	*June 1967*	*56*	*4*	*25*	*15*
United Kingdom	June 1967	59	4	22	15
United States	*January 1969*	*50*	*5*	*28*	*17*
United Kingdom	January 1969	41	8	29	21
United States	*March 1970*	*44*	*3*	*32*	*21*
United Kingdom	March 1970	46	6	24	24
United States	*October 1973*	*47*	*6*	*22*	*25*
United States	*December 1973*	*50*	*7*	*25*	*18*
Canada	November 1973	22	5	42	31
United Kingdom	December 1973	35	10	36	19
United States	*April 1975*	*37*	*8*	*24*	*31*
United Kingdom	July 1975	33	8	31	28
United States	*July 1982*	*41*	*12*	*31*	*16*
United Kingdom	August 1982	25	16	40	19
Canada	October 1982	17	13	70	a

SOURCES: Data for the United Kingdom from Gallup 1976, 932, 1032, 1092–1093, 1290, and 1425, respectively; 1982 figures for the United Kingdom and Canada from a Gallup poll, data from the Roper Center, University of Connecticut, obtained through the Center for Computer and Information Services, Rutgers University. Data for Canada for 1973 taken from Benesh 1979, 24. Data for the United States for 1967, 1970, and 1973 found in *Gallup Opinion Index* 1979a, 28; 1969 data from Gallup 1972, 2181; 1975 data from Gallup 1978, 458; 1982 data from *Gallup Report* 1982, 4.

NOTE: Some rows may not add up to 100 percent because of rounding errors.

[a]"Favor Neither" and "No Opinion" responses not separated.

224 POLITICAL POWER AND THE ARAB OIL WEAPON

What accounts for the difference? The United States has organized interest groups that build on such public opinion and, more important, possesses a set of political institutions unusually vulnerable to organized pressures at the level of the individual constituency. Both factors will be discussed later in this chapter.

Mass Media

The quantity and quality of American media coverage given to the Arab-Israeli conflict greatly increased as a result of the 1967 and 1973 wars. Even Kissinger's shuttle diplomacy helped, given that his entourage included fourteen American reporters who spent weeks reporting about the Middle East and talking to people in all of the relevant countries (Dunsmore 1980, 75). Terence Smith of the *New York Times* asserts that his paper alone had a bureau in Israel from 1947 to 1967 without a break; a number of other papers opened bureaus after 1973 (1983, 19–20). Smith estimates that before 1967 there were about twenty to twenty-five American correspondents in Israel; now there are normally more than fifty, and when the news is "heavy," there may be as many as five hundred. Moreover, the amount of television coverage has mushroomed. These developments have resulted in a larger number of reporters with more sophistication about the Middle East and a level of reporting that moves beyond the simple pro-Israeli positions typical before 1967.

Journalists find the Middle East difficult to cover for a variety of reasons. Chafets describes three ways the problem has been handled. First, UPI, AP, Reuters, the television networks, *Time, Newsweek,* the *New York Times,* the *Washington Post* and the *Los Angeles Times* adopt the "three-bureau" approach, each with full-time staff correspondents located in Jerusalem, Beirut, and Cairo. Second, the *Chicago Tribune, Baltimore Sun, Miami Herald, Christian Science Monitor, U.S. News and World Report,* and some other papers have one Middle East correspondent, usually living in Israel. The *Wall Street Journal's* Middle East correspondent is based in London. Third, other newspapers have no regular correspondent living in the area. They may use local journalists (stringers), send in their own reporters to cover hot stories, or rely on the coverage of others (Chafets 1985, 35–36).

Chafets, a former Israeli official, estimates there are only about thirty American reporters regularly living in the "Moslem Middle East." In the late 1970s the two American wire services together had only three American writers at their Middle East bureau centers in Beirut; North Africa was covered largely from Paris because of its French colonial background (Hudson 1980, 96). Their job is not easy.

> This, then, is the task that confronts a reporter in the Moslem Middle East—to cover, sometimes with one or two colleagues, often all alone, an area almost the size of Europe, where in most cases he cannot speak the language, knows little of the people, their history, geography, and culture, and cannot hope to get reliable and independent local helpers. Usually, he is unable to find citizens or officials who will speak to him frankly, and he has no access to any independent local media. There is no assurance that he will be able to get in to the country he wants to cover at the time he wants to cover it, and once he gets there, he may well be denied access to the area or individuals he wants to see. In certain places he can find himself the target of violence for unpopular reporting; and, if his story is too critical for the local authorities, may find himself permanently blackballed. (Chafets 1985, 47)

American television reporters have similar problems. Israel generally gives them a good bit of help in getting their stories (it had the first ground satellite station in the Middle East, for example), whereas Arab countries are much more difficult to cover.

It is a commonplace that American mass media tend to support Israel more than they support the Arabs (M. Hudson 1980; the most extensive analysis is Said 1981). Chafets, formerly director of the Israeli Government Press Office, argues that, on the contrary, the Western media have actually treated Israel much more harshly than they have treated the Arabs (1985), but this position is contradicted by a series of more specific studies in which the American press emerges as either pro-Israeli or reasonably well balanced. Spot analyses of syndicated columns and editorials in 1967 found strong support for Israel (Trice 1976b, 63–65; C. Wagner 1973, 314–315). A more systematic analysis of editorials in the American elite press from 1966 to 1974 found almost as many negative as positive comments about Israel, but also found that the Arab states and particularly the Palestinians received much more criticism. The pattern varied with the issue, however; in particular, Israel was heavily criticized for annexing Jerusalem and for its retaliatory raids. Although mass public opinion was skeptical of U.S. efforts to bring peace to the Middle East, the elite press strongly supported such attempts (Trice 1979, 312–320). An Israeli analysis of American television coverage of the October war concluded that all showed reasonable balance in using coverage from the Arab countries and Israel (Gordon 1975, 81–82).

Aside from Middle East coverage per se, the media also increased public concern about the oil crisis, according to one major study. American television news blamed the energy crisis primarily on OPEC (interestingly, this research does not distinguish between OPEC price rises and the Arab oil embargo) and the oil companies themselves (Theberge 1982, 1:27–38). In retrospect it seems clear that the press overemphasized conspiracy theories (perhaps in part because the oil crisis took place during Watergate, when conspiracy theories were understandably in vogue) and was unable to give the public a clear sense of how serious the energy problem was. A senior British journalist attributed this to qualities of American journalists: Their lack of training in economics (as opposed to British journalists); their tendency to ask congressional representatives, who also have no training in economics, for their opinions (in Britain, civil servants are more logical targets); and their skepticism toward explanations from the executive branch (Macrae 1983).

Interest Groups

The impact of interest groups on U.S. policy toward the Arab-Israeli dispute has been studied a great deal, in part because such organizations are widely believed to have more impact on this issue than on other foreign policy questions.

The Israeli Lobby.   The Israeli lobby in the United States is both professional and powerful, although hardly omnipotent. One authority sees it as composed of at least seventy-five groups; another says perhaps five hundred are involved at least occasionally (Trice 1976a, 79; Glick 1982, 96). These in turn are linked to two major coordinating groups: the Conference of Presidents of Major Jewish Organizations and the American-Israel Public Affairs Committee (AIPAC). Their historical roots can be traced back to the 1880s and pressure to encourage the U.S. government to support Jewish groups in Europe (Howe and Trott 1977, 284–285). The presidents' group tends to pressure the executive branch, whereas AIPAC is primarily concerned with Congress.

The Israeli lobby has much greater leverage in Congress than in the executive branch; this in turn dictates the issues on which it has the most influence. Trice makes a useful distinction between arms sales and peace-making issues (1981, 128). The Israeli lobby is more successful on arms sales, where Congress is involved and where the public can readily understand the issue, than on peace-making issues, where Congress usually has no particular role. Within arms sales, it is also easier to get more arms for Israel added to bills than to delete arms for the Arabs (S. Spiegel 1985, 159; Trice 1976b, 56–59; Feuerwerger 1979, 40–46). Actually, the relationship is more complex than this because congressional support increases the impact of interest groups within the executive branch as well—if only because of the access that senators and representatives enjoy even in the government agencies concerned with foreign affairs (Trice 1977, 466). Thus we can see some impact on diplomatic activities as well; Quandt argues that the negative reaction orchestrated by the Israeli lobby to the Rogers plan "virtually insured that no further pro-Arab initiatives would soon be taken" (1973, 274–275).

One rough indicator of the Israeli lobby's influence with Congress is that "about three-fourths of all senators, and a sizable majority of representatives, have put their names to pro-Israeli statements" (Quandt 1977, 22). Senator Fulbright reportedly put it more forcefully in 1973: "There are between seventy and eighty senators who will vote for anything Israel wants" (Howe and Trott 1977, 277). Conversely, during the 1967 war President Johnson thought Congress might not support strong measures backing Israel because of congressional fear of being drawn into a Middle East war (Quandt 1977, 70).

The reach of the Israeli lobby has been demonstrated in the American response to the Arab economic boycott. Along with its exertion of direct pressure on the national government, the lobby was instrumental in encouraging thirteen states to pass antiboycott statutes for banking, which is state-regulated. (For examples of such statutes, see N. Joyner 1976, 76–78.) The existence of varied state laws caused the banks themselves to press for a single, comprehensive federal law. The eventual result was a set of amendments to the 1969 Export Administration Act. The amendments were adopted in 1977 and incorporated wholesale into the Export Administration Act of 1979 (C. Joyner 1984, 259).

This law was written outside of Congress in a unique political process. The issue had become increasingly difficult as U.S. sales to Arab countries rose from 2.4 percent of all U.S. exports in 1973 to about 6 percent in 1976 ($6.9 billion). Lobbying from both supporters of Israel and business groups was intense, complicated by divisions within them, the reluctance of many adversaries to engage in a major confrontation, and significant differences during the 1976 presidential campaign between President Ford's opposition to such legislation and Jimmy Carter's campaign commitment to support it.

As a result of a chance meeting between the heads of the Anti-Defamation League and the Business Roundtable, these two organizations tried to negotiate an agreement between themselves. During the initial set of negotiations, there were twelve representatives of each group; corporate officials included the heads of Dupont, Exxon, General Motors, General Electric, Bechtel, FMC Corporation, and Federated Department Stores, along with senior executives from Citicorp, Chase Manhattan, and Mobil. The result was a statement of principles, which was then taken to Congress. The tentative agreement collapsed when its creators differed about its meaning before a congressional committee and when some other Jewish groups dissented.

At the request of President Carter, a new round of negotiations, involving the Business Roundtable, the Anti-Defamation League, the American Jewish Congress, and the American Jewish Committee got under way. After a week agreement was reached. President Carter announced the result and declared his support for it. The text was added to a pending Senate bill, passed without dissent by the Senate, adopted by the Senate-House conference committee, and became law as written (Franck and Weisband 1979, 200–209). The result was "the most extensive set of antiboycott provisions enacted in any jurisdiction in the world" (Stanislawski 1981, 163). Under the law, American firms cannot promise the Arabs not to deal with either the Israelis or companies that trade with Israel, cannot state that their goods do not contain Israeli components, cannot reveal their links to companies that have been blacklisted, and are required to report the receipt of any Arab boycott questionnaires (Phillips 1979, 6; see also Stanislawski 1981, 141–170; Nelson and Prittie 1977, 182–203; Stanislawski 1984, 140–142; C. Joyner 1984, 259).

Why is the Israeli lobby so powerful? The primary reason is that the lobby asks Congress to do what it would prefer to do anyway—support Israel. In addition, although the lobby has been able to effectively coordinate many groups under umbrella organizations, its central resource is clearly the American Jewish community. Although only about 3 percent of the American population is Jewish, this group is disproportionately involved in politics; the pollster Pat Caddell estimated that Jews constitute perhaps 7 percent of the voters outside of the south (Lipset and Schneider 1977, 29). More important, although a very diverse group, American Jews were united in support of Israel after the 1967 war (there had been a fair amount of dissent from Zionism before that time [Reich 1977, 366–367]) and have remained so, at least until the Israeli invasion of Lebanon in 1982. As Nathan Glazer reportedly said, "Israel has become the religion of American Jews" (Lipset and Schneider 1977, 27). The American Jewish community has been willing to expend time, money, and energy on the issue and has communicated the impression that its members will change their votes based on candidates' stands on the question.

As an organization, AIPAC is unusually competent and professional. Avoiding identification with either political party, it has also taken advantage of the geographic diversity of American Jews to develop a major resource, the capacity to mobilize thousands of people in most congressional districts for letter-writing and telephone barrages. One congressman is quoted as saying: "You make a statement one afternoon and within a couple of hours you get a call from their branch in [a city within the district]—not from their own man, but from a close friend of yours. It's amazing. It usually takes labor or other groups a few days. AIPAC's the quickest one at relaying the message" (Feuerwerger 1979, 95, 90–96). A backhanded tribute to AIPAC's reputation was a special meeting, attended in June 1977 by the senior officials of the Carter administration, including the president, "to consider our tactics in response to a large-scale campaign launched by the America-Israeli Political Action Committee" (Brzezinski 1983, 97)—not every interest group can trigger a rump National Security Council meeting. The lobby has also developed information sources within the Congress and the executive branch that allow it to intervene in issues before they are decided (Franck and Weisband 1979, 187–188).

The Israeli lobby exhibits the common tendency of pressure groups to talk to those who will be sympathetic rather than to those who are influential. One scholar found that its members interacted much more with Congress than with the State Department, even when the issues were diplomatic (Trice 1976b, 58–59), although another found that within

Congress unsympathetic representatives were approached, with some positive results. (Pro-Arab groups, reflecting their general skepticism regarding government receptivity, interacted relatively little with government representatives, whether sympathetic or not.)

Trice argues persuasively that one major strength of the Israeli lobby has been the ability of its member groups to form coalitions with non-Jewish groups:

> The consistent ability of pro-Israel groups to enlist the support of significant numbers of non-Jewish groups has been one of the major determinants of Congress's receptivity to pro-Israel policy preferences. . . . More important than the Jewish bloc vote and Jewish campaign contributions has been the capacity of pro-Israel groups to align themselves with more representative segments of the American population at the district and state levels. (1976b, 65–66)

He cites the examples of the Jewish War Veterans, which works closely with the American Legion and the Veterans of Foreign Wars on common concerns and is therefore able to influence them on Israeli issues; similar patterns manifest themselves in labor circles (see also Franck and Weisband 1979, 188).

Moreover, AIPAC is not the only instrument of the Israeli lobby. It responded quickly to the new opportunities of the political action committees (PACs); by 1982 thirty-one Jewish PACs not only contributed amounts of money equal to Washington's largest interest groups such as the American Medical Association (about $1.67 million), but also targeted these funds for congressional races that would affect the makeup of the committees working on policy toward Israel: the Senate and House foreign affairs committees and especially their foreign operations subcommittees. The PAC money became a kind of substitute for Jewish voters in certain districts (Novik 1986, 61–63).

The lobby is not invincible, however. It risks the possibility of a backlash. Richard Nixon provides an interesting example, although his antipathy for Jewish pressure groups makes it hard to generalize from the case. At a trivial level, in 1969 and 1970 the Nixon administration viewed congressional pressure to sell Israel more Phantom jets as coming from the Jewish lobby. The administration's "retaliation" came in the form of "a temporary halt to routine congratulatory messages for Jewish dinners, yearbooks, and bar mitzvahs." The conflict escalated, however, when President

Georges Pompidou of France visited the United States in March 1970. A group of Jewish war veterans harassed him, protesting French arms sales to Libya. Pompidou threatened to leave the country. Nixon, for whom renewed relations with France were quite important, was "incensed." When New York Governor Nelson Rockefeller and Mayor John Lindsay boycotted a dinner for Pompidou, Nixon himself attended in place of Vice-President Spiro Agnew (Nixon 1978, 479–480). More significant, the incident has been linked to a decision not to sell Israel any more Phantoms, even though the Soviets were sending sophisticated antiaircraft missiles and military personnel to Egypt to counter Israel's deep-penetration bombing. "These moves were unprecedented for aid to a non-communist country" (S. Spiegel 1985, 170, 190–191; see also Shadid 1981, 171–172; Hersh 1983, 214, 223n–224n).

More recently, resistance to AIPAC's lobbying efforts also seems to be increasing in Congress; one insider observed that "AIPAC often does with a sledgehammer what should be done with a stiletto," and some concern was expressed over the abrasive style of its second director, who resigned in 1980 (*Congressional Quarterly* 1979, 91, 95; see also Franck and Weisband 1979, 190; Curtiss 1982, 118). Senator Charles Percy (R-Ill.) refused to sign the "seventy-six letter" opposing the administration's "reassessment" of policy toward Israel in 1975 and received twenty thousand pieces of mail in consequence; he was outraged and refused to knuckle under. (He also lost the next election.) Since 1973 considerable dissension has developed in the Israeli lobby about tactics, some of which are less than effective (Howe and Trott 1977, 275–277, 312–319); Quandt notes that many letters protesting the U.S. condemnation of the 1968 Israeli attack on the Beirut airport "had an identical text, the same misspelling of the assistant secretary's name, and suspiciously similar signatures" (1973, 280).

Even more threatening to the Jewish lobby's effectiveness is a decline in Jewish unity on the issue of Israel, as evidenced by the decrease in the money given to Jewish causes when adjusted for inflation and, more important, by the fact that perhaps 25 percent of the American Jewish community may now publicly refuse to support some Israeli policies. A plurality of American Jews supports a more flexible Israeli policy toward the Palestinians, for example (Novik 1986, 68–82). This tendency is fairly recent, dating to the policies of the Begin regime and especially the invasion of Lebanon in 1982. But it is potentially an important shift, because the defection of even a significant minority could paralyze the community politically, especially in any sort of conflict between the U.S. and Israeli governments over the terms of a Middle East settlement (Novik 1986, 84).

The oil crisis also weakened the Israeli lobby by undercutting its basic rationale, that a strong Israel could ensure peace and defend American interests in the Middle East, and by legitimizing the argument that the Arab oil producers had to be appeased (S. Spiegel 1985, 221–222). Precisely because of the increasing strength of the opposition, however, the Israeli lobby was galvanized to greater activity, and many of its political strengths remained intact. (For example, in 1973 it raised $1.8 billion from private funds to send to Israel [Safran 1978, 573].) One of its most impressive efforts was in March 1975, when Israeli-American relations reached a low point, with Kissinger threatening a "reassessment" of U.S. Middle East policy. The letter from seventy-six senators (mentioned earlier) "shockingly undercut the entire thrust of the Administration's 'reassessment.'" Signatories included individuals with widely different political beliefs, such as Ted Kennedy, Barry Goldwater, Frank Church, Paul Laxalt, Walter Mondale, and Strom Thurmond (S. Spiegel 1985, 222–223, 295–297), although, as noted earlier, not Charles Percy.

Later that year the lobby was involved in altering a planned sale of Hawk antiaircraft missiles to Jordan, although the testimony of the Joint Chiefs of Staff is reported to have been more important; several subsequent attempts to stop arms sales to Arab countries failed without such support (Howe and Trott 1977, 294–297; S. Spiegel 1985, 303–304, 308–310, 347–348). Efforts to stop the sale of missiles to Saudi Arabia were weakened when Kissinger argued that this might cause higher oil prices; "the influence of the Israeli pressure groups appeared to be clearly reduced" (Roehm 1980, 479). Although AIPAC and the other organizations consistently opposed Kissinger's "step-by-step" diplomacy (Trice 1978, 244), they could not influence it. In 1975, when Congress was expressing concern about the very high levels of foreign aid to the Middle East after the Sinai agreements, AIPAC lobbied in favor of aid for Israel and did not oppose aid to Egypt, suggesting a limited view of its own political capabilities (Roehm 1980, 370).

The Israeli lobby was especially upset by the Carter administration, a collegial, Democratic administration in which no major figure supported Israel. The lobby was particularly distressed because it had never had good relations with the national security bureaucracies; contacts at the White House were therefore essential if it were to be involved in the political process. Steven Spiegel suggests that the Carter administration was not prepared to tolerate conflicting opinions on the subject precisely because the Middle East held the highest priority in Carter's foreign policy and that, on such issues, internal dissent could not be tolerated (1985, 320–329).

Like everyone else, the Israeli lobby was thrown into confusion by Sadat's trip to Jerusalem in November 1977. After some hesitation, it decided to encourage a separate Egyptian-Israeli peace as the United States' primary goal in the Middle East. This would seem to imply support for the Camp David accords. In fact, however, the U.S. government seemed to align itself with Egypt against Israel in preparing for the Camp David meeting; Carter felt that the Israeli lobby had opposed the Camp David initiative (1982, 315–316). In 1978 AIPAC led an unsuccessful attempt to block the sale of advanced fighter aircraft to Saudi Arabia, which the Carter administration had included in a package deal along with similar planes for Israel and Egypt. Carter's "Jewish aide" Mark Siegel resigned over the issue. "This hard-fought campaign . . . severely damaged the president's Jewish support. . . . Senior administration officials were reported to have boasted that they had 'broken' the pro-Israeli lobby. The president, who was bitterly irritated by the debate, never fully recovered politically from this episode" (S. Spiegel 1985, 349).

The coup de grace came after the Egyptian-Israeli peace treaty was signed, when a reconciliation between Carter and the Israeli lobby might have been expected. United Nations Ambassador Andrew Young had been so close to Jimmy Carter that it was hard to believe he had acted entirely on his own in initiating contact with the PLO. After he was forced to resign over the issue, black leaders rallied to his support, suggesting that the United States should at least talk to the PLO. This in turn outraged the Israeli lobby even more. Shortly thereafter the United States "by mistake" voted in favor of a U.N. resolution condemning Israeli occupation of Jerusalem. "There was little trust left between the White House and American Jews" (S. Spiegel 1985, 378, 375–379). In the 1980 presidential election less than half of the Jewish voters supported Carter, resulting in the lowest

Jewish vote for a Democratic presidential candidate since 1924. Ronald Reagan's 39 percent Jewish vote was the highest figure for a Republican since Abraham Lincoln (Novik 1986, 67–68).

Despite this support, the Reagan administration provided the Israeli lobby with some unpleasant surprises. Its initial rhetoric had been quite sympathetic toward Israel, so the lobby was caught off guard when the AWACS sale to Saudi Arabia was proposed, in part because Secretary of State Haig had told them that no new weapons would be included in the package (personal interview with a former senior official, February 1985). The result was a bruising political battle that the president won, but at a considerable political cost. Similarly, the lobby was not enthusiastic about the Reagan plan.

The Israeli lobby is undoubtedly powerful. It has significant resources of money, expertise, and organization. But these resources are neither unique nor unlimited. Although it may influence elections in certain key states, for example, the Jewish vote is not overwhelming in the United States. As noted earlier, Jews make up only 3 percent of the American population, constitute more than 10 percent in only one important state (New York, where they are about 12 percent of the electorate), and have not always voted in favor of candidates supporting Israel (Glick 1982, 105). Trice found that the percentage of Jewish voters in a senator's district and the amount of money he or she received from Jewish groups accounted for only a small percentage of the voting variance on Arab-Israeli issues; in another analysis he found little relationship between activities of the Israeli lobby and support for Israel in the general public (1977, 456–462; 1978, 247–250). Moreover, the lobby faces the same structural problems of any interest group in foreign policy: Decisions are made primarily in the executive branch, access is difficult, and lobbyists' arguments are discounted because of their obvious bias.

On balance, the Israeli lobby has had the most influence on issues where Congress is involved. It has demonstrated the ability to get remarkable amounts of U.S. government funds for Israel, and it can make weapons transfers to Arab states politically expensive, although in the end it cannot usually prevent them. It has had relatively little direct impact, however, on diplomacy and strategy formulated within the executive branch, the most important aspect of American policy toward Israel. "Pro-Israeli groups are often most influential when they do nothing at all to influence policy," when decision makers anticipate their response and act to avoid trouble with them (Quandt 1977, 20).

The Arab Lobby.   The Arab lobby is clearly much smaller, composed of a number of groups, all organized fairly recently. The leading group is the National Association of Arab Americans (NAAA), a formal lobbying group organized in 1972 and modeled after AIPAC. Other organizations include the Association of Arab American University Graduates (AAUG), a more academic organization of about the same vintage; the Action Committee on American Arab Relations (ACAAR), organized in 1964 (of these groups, probably the organization most committed to direct opposition to Israel but oriented more toward influencing public opinion than lobbying); the revived Federated Organizations on American-Arab Relations; the Palestine Congress of North America, formed in 1979; the American Arab Anti-Discrimination Committee; the Foundation for Middle East Peace; the Center for U.S.-European Middle East Cooperation; the American Arab Affairs Council; and the American Educational Trust. A number of separate Arab governments have also been involved in the process at various times, including Kuwait and Saudi Arabia, as well as the Arab League Information Office (Trice 1976a, 88–89; Trice 1976b, 37; Shadid 1981, 161–162; Howe and Trott 1977, 337–338; Barberis 1976, 213–220; Curtiss 1982, 127–130, 141–144).

There are perhaps about two million Arab Americans (although Trice [1981, 125] estimates half a million to one million), as opposed to about six million American Jews. Moreover, until recently Arab Americans have stressed assimilation. They have difficulty representing a divided Arab world and are divided among themselves. In 1976, when they might have been expected to be strong because of the oil embargo, the civil war in Lebanon split the Arab American community, which is mostly of Lebanese Christian descent. Politically, most of them are Democrats, but many of the more successful are Republicans (Howe and Trott 1977, 337–341). Divisions between native- and foreign-born Arab Americans were important as well (Trice 1976b, 54–55; Curtiss 1982, 126–127).

Arab Americans lack two major resources of their domestic opposition—unity and willingness to expend resources on the issue. The NAAA has gained considerable influence in Washington, however, and has tried to increase political awareness among Arab Americans in a long-term effort to counterbalance the political leverage of American Jews (*Congressional Quarterly* 1979, 96–100). This process seems to have borne some fruit; Jimmy Carter noted in his diary in December 1979 that Arab Americans were giving "all the staff, Brzezinski, Warren Christopher and others, a hard time," which would have been difficult to imagine ten years before (1982, 299).

Whereas the NAAA has primarily targeted Arab Americans, the Action Committee on Arab American Relations has been more concerned to press the Arab case among leftist intellectuals. Not surprisingly, it devotes a good deal of effort to college campuses. This activity has not been without risk; in 1974 the New York office of its newsletter was burned, and its director, Dr. M. T. Mehdi, was beaten, apparently by Jewish Defense League members (Howe and Trott 1977, 343–344).

Like supporters of Israel, the Arab lobby has worked on building coalitions. One interesting example is the Palestine Human Rights Campaign, bringing together "peace, church, civil rights, and Middle East–related" groups; another is the Middle East Peace Action Coalition (F. Spiegel 1981–1982, 70).

Corporations.   There has been a good deal of debate in the United States about the ability of major corporations to affect foreign policy. The role of the major oil companies has been particularly controversial. Without delving extensively into the issue, it seems evident that the oil companies have had relatively little success in altering U.S. policy toward the Middle East, particularly its continuing support of Israel. Essentially, they have had to adapt to a policy they could not change (Wilkins 1976, 164).

> *Pressure* is hardly the word to describe the efforts of oil companies with respect to Arab-Israeli issues. Rather, appeals to interests and attempts to explain problems of the oil industry highlight the efforts of oil representatives. Without a public voice, the oil companies do not carry as much weight as the pro-Israelis. In addition, they are less ambitious, trying merely to temper the United States' position on the Arab-Israeli conflict and pushing for particular favors from the Administration when oil problems arise. (Quandt 1973, 269, emphasis in the original)

American corporations, including the oil companies, have preferred to deal directly with the government rather than going to the public. The major lobbying group for big oil is the American Petroleum Institute, with very impressive resources and access. But it focuses primarily on U.S. energy policy rather than on Middle East strategies. Similarly, John J. McCloy, the "prince of oil lobbyists," who "operates at a level with which even the Israeli lobby cannot compete" (Howe and Trott 1977, 356–358), is not primarily concerned with Middle East affairs, and his efforts in this area in 1973 were notably unsuccessful, as we shall see.

Thus, much of the public-relations work has been done by individual oil companies, which have generally focused on explaining Arab positions and the corporations' own behavior (Trice 1976a, 91–92). For example, between May 1972 and April 1973 Mobil sponsored a rather expensive series of advertisements in American newspapers, arguing that the energy crisis was real and must be dealt with. Rustow writes: "When else in the history of institutional advertising did a major corporation choose to spend hundreds of thousands of dollars to inform the public, not of the availability and attractiveness but of the imminent disappearance of its major product?" (1982, 165–166). Rustow, in fact, argues that this publicity campaign, together with activities by the State Department, created a public mood of concern over oil supplies that strengthened Saudi Arabia's bargaining position.

It is more difficult, however, to trace any impact the oil companies may have had on U.S. policy toward Israel. In 1967 the chairman of Continental Oil Company urged Walt Rostow, Lyndon Johnson's national security advisor, to show that the United States was evenhanded in the Middle East, without much notable effect: "Oil company influence on Nixon in the foreign policy arena was almost non-existent . . . the companies were consistently unable to affect the Administration's foreign policy" (Novick 1986, 126). In December 1969 senior oil company officials met with Nixon and Kissinger about the Middle East, but the oil executives felt they had no influence at all (S. Spiegel 1985, 171). In May 1973 senior officials of ARAMCO and the ARAMCO companies (Exxon, Texaco, Mobil, and Standard of California) were told by King Faisal that American policy must stop favoring Israel; hints were passed that ARAMCO oil supplies might be jeopardized. They first passed the information on to the government—Joseph Sisco at the State Department, William Clements at the Department of Defense, and members of the presidential staff. It was received skeptically (Kelly 1980a, 389; S. Spiegel 1985, 242–243; Sampson 1974a, 1979, 292–293).

The oil companies did not stop there. The chairman of Standard of California sent a letter to his 300,000 stockholders, asking for "understanding" of Arab aspirations; the resulting Jewish protests and the return of thousands of credit cards were used as evidence of their efforts when ARAMCO officials spoke to the Arabs. Mobil sponsored an advertisement in the *New York Times* on June 21, 1973, calling for a Middle East peace settlement, which the *Times* felt was too political to appear on Mobil's usual spot opposite the editorial page (Rustow 1982, 154–155; S. Spiegel 1985, 459–460; Sampson 1979, 293–295; Emerson 1985, 27–35). None of these activities seemed to have any perceptible impact on U.S. foreign policy.

On October 12—after the decision to rearm Israel but before the embargo—John McCloy, chief lawyer for ARAMCO, hand-delivered a letter to Alexander Haig, Kissinger's chief of staff. He enclosed a memorandum to be passed on to Kissinger and Nixon from the ARAMCO company chairmen recommending against military aid to Israel and asserting that the Saudis would reduce production as a result of acts already taken, a letter one observer sees as "the Saudi Government's last warning to the U.S. government" (Bouchuiguir 1979, 223–224). It seems to have had no effect (Sampson 1974b and 1979, 300; S. Spiegel 1985, 258–259; Kassis 1981, 53–54).

On October 25, after the embargo had begun, James Akins, U.S. ambassador to Saudi Arabia, asked ARAMCO to direct its member companies to inform Washington officials that the restrictions would not be lifted unless a political resolution satisfactory to the Arabs was reached (Sampson 1979, 303); the idea of an American ambassador using a private corporation to communicate with his own superiors is intriguing, but the act does not seem to have led to anything: In Kissinger's discussions with the oil companies during the embargo, he seems to have followed his own policy star, although the results did not clash with company interests (Nau 1980a, 42).

ARAMCO and its parent companies were clearly in a difficult position during the embargo. On the one hand, they enforced the embargo against the United States, even giving the Saudis detailed information about planned shipments to the U.S. military and canceling such shipments on order. Moreover, the embargo was more feasible because these few companies controlled so much oil, making it easier for the Saudis to control the flow (Sampson 1974b; 1979, 304, 316; Emerson 1985, 38).

On the other hand, the oil companies fundamentally undercut the Arab embargo of the United States and Holland by diverting non-Arab oil from other destinations to these countries. Apparently, the Arabs brought no pressure on the companies to alter this policy (U.S. Congress 1975b, 10), and we do not know how the oil companies would have reacted to such pressure. In a congressional hearing, John Sawhill, administrator of the Federal Energy Office, was asked by Senator Charles Percy about the conflict for the companies between high profits and the American national interest in getting oil during the embargo. Sawhill said the corporations had seen such a conflict but "almost uniformly they felt the national interest came first" (U.S. Congress 1975a, 170, 34). Interestingly, once they decided to spread the pain, the companies do not seem to have favored their home country particularly: "As a percentage of projected current demand, during the period January to April 1974 the five U.S. majors reduced supplies to the United States by 17 percent, to Europe by 18.6 percent, and to Japan by 16 percent" (Schneider 1983, 245). At the same time, it is interesting to speculate about American public and government reaction if, say, France had been embargoed and the companies had cut shipments to the United States to equalize the impact around the world.

Before the oil crisis, the oil corporations were almost alone in their concern with the Middle East. Higher oil prices greatly increased the number of American corporations with serious economic stakes in the Arab world. U.S. exports to the Arab states went from $1 billion in 1971 to $5.4 billion in 1975, $6.9 billion in 1976, $7.15 billion in 1977, $8.36 billion in 1978 (Stanislawski 1981, 146), $15.3 billion in 1980 (Rubin 1984, 69), and $14 billion in 1983 (Emerson 1985, 410). "More U.S. corporations than ever before had incentive to question U.S. support for Israel" (S. Spiegel 1985, 220; for particular examples, see Emerson 1985, 44–57). In particular, a new coalition of arms manufacturers looking for new markets, Defense Department officials favoring longer production runs to reduce weapon unit costs, State Department personnel interested in improving relations with the Arabs, and other government officials wanting to recycle petrodollars through the United States became particularly effective in supporting massive arms sales to Arab countries (S. Spiegel 1985, 220–222, 309–310, 398).

American corporations therefore assembled an informal lobbying group to support the AWACS sale to Saudi Arabia in 1981, including Boeing, Pratt and Whitney, Exxon, Mobil, Brown and Root, Bechtel, and NL Corporations (Curtiss 1982, 134; Emerson 1985, 185–214; Novik 1986, 127–129). There are contradictory reports on the degree to which these companies were actually pressured by the Arab governments to use their influence on the Arab-Israeli dispute (Shadid 1981, 160–161; Emerson 1985, 61–74, 189–191, 212–214). Another interesting corporate tactic involved funds to help the public image of Arabs by, among other things, making grants to American universities for Arab studies programs.

But Shemeul Meir argues that the AWACS issue was the high point of business influence on Arab-Israeli policy (1986, 86–87). He notes that the decline in oil prices reduced expenditures by the oil-producing states, which greatly reduced the lobbying incentive for non-oil corporations. Many of the corporations have either stopped doing business with the Arabs or even gone out of business.

Other Pressure Groups.   Another interesting type of pressure group, the anti-Zionist Jewish group, is epitomized by the American Council for Judaism. But, though this organization still exists, it apparently has not been a significant actor ever since the 1967 war, when its attacks on Israeli policy severely reduced its constituency (Trice 1976b, 39). Its role is carried on, however, by the American Jewish Alternatives to Zionism, which periodically purchases full-page newspaper advertisements. Breira was another group that sought "alternatives" (*breira* in Hebrew) for policy toward Israel and the Palestinians. Founded after 1973, Breira became inactive later in the decade (Curtiss 1982, 122–123). The Foundation for Arab-Israeli Reconciliation includes both Israelis and Palestinians (Howe and Trott 1977, 347–350).

The Impact of Pressure Groups.   What difference does all this activity make? One commonly hears that interest groups determine U.S. policy toward the Arab-Israeli dispute; according to political mythology there are few more powerful groups in this country than the Israeli lobby on one side and major corporations, particularly the oil companies, on the other. Ironically, each side assumes that the other's lobbyists are overwhelmingly powerful. Lobbyists for Israel speak of millions of petrodollars funding propaganda campaigns in the United States and Western Europe (Howe and Trott 1977, 359–360); Steven Emerson refers to an American "petrocorporate class" (1985, 111). Senator James Abourezk (D-S.D.) said the Israeli lobby could "accomplish virtually any legislative feat involving military or economic assistance to Israel" (*Congressional Quarterly* 1979, 89). Clearly the lobbies have some impact on policy. A more plausible view, however, documented in Steven Spiegel's massive study of American Middle East policy decisions from Roosevelt to Reagan (1985), is that neither set of lobbyists is dominant. Instead, the executive generally makes decisions, then packages them to limit damage with both groups. (For similar arguments, see Glick 1982, 102–106; Trice 1978, 240.) This was particularly true during the oil crisis itself where, owing to Kissinger's extraordinary dominance of the policy process, none of these outside groups seems to have had much leverage. "The decisions made by Nixon and Kissinger can be explained almost without reference to inputs from outside the bureaucracy, even though the White House was bombarded by Congressional and other figures during the first week of the war" (Dowty 1984, 240n; see also S. Spiegel 1985, 235–236; Quandt 1977, 203).

Why, then, does practically everyone assume the lobbies have so much impact? Spiegel argues that everyone in the process has a vested interest in maintaining one version or another of the central myth (1985, 388–390). The idea of U.S. policy being shaped by the Israeli lobby, for example, is useful to a number of different participants. The lobby itself can demonstrate what good work it is doing. Presidents and Congress show their responsiveness. Arab sympathizers in the United States use it to explain their failure to influence policy and to suggest that the Arabs not blame the U.S. government, corporations, or people for irrational policies. Bemused foreigners see it as explaining why the American government is so resistant to their arguments; without the myth of the Israeli lobby, they might be forced to question the persuasiveness of their own logic. Conversely, the myth of the Arab lobby, usually tied to major corporations, is equally useful for Israel's supporters in explaining past failures and justifying renewed future efforts.

Governmental Influences on U.S. Policy

Congress

The Constitution gives the United States the strongest and most independent legislature of any country in this study. The separation of powers (the United States is the only nonparliamentary system studied here) produces formal congressional independence from the executive and allows different parties to control the institutions at the same time. Perhaps more important, the lack of party discipline means individual legislators can ignore party mandates as long as they can maintain strength in their own constituencies; this in turn greatly increases the power of any group that can influence votes in these constituencies. In addition, the rise of the committee system has given Congress specialized knowledge both in its staff and among its members, which has been particularly important in the relatively esoteric area of foreign affairs. In the classic phrase of Edward S. Corwin, "The Constitution . . . is an invitation to struggle for the privilege of directing American foreign policy" (1957, 171).

Nevertheless, it is also clear that the executive enjoys much greater dominance over Congress on foreign policy issues than it does on domestic questions; one prominent theorist has described this situation as "the two Presidencies" (Wildavsky 1975, 448–471). Again, this stems in part from the Constitution, which gives the president command of the armed forces and responsibility for foreign affairs, greatly strengthening this position. Executive prerogative is reinforced by the perception of foreign affairs as an area where speed and secrecy are often crucial, with high stakes and uncertain electoral impact; Congress is thus often reluctant to intervene. The result is that the executive initiates foreign policy; Congress chooses whether or not to react.

The executive (in practice, Henry Kissinger) faced particular problems with Congress at the time of the oil embargo. Both houses were dominated by the Democrats, while presidents Nixon and Ford were Republican. Middle East policy involved the extensive use of foreign aid, over which Congress potentially exercises a good deal of influence. The Ninety-third and Ninety-fourth Congresses had been particularly concerned to establish a separate congressional role in foreign affairs, in part because of Vietnam. Congress was engaged in the elaborate process that eventually would force President Nixon to resign, and the complex committee structure and the weakened congressional leadership made it difficult to negotiate with Congress (Roehm 1980, 167–199; Roehm 1981, 27–29; Franck and Weisband 1979, 6–9).

Aside from these general problems, Congress has a long tradition of supporting Israel (it passed a resolution in 1922 supporting a Jewish home in Palestine). A quantitative study found that from 1969 to 1976 "an average of about 80 percent of the Senate and 86 percent of the House cast votes favorable to Israel" (Feuerwerger 1979, 28). Even more impressive, another study found that "*none* of the senators who served during the entire 1970–1973 period voted against Israeli interests on all seven votes" (Trice 1977, 448–451, emphasis added).

This support extended to opposing the administration at times. In 1968, for example, when President Johnson announced he would not sell Phantom jets to Israel, Congress inserted a section in the foreign aid bill recommending such a sale (Sorley 1983, 77–78). The sale was made, and the delivery of the Phantoms led to Israel's devastating deep-penetration bombing of Egypt in 1970 and Soviet intervention in Egypt. There was significant conflict between Congress and the executive during the 1973–1976 period over the issues of arms sales to Arab states, the balance of aid given to Israel and the Arabs, rhetorical commitment to Israel versus a more balanced approach, and the degree of flexibility in negotiations accorded to the executive (Roehm 1981, 29–30).

One sophisticated study explains Congress's consistently pro-Israeli attitude as a case where personal beliefs of the members and the perceptions of constituent beliefs coincide. These factors are reinforced by the dominance of pro-Israeli interest groups, the behavior of fellow representatives, and the pro-Israeli positions of both political parties. Members of Congress identify with Israel because it is a friendly country similar to the United States, is a traditional U.S. commitment, is opposed to the Soviet Union, and holds a special moral standing because of the Holocaust. In addition, many of them have visited Israel and been impressed, whereas few have been to Arab countries (until recently), in part because of the absence of diplomatic relations with most Arab states from 1967 to 1973. In electoral terms, support for Israel is seen as helping members of Congress with their Jewish constituents and not hurting them with others (Feuerwerger 1979, 76–115, 122–123).

It is worth noting that Congress's pro-Israeli proclivity has limits. A detailed study of the activities of the House Foreign Affairs Committee during 1975 found that for a variety of reasons the committee was unusually willing to oppose the executive in support of Israel. But the committee also refused to cut administration requests for aid to Arab states, did not veto arms sales to Arab states, and rejected amendments that would have limited the administration's ability to give aid to the Arabs. The author deduced the following rule: "*Approve all reasonable measures which directly benefit Israel regardless of executive branch attitudes, but seek compromise on all controversial 'anti-Arab' measures which threaten executive branch flexibility"* (Feuerwerger 1979, 170, 141–176; emphasis in original). Congress also became the forum for discussion of opening relations directly with the PLO—not exactly an Israeli goal—in a series of hearings by the House International Relations Committee in 1975. At about the same time senators Howard Baker (R-Tenn.), George McGovern (D-S.D.), and Ted Stevenson (R-Alaska) met with PLO chairman Yasser Arafat during visits to the Middle East, even though Kissinger had refused to initiate any formal contact with the PLO (Roehm 1980, 383, 386).

Congressional pressure on the Middle East has varied somewhat over time. Before the oil embargo, Nixon found himself constantly pressured by Congress to support Israel. In particular, Congress opposed the Rogers plan as requiring excessive concessions from Israel, pushed the executive to sell Israel more Phantoms, and urged it to react more strongly to the Soviet troop commitment in Egypt (S. Spiegel 1985, 169–170, 183, 192–193, 210).

Congressional opinion was more divided after the oil crisis, but support for Israel remained widespread, if not necessarily deep. The 1975 letter signed by seventy-six senators opposing administration pressure on Israel was reportedly drafted by the head of AIPAC (Shadid 1981, 172). From 1970 to 1977 Congress increased administration requests for economic aid to Israel 30 percent, while cutting overall foreign assistance by 23.5 percent. Indeed, support for Israel was so strong that the administration deliberately put such measures in controversial military aid bills to get them through Congress (Feuerwerger 1979, 29–30, 126–140).

Despite these differences, Congress essentially supported Kissinger's policies from 1973 to 1975 (Kassis 1981, 196–197): It approved the $2.2 billion aid request for Israel, which triggered the oil embargo (indeed, there is evidence to suggest that the large amount was requested in part to keep pro-Israeli congressional groups happy); the cease-fire agreements; the first-stage disengagement agreements between Israel, Egypt, and Syria; and the foreign aid bill of 1975. One exception was Nixon's proposal to sell nuclear reactors to Israel and Egypt; Congress did not actually veto the proposal, but it delayed it for two years with hearings before no fewer than three committees. Watergate may actually have helped Kissinger, because some members of Congress wanted to avoid a total collapse of American foreign policy and may therefore have been more tolerant of him than they otherwise would have been, at least until Nixon resigned (Roehm 1980, 201–204).

In 1975 and 1976 congressional support for Israel declined, ostensibly over the issue of arms sales to the Arabs. Concern over proposed arms sales to Saudi Arabia produced the Nelson-Bingham law, which requires the executive to inform Congress of any arms sales exceeding $25 million (later reduced to $7 million) and allows Congress, by concurrent resolution, to veto it. The first test of this procedure came in 1975 when congressional pressure forced the Ford administration to alter a proposal to sell Jordan Hawk antiaircraft missiles (S. Spiegel 1985, 303–304). The sale eventually went through, but the mobility of the missiles was severely restricted (Franck and Weisband 1979, 98–103, 183–184; Lucas 1981, 80–81).

In order to sell six transport aircraft to Egypt, the administration had to promise not to ask for more arms for Egypt in 1976. A request to sell air-launched missiles to Saudi Arabia was also cut drastically after a compromise between the administration and senators Humphrey and Javits did not hold up in the Foreign Relations Committee; Kissinger had to appear to persuade the committee to reverse its vote (Roehm 1980, 333–353). In 1977 congressional response delayed the proposed sale of AWACS aircraft to Iran until 1979, whereupon Iran's new revolutionary leaders canceled the sale. In 1978 Congress and the Carter administration deadlocked over a proposal to sell F-15 fighters to Saudi Arabia; the eventual solution, suggested by Kissinger, was to sell more F-15 fighters to Israel as well (Franck and Weisband 1979, 105–111, 184–185; Lucas 1981). In 1982 a Reagan administration proposal to sell AWACS aircraft to Saudi Arabia was initially opposed in a letter by fifty-five senators; it passed only after Saudi Arabia agreed to limitations on the AWACS' capabilities and after Israel was promised still more F-15s (Haig 1984, 167–194). At the end of 1982, after the Israeli invasion of Lebanon and despite a major effort by the Reagan administration, Congress once again increased aid to Israel (S. Spiegel 1985, 422–423).

Aside from these concrete acts, Congress also became increasingly involved in rhetoric on the Middle East during this period, sometimes in direct opposition to administration policy. Perhaps the clearest example—mentioned earlier—was the March 1975 "seventy-six letter" opposing the administration's threatened "reassessment" of Middle East policy at a time when Kissinger was trying to pressure Israel to negotiate the Sinai agreements. Congress also inserted limitations in the joint resolution authorizing the Sinai observation force and eventually published the "secret clauses" of the agreements (Franck and Weisband 1979, 141–143). As Roehm describes it:

> Congress and Kissinger were clearly at cross purposes concerning the U.S. approach to the Sinai Agreement and future negotiations. Kissinger sought, through the intentional use of ambiguous language and secret executive agreements, to maximize U.S. (and his own) flexibility, in order to move Egypt and Israel . . . closer to agreement. Congress, on the other hand, sought to reduce the ambiguity surrounding U.S. commitments to the Middle East . . . and to reduce sharply the amount of executive discretion to be permitted in U.S. Middle East policy. (1981, 37; see, generally, Roehm 1980, 327–333, 353–363)

A related issue, the American response to the Arab economic boycott against Israel, reflects the Israeli lobby's strength in Congress. Because of domestic pressure, Congress went on record as early as 1956 in opposition to the boycott. In 1965 an amendment to the Export Administration Act required companies to report requests for information in connection with the boycott. However, by 1974 such reports had decreased, even though U.S. trade with the Arab states had increased and American investment in Israel had dropped from $185 million in 1973 to $43 million in 1974 and was "negligible" by 1975 (N. Joyner 1976, 43). In 1975 Congressman John Moss (R-Calif.) began an investigation into the domestic effects of the boycott; this was followed by similar investigations by Senator Frank Church's Subcommittee on Multinational Corporations. Late in 1975 President Ford tightened the reporting requirements of the Export Administration Act, and in 1976 Congress passed an amendment to the tax bill denying a variety of tax benefits to corporations participating in secondary and tertiary boycotts. The process culminated in the 1977 amendments to the Export Administration Act.

As noted earlier the amendments were drafted by an unprecedented team of Jewish pressure groups and the Business Roundtable and are "the most extensive set of anti-boycott provisions enacted in any jurisdiction in the world" (Stanislawski 1981, 163, 141–170; Nelson and Prittie 1977, 182, 103; Roehm 1980, 380–382; Stanislawski 1984, 140–142; C. Joyner 1984, 259), although it is worth noting that the law allows American corporations to participate in the primary boycott (C. Joyner

1984, 260–263). These laws were all enacted despite the desire of the Ford administration to handle the issue through diplomacy (Feuerwerger 1979, 36–37).

As often happens in Congress, these votes seem to have reflected a number of different issues. Certainly the impact of the Israeli lobby was a major factor; during the earlier period, the lobby's goals had generally coincided with those of the administration, while they now began to diverge. There was also growing concern about Kissinger's tendency to conduct secret diplomacy and commit the United States without consulting Congress; this was particularly true of the Sinai agreements of 1975, which included a number of important secret clauses. Moreover, Kissinger's policy of large arms sales to the Arab frontline states was seen as risky and expensive. At the same time he ceased to benefit from Watergate; now that the presidency was secure, there was less incentive for Congress to refrain from attacking his policies (Roehm 1980, 207–218, 390–399).

Kissinger realized the special concern of Congress for Middle East policy. During his thirty-nine months as secretary of state, he appeared eighty-four times before congressional committees. Fifty-four of these appearances were before the foreign relations committees of either the Senate or the House; thirty-nine of these appearances (72 percent) concerned the Middle East. These figures apply only to formal sessions; Kissinger also devoted a great deal of time and effort to informal contacts, particularly with members of the Senate Foreign Relations Committee. These sessions, however, tended to involve presentations by Kissinger rather than solicitations of advice—this became a serious problem when Kissinger undertook major commitments without prior consultation of Congress, as in the nuclear reactor, Hawk missile, and Sinai agreement issues. Kissinger and other members of the executive branch argued that problems inherent in Congress itself hindered such consultation, in particular the many overlapping committees and subcommittees in foreign policy, the decline of congressional leadership, and problems in keeping classified information secret (Roehm 1980, 221–244, 255–259).

In some ways congressional input was not particularly noteworthy. The arms sales to Arab countries were in fact completed, and the original requests may well have been deliberately inflated to allow Congress to cut them without doing much damage. The Sinai agreements were signed and put into effect. Congressional appropriations for additional arms as-

sistance for Israel were often not expended, and Congress has generally
been unable to impose its diplomatic preferences (requiring direct negotiations between the Arab states and Israel, for example) on the administration (Feuerwerger 1979, 41–46).

At the same time, this high level of congressional activity clearly made it
more difficult for Kissinger to develop a coherent Middle East policy. The
single most important difference between Congress and Kissinger was the
pro-Israeli position of the former; Congress, in general, expedited executive actions that aided Israel and hindered those that did not. The public
expression of this position may have negated the goodwill that some of the
Arab arms sales were designed to produce and almost certainly denied the
administration the plausible threat of cutting off aid to bring the Israeli
government to the conference table (Roehm 1981, 37–39). Substantively,
Congress increased total aid to Israel 8.7 percent over administration requests from 1970 to 1977, while cutting total foreign aid expenditures
about 25 percent (Feuerwerger 1979, 40).

Deeper congressional involvement in Middle East policymaking certainly complicates life for the executive. Nevertheless, there is an argument
in favor of such participation. It may be easier, for example, to conclude
secret treaties than public ones, but such documents are much less likely to
commit the U.S. government than those that have to undergo the tortuous
ratification system. This is not news abroad. Israel apparently insisted that
the Sinai observation force, a critical part of the Sinai II agreements of
1975, be established by legislation, not by presidential policy statement or
written agreement, as the price for Israeli troop withdrawal (Franck and
Weisband 1979, 160–161).

Roehm contends, reasonably, that formal congressional acts on Middle
East policy tended to focus on marginal issues, such as the size of arms
sales to Arab states, rather than on more fundamental questions such as the
advisability of excluding the Palestinians from negotiations, the wisdom of
separating Egypt from the other Arab states, or the advantages and disad-
vantages of "step-by-step" and "comprehensive" approaches to negotiation (1980, 484–486). Congress's major impact was probably in influencing Kissinger's sense of what was possible domestically, limiting any "tilt
toward the Arabs," and probably reinforcing his personal preference for
not tackling the Palestinian issue directly.

What of the future? Novik argues persuasively that shifts within Congress itself, its increasing youth and conservatism, will affect its support of

Israel (1986, 38–55). He contends that there will be general support but that it will be much less automatic, with the Congress more likely to turn on Israel punitively in response to short-term policy differences.

The State Department and the White House

Before the 1973 oil embargo William Quandt endeavored to explain how U.S. policy in the Middle East is made:

> It is necessary to break down the image of the United States as a unitary rational actor. In fact, the number of relatively high-level officials engaged in writing memos, reports, speeches, and analysis of the Arab-Israeli problem probably exceeds fifty in Washington alone. At least four major departments within the executive branch deal with Arab-Israeli affairs and each of these contains subdivisions which differ among themselves as to appropriate policies. (1973, 264)

The numbers are now undoubtedly higher.

Within the U.S. government, the White House has often been more supportive of Israel than has the State Department, presumably because the Israeli lobby made it good domestic politics. The Defense Department's position is less fixed. Before and after the oil crisis, various bureaucratic organizations had significant impact on Middle East policy. From 1973 to the end of the Ford administration in 1976, however, Kissinger dominated foreign policy in a quite unusual manner. As secretary of state and national security advisor, he was delegated almost total powers by both Nixon and Ford, albeit for different reasons. As a result, organizations other than the State Department had relatively little impact on policy (S. Spiegel 1985, 236).

Steven Spiegel argues that the State Department's tendency to support Arab positions is in part the result of its heavy regional focus as contrasted with the global focus of the Defense Department (1985, 5–6). The regional focus pulls the analyst toward a pro-Arab position because almost all countries in the area are opposed to Israel. So, if one's concern is to facilitate relations in the region, support of Israel makes no sense. The Pentagon, in contrast, has been concerned with other regions since 1945, so its interest in the Middle East tends to be global.

Robert Trice argues that the State Department resists pressures from any domestic group—either pro-Israel or pro-Arab—and that this desire for independence is more important than any Arabist tendencies in the department (1976b, 57–59). (Some State Department officials reportedly believe the Israeli embassy in Washington routinely has access to its internal communications on the Middle East, suggesting Israel is not without friends in the institution [Hersh 1983, 224–225].) Trice also notes that the department was not receptive to pro-Arab pressure-group demands, which may explain why Arab American groups feel equally rejected by all sectors of the U.S. government. The ultimate test, however, is the policies advocated over time; although the State Department may not have been pro-Arab, it has generally recommended a more evenhanded policy in the Middle East, probably for the reasons suggested by Spiegel.

     The State Department does have Arabists (that is, people who speak Arabic with considerable field experience in the Middle East) in its Bureau of Near Eastern and South Asian Affairs, but they are much less influential than their counterparts in Britain. They do not always agree with one another (about how to respond to radical Arab regimes, for example), and their influence within the department seems to have declined as the Middle East became a more crucial issue. When Richard Nixon became president, he appointed Joseph Sisco, a Soviet specialist, as assistant secretary of State for the Near East and transferred a number of Arabists out in order to define the Arab-Israeli issue more in terms of the Soviet-American competition (Pollock 1982, 58). The bureau has not been headed by an Arabist since. State has increasingly required its Middle East specialists to serve in both Israel and the Arab countries, presumably offsetting pro-Arab biases. Indeed, in early 1985 the two Middle East staff members on the National Security Council were both career Foreign Service officers with more field experience in Israel than in Arab countries (personal interview with a former senior official, February 1985).

     State's tendency toward pro-Arab positions is often offset by other advisors and departments (Quandt 1977, 25–26). One study found State Department people from Intelligence and Research allied with the Pentagon against Near Eastern desk officers and policy advisors (Bonham, Shapiro, and Trumble 1979, 21–43).

     Nonetheless, the direction of State Department policy seems to have been fairly clear. In 1948 President Harry Truman recognized Israel over the objections of the secretaries of state and defense. "Defense worried about strategic access to oil and bases; State was concerned with political relations between the United States and the Arab world, as well as with oil" (Quandt 1977, 25). A member of the Foreign Service, having been transferred to the Middle East from Europe in 1956, found himself in a new environment:

> In my previous experience, colleagues had generally approved of American policies in the areas in which they served. . . . In the Middle East, however, the atmosphere was totally different. Patriotic American foreign service officers there considered our developing policies a prescription for disaster, and their best efforts were devoted to compiling evidence that U.S. policy should be reconsidered and revised in Washington. . . . There was general agreement that knee-jerk American partisanship on the side of the Israelis was complicating efforts to settle the problems. (Curtiss 1982, iii–iv)

Similarly, early in the Nixon presidency, the Israelis learned to bypass Secretary Rogers and deal directly with Nixon and Kissinger, who were more sympathetic (Quandt 1977, 25, 81). State pushed strongly for immediate negotiations between the great powers for a Middle East settlement, despite Israeli opposition; this move had considerable impact because Nixon had delegated the Middle East to Secretary Rogers, who in turn adopted the general thrust of his department. The result was eventual negotiations between the Soviets and the Americans; when these failed, Rogers went public with his plan. Kissinger contends that Rogers did so before the National Security Council had approved it; Steven Spiegel cites this as a rare case where a national security bureaucracy essentially made policy during the Nixon administration (1985, 187).

Later, when the Soviets committed weapons and troops to Egypt to halt Israeli deep-penetration bombing, the Nixon administration's response was "uncharacteristically muted." Spiegel attributes this to Nixon's increasing preoccupation with other things—the national security bureaucracies, left to themselves, therefore tried to use Soviet moves to pressure the Israelis to make more concessions (1985, 191–192). These same organizations were extremely reluctant to admit that the Egyptians had violated the cease-fire agreements on the Suez Canal in August 1970. Similarly, the State Department's reaction to the Jordan crisis of 1970 differed from that of the White House. Nixon and Kissinger saw the problem as one of repelling a Soviet thrust in the Middle East, whereas State saw the crisis as a diversion from the search for an Arab-Israeli peace. The result was incoherent policy, with State unable to deliver on its implied promises to the Arabs, and Israel incorrectly assuming the White House would intervene on its behalf when necessary. Rustow argues that the State Department's Arabists, by warning of an embargo before it occurred, increased the embargo's effect by making the American public and government overreact (1982, 164–166 and 171–173). The major lesson of the 1973 war for the Middle East specialists at State seems to have been increased distrust of the Soviet Union, but their commitment to negotiations as the way to solve the Arab-Israeli conflict did not waver (Bonham, Shapiro, and Trumble 1979, 40–43).

In 1973 Kissinger became both secretary of state and national security advisor. Until his departure in 1976, interdepartmental battles were minimized because he controlled the process so rigorously. Melvin Laird, secretary of defense, was concerned with Vietnam and played little role in the Middle East. The same was true of the chairman of the Joint Chiefs of Staff, Earl Wheeler. CIA Director Richard Helms "was not . . . a strong advocate of specific policies in the Middle East" (Quandt 1977, 75). State Department influence probably increased a bit after Kissinger became secretary, but the process remained tightly centralized; in general, area specialists became involved in the decision process (as opposed to being consulted about specifics) only if they were in the Kissinger retinue. This centralization of decision-making authority is common during crises; but unlike other crises in 1958 and 1970, normal processes never resumed after the resolution of the oil crisis, and indeed this pattern lasted until Jimmy Carter became president (Dowty 1984, 204n, 309–312).

Even though it adopted a different policy, the Carter administration was able to maintain the same tight control over interdepartmental differences on the Middle East. Its stress on building relations with the Arab world, particularly the Palestinians, meant that the traditional State Department view was dominant for the first time. Carter's appointees all shared this position; there was no strong supporter of Israel among them, and personal amity between State Department and National Security Council personnel was high. Administration unity was therefore maintained to an unusual degree (S. Spiegel 1985, 320–321). Essentially, then, bureaucratic conflicts had little or no direct influence on U.S. policy toward the Arab-Israeli dispute for the decade following the oil crisis.

The situation changed again under the Reagan administration, which had a great deal of internal disagreement, little central control, and major policy shifts. It hardly seems coincidental that its secret contact with the PLO was made through the State Department. Indeed, while Secretary

Haig was certainly aware of the initial contact (and informed the president of it), further talks seem to have been conducted by the head of the Near East Bureau, Nicholas Veliotes, on his own authority. Later the bureau opposed closer relations with Israel, even after the president had decided to implement them (S. Spiegel 1985, 418, 426). Once again the State Department was involved in the coalition pushing for a less pro-Israeli policy.

The importance of the White House in these questions makes William Quandt's analysis of American Middle East policy in terms of the U.S. electoral cycle particularly interesting (1986, 8–29). He contends that in the first year of his first term, the president learns about the Middle East and about the domestic political aspects of the issue; any initiatives undertaken during this period are likely to fail. By the second year, however, this experience means policy is more in tune with reality and therefore is more likely to succeed. In the third year, the president becomes concerned to demonstrate success and may either pay a disproportionately high price to get some tangible product, like an agreement, or may drop the entire issue. Nothing much is likely to happen in the fourth year because presidents must avoid controversy for the election and because neither the Arabs nor the Israelis are willing to make major concessions for an American president who may shortly leave office.

The Influence of Individuals on U.S. Policy

How important are individuals in the making of U.S. policy on the Middle East? One sophisticated observer is forthright:

> The critical factors determining the content of American policy are: the basic assumptions of the president, the individuals on whom he relies for advice, and the resulting decision-making system which converts ideas into policies. . . . All Presidents enter the White House with specific assumptions that prove remarkably resistant to the effect of outside forces—interest groups, events and crises in the area, the bureaucracy. . . . In the end policy can be understood only by examining an administration's foreign policy priorities, philosophical assumptions, decision-making system, and key personalities. (S. Spiegel 1985, 10)

Richard Nixon

As president, Richard Nixon was able to dominate U.S. foreign policy if he chose to do so. He influenced both the substance of policy toward the

Middle East and the process by which that policy was made. The initial priorities of Nixon's foreign policy were Vietnam, the Soviet Union, and China, endeavors that were controlled by Kissinger from the White House. Other areas, including the Middle East, were left to the State Department, at least early in his first administration (Quandt 1977, 77). As one author describes it: "The administration gave high priority to dealing with the instabilities of the Middle East, except when other issues were perceived as more critical. No administration . . . had entered office with a conception of the Middle East as being more important to American interests" (S. Spiegel 1985, 168).

The central idea of the Nixon administration in the Middle East was that progress on the Arab-Israeli conflict was essential to improve the American position in the Arab world. The central goal was largely to bar Soviet influence in the region. Two somewhat separate positions emerged in the administration. Those concerned primarily with U.S.-Arab relations felt resolving the conflict was the only strategy to follow. In contrast, those primarily concerned with the Soviet Union advocated a secondary strategy of relying on proxy states in the area, particularly Iran and Israel, in case a settlement could not be reached. Not surprisingly, this produced somewhat contradictory policies: Israel was to be pressured into an agreement (to be saved despite itself) and at the same time was to be an American proxy. This duality accurately reflected Nixon's personal ambivalence toward the situation, and particularly toward Israel—on the one hand, he was contemptuous of American Jews, whom he saw as political enemies; but on the other hand, he greatly admired the Israeli military because they were able to use force successfully (Hersh 1983, 214). The result was a convoluted U.S. policy, as when, in response to protests over the Rogers plan, Nixon did not disavow it but sent a series of messages to both American Jews and the Israeli government strongly implying he did not support it (S. Spiegel 1985, 171–173, 178–181, 187–189, 215–218; Hersh 1983, 220–221).

In 1969 Nixon and Kissinger worried about the Middle East primarily because of the possibility of a clash with the Soviet Union; the State Department was mainly concerned about threats to American interests in the area, although a threat to U.S. oil supplies was not thought likely (Quandt 1977, 79, 79n). Interestingly, despite large campaign donations from the major oil companies, including a secret gift of $100,000 from Gulf Oil, Nixon remained impervious to their desire to alter Middle East policy

(S. Spiegel 1985, 170–171). Hersh argues that Nixon encouraged Israel's deep-penetration raids in 1970, which led to Soviet intervention (1983, 217–218, 223–224). This argument ignores the fact, however, that the United States eventually failed to support Israel; a more plausible interpretation is that Yitzhak Rabin, the Israeli ambassador to the United States, exaggerated U.S. support in his reports to Jerusalem (Shlaim and Tanter 1978, 493–494, 501–502, 514). Quandt writes that "Nixon's great hope during his first term was that he would be able to recreate a domestic consensus on behalf of his foreign-policy goals. The style and timing of each major foreign-policy step was taken with an eye toward domestic public opinion" (1977, 79).

Thus, one reason for the "standstill diplomacy" from 1970 to 1973 was that Nixon was focusing on the Soviet Union, China, and Vietnam in preparation for the 1972 elections and, after they were over, had no incentive to work quickly (Quandt 1977, 147). Kissinger himself sees a similar effect in the second term: "Lifting [the embargo] turned almost into an obsession for the next five months, partly because Nixon thought that it lent itself to a spectacular that would overcome Watergate" (1982, 873).

When the Jordanian crisis broke out, Nixon saw an opportunity to crush the Palestinian guerrillas, while both the State Department and the military saw a crisis to be avoided, and even Kissinger wanted to back possible Israeli intervention rather than commit American troops. Hersh asserts that Nixon ordered Secretary of Defense Laird to bomb PLO hideouts with U.S. Navy aircraft after the series of hijackings, but Laird reportedly used the excuse of "bad weather" to avoid executing the order (1983, 235–236). Both Kissinger and Nixon saw the subsequent Syrian invasion of Jordan as directly linked to the Soviet Union rather than to internal Syrian politics (S. Spiegel 1985, 198–202; Hersh 1983, 234–249).

Nixon's personality also clearly influenced the structure by which decisions were made. He was unusually concerned to rigorously restrict discussion and information about his own attitudes: "His experience in foreign affairs, such as it was, came largely through his own first-hand experience and discussions. He had little patience with academic studies or lengthy briefing materials" (Quandt 1977, 73). Moreover, Nixon decided where Middle East policy would be made in his government—originally in the State Department and later, after the Jordanian crisis, in the White House itself (Quandt 1977, 128; S. Spiegel 1985, 197).

Nixon was uncomfortable with personal confrontations and disagreements, and this was a particular problem in Middle East policy, where Secretary of State Rogers and Henry Kissinger clashed repeatedly. Kissinger's preoccupation with the Soviet Union inclined him toward strengthening Israel; Rogers favored conciliation and reflected the regional orientation of the State Department, which gave more weight to propitiating the Arabs. Nixon's solution was to refuse to choose, in part because he himself was ambivalent. Thus, before the 1973 war there were actually two independent teams carrying out semi-independent foreign policies, a degree of division unprecedented in previous administrations (S. Spiegel 1985, 174–177, 215–218).

This situation ended after the 1972 election, when Nixon moved Kissinger to the State Department just in time for Kissinger to handle the 1973 war and the oil crisis. From 1973 to 1975 Kissinger was both secretary of state and national security advisor, a uniquely powerful position brought about by Nixon's personal decision. Steven Spiegel (1985, 228–230) argues that it drastically reduced the ability of outsiders to influence the policy process and that this in turn encouraged them to try to bring public pressure to bear early in the process. He cites both the oil companies before the embargo and the pro-Israeli lobby afterward as evidence of this.

Kissinger's leverage was strengthened by Watergate, which had made it more difficult for the president to attend to foreign affairs. Despite Nixon's rather cool personal relationship with Kissinger, the president "abdicated American diplomacy to Kissinger" (S. Spiegel 1985, 219). By the time of the 1973 war Kissinger was clearly the central figure in U.S. foreign policy and "forged the American response to the Arab-Israeli conflict almost single-handedly" (Maghroori and Gorman 1981, 4).

Interestingly, Kissinger's unusual power continued under Gerald Ford; President Ford had little experience in foreign affairs and possessed a warm relationship with Kissinger. "When Gerald Ford became President in August 1974, foreign policy changed less than between any two administrations in the post-1945 era. Kissinger pursued the same strategy under both Presidents" (S. Spiegel 1985, 230–234). The 1973 Arab-Israeli war broke out in the middle of Watergate, the crisis that eventually forced Nixon to resign the presidency. In October alone, while the war broke out and the embargo was declared, Spiro Agnew resigned as vice-president, Gerald Ford was selected to replace him, the Saturday Night Massacre occurred, the War Powers Act was passed, and the House of Representatives began its impeachment inquiry. Nonetheless, Watergate seems to have had relatively little impact on Middle East policy other than strengthening Kissinger's own position somewhat (Kissinger 1982, 195; Quandt 1977, 250–252), although Dowty suggests it made Nixon more likely to overreact (1984, 261).

There is some fragmentary evidence to support Dowty's suggestion that Watergate made Nixon more prone to extreme measures. Kissinger asserts that during the week before Nixon resigned, the president requested that papers be prepared to "cut off all military deliveries to Israel until it agreed to a comprehensive peace" (1982, 1205). The papers were drawn up but never signed. However, this was not unprecedented behavior. Nixon had often issued exotic orders that subordinates would quietly ignore; we have already noted a report that he wanted to bomb PLO positions in 1970. A more interesting argument is that the magnitude of the $2.2 billion aid bill for Israel—which seems to have triggered the embargo—may have been dictated by Nixon's need to appease a pro-Israel Congress because of Watergate. Administration witnesses were apparently unable to explain what they were going to do with funds in excess of the $850 million recommended by the Pentagon; evidently the Israelis had not requested assistance of this magnitude (Kassis 1981, 121–122; Roehm 1980, 297–298). Kissinger also was apparently involved in this decision, however, so it is unreasonable to attribute it entirely to Nixon's overreaction at the time.

Quandt, among others, has found it difficult to determine how much influence Nixon had on U.S. foreign policy in the Middle East during the 1973 war (1977, 250–252, 183, 192, 196). Nixon was clearly preoccupied with Agnew's resignation on October 10, and he left details to others, but he is reported to have decided on a massive resupply of arms to Israel on October 13. The Saturday Night Massacre took place on October 20 and by that time Kissinger had the authority to negotiate a ceasefire without further approval by Nixon. A few days later Kissinger was also empowered to order a military alert in response to threatened Soviet intervention to save the Egyptian Third Army.

Henry Kissinger

Henry Kissinger was clearly the dominant figure in U.S. policymaking on the Arab-Israeli conflict from 1973 to 1976. We can trace some reasons for this to the particular needs of presidents Nixon and Ford, to Watergate, and so forth, but much of Kissinger's power stemmed from his singular intellectual gifts and his personality; it is hard to imagine another person in the same situation who could have converted the clear potential for power into the foreign policy supremacy that Kissinger enjoyed during this period.

Kissinger had not been deeply involved in Middle East or energy policy before 1973, although one unsympathetic observer argues that he was largely responsible for sabotaging the Rogers plan initiatives (Hersh 1983, 213–233, 402–414). He had not felt free to concentrate on the Middle East until the 1972 elections were over and a Vietnam agreement was reached in January 1973. Even then, however, he felt no real urgency to act, suggesting at one point that nothing much could be accomplished until after the Israeli election in October. Indeed, Kissinger believed one of the lessons of the failure of the Rogers plans was that the United States should not be involved too early in the negotiations. Moreover, Watergate began to distract the administration's attention; on April 30 White House Chief of Staff H. R. Haldeman and presidential assistant John Ehrlichman were forced to resign. It took the 1973 Arab-Israeli war to bring the issue to the center of Kissinger's concerns (Quandt 1977, 155–157, 207–208; Quandt 1975, 35).

Kissinger was the paramount figure during the October war because he had the support of the president and was in fact the only person engaged in foreign policy who had routine contact with him. Personality and bureaucratic differences grew less important during the crisis, and Kissinger's personal dominance resulted in an unusually coherent U.S. policy. He certainly altered policies during this period, but at any given time he spoke for the United States and was largely able to mute domestic dissent. This unique position of power lasted until he left office in 1976 (S. Spiegel 1985, 312–314; Pollock 1982, 201–203). Similarly, he dominated Congress, at least up to 1975, although, as noted earlier, he had some problems later on (Roehm 1980, 207–218). Kissinger even altered Gerald Ford's position on the Arab-Israeli dispute. As a congressman, Ford had been a strong supporter of Israel; indeed, in 1972 he had advocated recognition of Jerusalem as the capital of Israel. He rejected this position in his first news conference, however (an interesting contrast to Joe Clark of Canada). In general, he quickly altered his verbal positions to conform with Kissinger's diplomacy (S. Spiegel 1985, 230–234).

What real difference did Kissinger make? How would U.S. policy have been different if someone else had been in his position? (The best introduction to the literature on Henry Kissinger is Starr 1984.) Several possibilities suggest themselves. First, he chose to use this dominance to work on the Middle East. Quandt asserts that between October 1973 and May 1974 no issue ranked higher in U.S. foreign policy than the Middle East negotiations. During this time, Quandt estimates, Kissinger spent one-half to two-thirds of his time and energy on this issue (1977, 202–205; Quandt 1975, 38), presumably because he felt he could make a difference; it is not at all clear that another secretary of state would have expended as much effort on the issue.

Second, despite Kissinger's skill at working with regional actors, much of his concern always stemmed from his central focus on Soviet-American relations. For instance, the decision to rearm Israel during the 1973 war was meant in part as a rejoinder to the Soviet Union's high level of support to the Arab countries (Dowty 1984, 246–248). Indeed, Kissinger's willingness to concentrate on the Middle East for so long seems to have stemmed more from his linkage of the problem to U.S.-Soviet relations than from an analysis of regional factors, including the oil embargo itself. Third, the process itself may have stemmed from experience. Kissinger seems to have based his step-by-step diplomacy in the Middle East in part on his successful experience with a similar technique in negotiating with the Chinese, Russians, and Vietnamese (Quandt 1977, 33; Starr 1984, 69).

Finally, it is quite possible that only Kissinger could have made shuttle diplomacy work; it remains the premier example of personal diplomacy of our times. He was walking a number of tightropes at once—not only managing the substantive problems of several different sides but also dealing with his own delicate problem of not *seeming* to negotiate because of the embargo, while trying to persuade the Arabs to end the embargo because of his negotiations. "By using its peacekeeping efforts to get the embargo lifted, the United States, in fact, was trying to take advantage of the Arab policies which sought to pressure Israel through the United States" (Ramazani 1977, 9).

Kissinger played the role of outsider in the Middle East negotiations:

> Unlike his other negotiations, the disengagement talks found Kissinger in the role of mediator, moderator, message-carrier, with relatively few cards of his own to play. . . . Basically his task was to urge the parties to recognize their own interests in reaching limited agreements. A major part of Kissinger's time went

into educating each party about the constraints operating on the other. As the only person in intimate contact with Sadat, Assad, and Prime Minister Meir, he was able to offer his judgment on what each party could realistically afford to accept in light of its domestic political situation. (Quandt 1975, 39)

His strategy was to break issues into as many pieces as possible, on the theory that, when some issues had been settled, a momentum would build up, allowing the more difficult ones to be attacked. He also felt strongly that the negotiations had to be kept secret; on the American side, that meant that only he would know everything that was going on. He even refused to use an American interpreter in the high-level negotiations for this reason. He used this control of information to selectively inform the press and Congress as the negotiations went forward to build American public support (Quandt 1975, 46; Dowty 1984, 305–306).

His negotiating tactic was to take proposals from one side to the other, adding, if requested, a judgment about domestic political pressures on the other side and sometimes encouraging new or modified proposals. He would add his own proposals only when agreement seemed near. Kissinger also made it a point to talk to other Arab leaders about the negotiations to avoid isolating Sadat or Assad if they did reach an agreement with Israel; President Houari Boumediène of Algeria and King Faisal of Saudi Arabia were particularly important, Boumediène for Arab legitimacy and Faisal because of money and possible influence on the use of the oil weapon (Quandt 1975, 39–47).

One critical element in this process was Kissinger's ability to establish personal relationships with different leaders. It is hard to think of another person, let alone another American, who has done this so well with so many very different people on both sides of the Arab-Israeli dispute. His relationship with Sadat was particularly notable. At times Sadat would give Kissinger both his bargaining position and his fallback positions, confident of Kissinger's willingness to negotiate in Egypt's interest. This linkage built up political capital for the United States, which Jimmy Carter drew on later during the Camp David negotiating process (personal interview with a former senior official, February 1985).

Kissinger's impact was not limited to process; it also involved substance. He defined the problem as one of politics rather than economics or resources, a definition that suggested solutions should focus on political rather than economic tools. Thus, Kissinger was not really interested in altering U.S. energy policy; in fact, he basically ignored the embargo, focusing instead on using it to bring about a political negotiating process in the Middle East (Nau 1974–1975, 436; Morris 1977, 259).

In general, however, as Steven Spiegel points out, Kissinger was reluctant to get overly involved in substance (1985, 313–314). He never stated a substantive *concept* for a Middle East political solution but developed a negotiating *process* instead. The process was to solve separate, smaller problems, presumably developing momentum that would carry over to others. This worked as long as military disengagement from the 1973 war was the central issue. No momentum developed when issues shifted to a political settlement, however, and Kissinger had nothing to substitute on his own. It is interesting to speculate whether such a new concept would have been developed if Gerald Ford had been reelected in 1976; Kissinger and Ford told Sadat they planned to seek a comprehensive settlement after the American election (Burns 1985, 184, 187).

Jimmy Carter

This pattern of high-level political involvement in Middle East policy continued after Kissinger left office. Indeed, the Middle East became a central foreign policy issue for the Carter Administration, perhaps more than for any other administration (S. Spiegel 1985, 317–318); Carter says he spent more time on this issue than any other (1982, 429). Carter's concern stemmed from a variety of sources, including his strong Baptist convictions regarding the biblical significance of the region (Carter 1982, 274). Politically, he was concerned about the energy problem and harbored a general commitment to improve human rights and relations with the Third World. "The oil question now replaced the U.S.S.R. and anticommunism as Washington's preeminent concern" (S. Spiegel 1985, 320; see also Quandt 1986, 4, 30–32; Carter 1982, 277–278).

Carter was personally more deeply involved in Middle East policy than any other president (S. Spiegel 1985, 324). Spiegel argues that no high-level officials in the Carter administration supported Israel; one reason might have been that Carter, because of his inexperience in foreign affairs, did not understand the depth of disagreement among Middle East experts and therefore assumed there was no need to ensure that he would be exposed to different views. Moreover, because the issue was so important, the administration could not tolerate internal dissent (Spiegel 1985, 318–329). In contrast, another observer notes that a number of Carter advisors who were supportive of Israel were involved in many, although not all, decisions of the period, including Walter Mondale, Stuart Eizenstat, Robert Lipshutz, David Aaron, and Harold Brown (personal interview with former senior official, February 1985; an interesting book by a member of this group is Charney 1984).

In general, Carter often had difficulty in gaining political support for his policies, regardless of how reasonable they were. In the Arab-Israeli dispute, he had problems dealing with potential allies who did not share his own particular motives, including the Arab states, Congress, and the Israeli lobby. This weakened his position significantly (S. Spiegel 1985, 328–329; Quandt 1986, 9–10, 30–32, 336). Nevertheless, Carter is largely responsible for the Camp David accords. When the State Department looked for a precedent of a U.S. president directly mediating a foreign conflict, it had to go back to Theodore Roosevelt working to end the Russo-Japanese War, and even Roosevelt had not participated in the talks himself (Vance 1983, 218). Carter decided to call the Camp David meeting, despite opposition from his bureaucracies, which feared it would be a disaster, and during the talks themselves "the central figure remained Jimmy Carter. His mastery of detail and great stamina saved the negotiations time and time again" (S. Spiegel 1985, 353). Indeed, the real negotiations were handled through the Americans, with the Egyptians and Israelis not meeting face to face. Carter suggested that the Sinai issue be separated from the West Bank–Gaza Palestinian issue, which in turn was central to the success of the talks. He also made the major error of not clarifying Begin's commitment on Israeli settlements; the letters of interpretation revealed that the two sides had very different interpretations of the agreement. He also assured Sadat that Saudi Arabia and Jordan would support Camp David; in fact they did not do so (S. Spiegel 1985, 353, 356, 359–360, 362, 370–373, 379–380; Quandt 1986, 206–258, 323; Vance 1983, 213–231; S. Brown 1983, 494–504; Carter 1982, 319–403).

Later, when major differences appeared, Carter decided to go to the Middle East himself and commit his personal prestige to hammer out a treaty, which he was able to do:

The president's decision was a breathtaking gamble and an act of political cour-age. The unresolved issues made final agreement unlikely without further nego-tiating rounds. Failure in personal presidential diplomacy, coming on the heels of the fall of the shah, could have sapped the administration's political strength as we were reshaping our security policy in Southwest Asia and the Persian Gulf, as well as girding for a difficult ratification fight over the SALT II Treaty. (Vance 1983, 245)

After Camp David President Carter apparently decided to disengage him-self personally from Middle East questions, presumably for domestic po-litical reasons (Brzezinski 1983, 437–438).

Zbigniew Brzezinski and Cyrus Vance

Zbigniew Brzezinski and Cyrus Vance were clearly the other two major players in the Carter foreign policy process. Brzezinski had had some inter-est in the Middle East before taking office, although his primary concern had been with the Soviet Union and Eastern Europe (Quandt 1986, 35). After the oil crisis he wrote a series of articles arguing for American pres-sure on Israel, establishment of a Palestinian state, and a U.S.-Soviet guar-antee of Middle Eastern frontiers. His analysis assumed that the Arab-Israeli issue had to be settled in order to deal with the energy problem:

Kissinger had concentrated on the Russians more and on the Third World less, but Brzezinski shared with Kissinger a deep-seated fear of another oil embargo. Kissinger had considered Egypt the key state in the Middle East and saw block-ing Russian influence there as the central problem facing American policy. Brzezinski saw the key state as Saudi Arabia and energy as the central problem. Kissinger had a role for Israel to play; Brzezinski had little use for Israel in his plans. (S. Spiegel 1985, 322).

In 1975 Brzezinski participated in a major Middle East study at the Brookings Institution that reached similar conclusions and enjoyed con-siderable currency, particularly among knowledgeable Europeans; this was not surprising, because it ran along lines favored there. The Brookings group included a number of people later prominent in the Carter admin-istration, such as William Quandt of the NSC; Robert Bowie, who be-came CIA deputy director; and Philip Klutznick, later secretary of com-merce. The Brookings group included a wider variety of views than later found in the Carter administration, and the resulting report showed this divergence. It stressed the need for a comprehensive rather than step-by-step approach but was vague about what territories the Israelis should relinquish and what kind of concessions would be required of the Arabs: "The Brookings Report was vague on key issues. . . . Except for the emphasis on normalizing relations between Arabs and Israelis, . . . Carter's policy would probably have been similar if the Brookings report had never been written" (S. Spiegel 1985, 323; see also Brzezinski 1983, 84–85).

Brzezinski spelled out his own position before taking office in an article called "Recognizing the Crisis." His two short-term policy recommendations were (a) to work for a settlement of the Arab-Israeli dispute and (b) to engage in crisis planning, "for there is high probability of renewed [oil] production curtailments or embargoes in early 1975, especially barring tangible progress toward a Middle Eastern settlement" (Brzezinski 1974–1975, 70). He therefore believed a Middle East settlement was central for the well-being of the international system as a whole and defined the issue as seeking a settlement rather than, for example, defending Israel or preventing Soviet entrance to the area.

Cyrus Vance had less of a public record on the Middle East before becoming secretary of state, but what he had said was "startlingly similar" to Brzezinski's position (S. Spiegel 1985, 322). He quickly became deeply involved in the issue and spent a good deal of time in direct negotiations with both sides; Quandt asserts that he was able to get and keep the confidence of both Arabs and Israelis better than any American who dealt extensively with both sides (1986, 35).

Nevertheless, it is difficult to trace the direct influence of either Brzezinski or Vance on the Middle East policies of the Carter administration. There are at least two major reasons for this. Although disagreements between Vance and Brzezinski became increasingly public during the Carter administration, the two men basically agreed on the main thrust of Middle East policy (S. Spiegel 1985, 320–321; Quandt 1986, 35). More important, in the Middle East as in other areas of foreign policy, Jimmy Carter was the dominant figure during his administration; Brzezinski quotes Carter as saying: "There have been presidents in the past, maybe the not too distant past, that let their Secretaries of State make foreign policy. I did not" (Brzezinski interview with James Reston, *New York Times*, December 31, 1978, cited in Moore 1984, 67).

George Shultz

If the Carter administration appointees agreed that an Arab-Israeli solution was critical for the United States, their counterparts in the Reagan administration shared a similar consensus that the major problem was U.S.-Soviet relations and that everything had to be subordinated to it. In the Middle East this produced the bizarre notion that the Arab-Israeli dispute could be settled by persuading all parties to join in a common alliance against the Soviet Union—a breathtakingly naive concept in light of the region's history. It is hard to imagine another foreign policy team in recent history that could have put such an idea forward seriously. Interestingly, of the top-level appointees, Haig seems to have been quite sympathetic toward Israel for strategic reasons, whereas Secretary of Defense Caspar Weinberger was much more interested in Saudi Arabia, perhaps in part because of his experience with Bechtel in the Middle East.

As might have been expected, this initial thinking did not last long. Just as the original idea was the product of the individuals Ronald Reagan selected to fill his foreign policy positions, so was the change. George Shultz seems to have taken charge of Arab-Israeli policy fairly quickly; at least one observer attributes the Reagan plan of 1982 largely to Shultz (Rubin 1983, 378). Before the oil embargo, Shultz had headed a task force on oil imports that had been skeptical of the Arab producers' ability to withhold oil for political purposes (Sampson 1974a). As secretary of the treasury in 1973 he had opposed concessions to the Arabs. He came into office as Reagan's secretary of state reportedly feeling sympathetic toward the Arabs; however, over time he seems to have been repelled by the inconsistency and deception of some Arabs and to have moved toward the idea that the Israelis were at least sane, albeit difficult—a process assisted by Lawrence S. Eagleburger, his undersecretary and an advocate of a strong Israel (S. Spiegel 1985, 404–405; personal interview with a former senior official, February 1985).

Ronald Reagan

Ronald Reagan's approach toward the Middle East has been dominated by his deep concern for the Soviet threat; otherwise, he seems to have had little in the way of specific ideas about the Middle East. Before his election, he strongly supported Israel as an ally against Soviet involvement in the area. The vagueness of his ideas, however, allowed for very different interpretations of an anti-Soviet coalition. Thus, Secretary of State Haig argued in favor of an alliance between the West, Israel, and the Arab states, while Secretary of Defense Weinberger felt Israel should be excluded (S. Spiegel 1985, 403). Reagan's unwillingness either to specify what he meant or to resolve quarrels between his subordinates increased the policy confusion. The result was at least three different foreign policy teams dominating foreign policy during the first three years of his first administration, including two secretaries of state, two NSC advisors, and three Mideast negotiators serving from mid-1982 to late 1983. "In no previous administration had the turnover of major foreign policy personnel been so rapid or extensive" (S. Spiegel 1985, 404, 423).

Ironically, the Reagan administration had a clear philosophy of foreign policy. But the implications of the philosophy were so unclear for particular policy areas that it required someone to make decisions. During his first term, Ronald Reagan was unwilling to do this himself or to appoint someone who would. The result was unprecedented levels of internal conflict and drastic policy shifts:

> Reagan's approach to the Middle East was a combination of emotion, ideology, lack of knowledge, and instinctive political acumen that made him so difficult for analysts to comprehend. He could proclaim a united pro-American Lebanon vital to U.S. interests in one month (January 1984) and then act as if the country did not exist the next. He presided over periods of Israeli-American tensions unusual even in the checkered history of relations between Jerusalem and Washington and yet later approved new levels of assistance to Israel. . . . He reveled in rebuilding American military might and failed to use arms effectively in Lebanon. He was preoccupied with the expansion of Soviet arms and influence and then watched helplessly as Russian involvement escalated in Syria.
>
> He reviled the PLO yet saved its leadership from destruction. Ronald Reagan seemed blind to these contradictions. (S. Spiegel 1985, 428–429)

How important have personal differences among these various individuals been? William P. Bundy argues that the differences were tactical and marginal rather than dominant: "Each of the four Presidents of the period—and indeed their predecessors—have come to practice similar policies toward the Arab-Israeli problem: if one were to bring together these four men today . . . the areas of disagreement would be minor and tactical" (1984, 1219). Kissinger and Carter can undoubtedly lay the greatest claims to influencing policy directly, but one can argue that both were more notable for their ability to carry out policy than to develop new ideas about what that policy should be. The ability to implement policy in an area as difficult as the Arab-Israeli dispute is not to be despised, but the undoubted changes that have occurred in U.S. policy should not obscure a remarkably strong element of continuity from 1967 to the present. Such continuity, in retrospect, has not been marred much either by different personalities or by external influences such as the oil weapon.

Impact of the Oil Weapon on U.S. Middle East Policy

It seems useful to analyze the Kissinger and Carter eras separately. There are a number of different interpretations of the oil weapon's impact on Kissinger's diplomacy, but they basically pose two different questions: (1) Did the United States change its policy toward the Arab-Israeli conflict? and (2) Was the change, if any, brought about by the oil weapon?

One view holds that no real change occurred in U.S. policy, only cosmetic shifts. Not surprisingly, this is the view of U.S. officials (Szyliowicz 1975, 183–184), supported by at least one outside analyst (Sicherman 1978, 12).

A second view is that, although policy may have remained constant, notable shifts took place regarding priorities. Thus, Ian Smart suggests that Kissinger's negotiation of Israel's withdrawal from some territories was "moved to a new urgency by the use of the oil weapon" (1976, 262; see also Stockholm International Peace Research Institute 1974, 29). Szyliowicz pinpoints the shift "from cautious diplomacy to shuttle diplomacy" and greater sympathy with Egypt as two changes in policy during this period; he summarizes his findings by saying that the embargo made the Middle East a top-priority issue and that Kissinger then applied his pre-embargo ideas to try to solve it (Szyliowicz 1975, 208–211, 230; also Szyliowicz and O'Neill 1977, 48, 51). Kassis argues that the administration delayed resupplying Israel with arms for several days because of apprehensions about a possible embargo as well as a number of other considerations but that it eventually decided to run the risk (1981, 112–118, 124–127, 191–194; see also Sorley 1983, 90). Sheehan (1976, 116) and

a former senior U.S. official (personal interview, February 1985) suggest
that the Syrian shuttle diplomacy was the price for ending the embargo.

Steven Spiegel notes that the agreements of 1974 and 1975 resembled
Kissinger's strategy in 1971 and that the administration had deliberately
not adopted at least two options suggested at the time: confrontation with
the Arab oil-producers, or pressuring Israel to accept an independent Palestinian state (1985, 224–227). But he also observes that, although
Kissinger publicly denied any linkage between the oil embargo and his foreign policy, the internal memos of the administration show that such a relationship did indeed exist. The overall strategy did not change, but the
embargo was important in influencing particular decisions, such as pressuring Israel to attend the Geneva Conference.

A third view holds that changes took place but that they were caused by
factors other than the oil weapon, such as the new balance of power in
the Middle East resulting from the partially successful Egyptian attack or
the involvement of the Soviet Union. William Quandt, for example, maintains that the war changed the assumption of short-term stability that un-
derlay U.S. policy, although "probably U.S. diplomacy between November 1973 and May 1974 would not have been significantly different even if
there had been no Arab oil boycott of the United States" (Quandt 1978,
288; see also Knorr 1976, 230; Smart 1977, 20–27; Sheehan 1976, 35;
Bouchuiguir 1979, 382–386; Schneider 1983, 243–244; AlRoy 1975,
36). Ali makes a similar argument:

> Had there been no oil embargo, there would have been a cease-fire between the
> Arabs and the Israelis, but no quick Israeli pullback from the Egyptian territory
> on the west side of the Suez Canal and the Syrian territory from the Golan
> Heights. Precisely, then, the Arab oil embargo accomplished only the initial Is-
> raeli pullback and exchange of prisoners of war. (1976, 156)

A final position is that the oil weapon forced the United States to fundamentally alter its foreign policy and abandon its single-minded support
of Israel. One piece of evidence is the United States' pressure on Israel to
accept a cease-fire when it seemed on the point of wiping out the Egyptian
Third Army in the Sinai (Rand 1975, 318–319). Other authors agree:
"Indeed the oil embargo may have been the most decisive part of that
war—the part that led to an unfavorable diplomatic outcome for Israel,
and that continues to have the most far-reaching consequences" (Fried-

man, Seabury, and Wildavsky 1975, 155; see also al-Sowayegh 1980, 222–223; Manoharan 1974, 95, 103; Itayim 1974, 4). Rustow (1977, 507) and Steven Spiegel (1985, 271–272) do not agree with this position but suggest that the Saudis and other Arabs may see it that way, particularly given later U.S. shifts on Middle East policy; Bouchuiguir, after reviewing the Arabic literature, agrees that this was the Arab consensus (1979, 9–19). Related opinions maintain that the oil weapon could have had such influence if the Arab states had only used it properly (Shadid 1981, 180–181) and that it can have such influence in the future (al-Sowayegh 1984, 138, 140–141, 151–154, 194–201).

Leaving aside the question of how it changed, how successful was Kissinger's policy? He apparently told the Arabs he was working toward a comprehensive political settlement while telling the Israelis he was buying time.

> If step-by-step diplomacy as carried out by Kissinger is to be judged on its own merits, it rates high as a tactic but fails to convey any sense of long-term purpose. . . . Kissinger's justification for his efforts was that without the disengagement agreements there would be another war, accompanied by another oil embargo and by a resurgence of Soviet influence in the Arab world. . . . A political process must begin that would offer the Arabs an alternative to war, but it must be carried on at a pace that the Israelis could accept. This was the extent of Nixon's and Kissinger's initial conceptualization. (Quandt 1977, 250–251)

On balance, it seems Kissinger was not fundamentally influenced by the embargo; his policy remained essentially the same. There were some changes of timing and in specifics but certainly much less than the supply theory of economic sanctions would predict.

Ironically, the embargo seems to have had more influence on the Carter administration than on Kissinger, who was in power when it occurred. The oil weapon was apparently one reason the Carter administration elevated the Arab-Israeli issue to its top foreign policy priority. It also made Carter feel that it was essential to propitiate the oil producers, particularly over the Palestinian question. There is no suggestion in the literature that European pressure had any influence on this policy shift.

Steven Spiegel suggests that it is not unusual for a new administration to have thought through the lessons of an earlier experience and incorporated them into its philosophy (1985, 317, 399). A broader, theoretical reason also leads us to expect such a pattern of behavior. Philip Burgess, in explaining the impact of the German invasion on Norway's willingness to abandon neutrality after World War II, contends that the experience did not really change the minds of individuals in power but, rather, brought to power different individuals with different ideas (1968, 143–164). Combined with the Spiegel argument, this suggests the intriguing hypothesis that the major effects of sanctions will occur well after their actual imposition, unless they produce immediate personnel changes in the target government. Steven Spiegel argues that this same effect explains the Reagan administration's concern for a Middle East peace, although he notes that, in fact, other administrations have adopted the same position (1985, 399). Moreover, no such "second-administration effect" seems to have occurred in the other countries we have studied.

But why, then, did U.S. policy toward the Middle East shift over time? Like other Western countries, the United States found the Arab oil-producing states much more lucrative markets after the embargo as a result of their new wealth. This was particularly true for arms exports. Indeed, Lewis Sorley argues that the new wealth of the Saudis and the Iranians was vital for the Nixon doctrine of building up regional powers, since Congress would not have voted sufficient foreign aid to do so (1983, 44–45).

Overall, the new Arab wealth probably had less direct impact on the U.S. economy than on the other states in this study because the American economy is much less dependent on exports. Similarly, Arab oil money seems not to have been an important independent factor in determining Arab-Israeli policy (personal interview with a former senior official, February 1985), although it may have had an indirect impact on the Reagan administration through Caspar Weinberger and George Shultz, who had both worked for Bechtel (a corporation that had been heavily involved in Middle East contracts) and who, at least initially, seemed somewhat more sympathetic to Arab positions than they might otherwise have been.

The Nixon-Ford administrations (that is, Henry Kissinger) seem to have been primarily occupied with the role of the Soviet Union in the Middle East. This, along with some apprehensions about the stability of the Atlantic alliance, motivated Kissinger to pursue a step-by-step settlement in the Middle East. The Carter administration, however, seems to have been more concerned with energy; indeed, the oil weapon's strongest claim of influence over U.S. policy is that it influenced Jimmy Carter to make the Middle East his major foreign policy issue. The Reagan administration's concerns were clearly similar to those of Kissinger, focusing almost exclusively on the role of the Soviet Union with no serious interest in oil or energy.

In historical perspective, however, the Kissinger and Carter foreign policies are much more similar to each other than to the Reagan administration's; certainly they agree much more than they disagree. Kissinger had indicated he would seek a comprehensive settlement in a second Ford term, as Carter initially did. Ironically, this initiative was torpedoed by Sadat's trip to Jerusalem, and Carter became identified with the Camp David process, the logical culmination of Kissinger's step-by-step diplomacy, showing both the strengths and the weaknesses of that process. Given the same circumstances, Kissinger would undoubtedly have attempted to do the same. Carter showed greater sympathy for the Palestinians than did Kissinger, but this is probably the result of Carter's greater concern for human rights and the Third World. Both, in sharp contrast to the Reagan administration, gave high priority to the Arab-Israeli dispute and actively attempted to resolve it, risking considerable political capital in the process.

In an interesting analysis, Blema Steinberg argues persuasively that American concern over oil did not have to be translated into the United States' deep involvement in managing the Arab-Israeli dispute (1983, 137–142). Such concern for oil (and petrodollars) does, however, relate to the willingness of the United States to make massive transfers of sophisticated weapons to both Israel and the Arab countries, particularly Saudi Arabia.

On balance, then, it is difficult to argue that the oil weapon significantly altered U.S. foreign policy toward the Arab-Israeli dispute. The strongest argument for such a position is its impact on Jimmy Carter, but his policy was not significantly different from that of Henry Kissinger, who seems to have been influenced only marginally. Shifts in U.S. policy certainly have been less severe than the supply theory of economic sanctions would have predicted.

7

The "Lessons" of the Oil Weapon for Theory and Policy

We have examined in some detail the responses of five governments that were targets of the oil weapon of 1973–1974. How do their experiences help us to answer the questions outlined in the first chapter? Did the use of the oil weapon mark a new era in international affairs when Third World countries successfully used economic sanctions to force industrial countries to alter their foreign policies, as a number of noted policy analysts had suggested? Or did the experience instead confirm the skepticism of the theorists who argued that such influence was almost certainly impossible?

Did the Target Countries Alter Their Policies?

Policy Change During the Period of the Oil Weapon

I have tried throughout this study to distinguish between foreign policy changes during the period of the oil weapon itself and changes that have occurred since. During the period of the oil embargo itself there was almost no policy change on the Arab-Israeli dispute by the countries discussed here. The November 6 statement by the EC countries represented only a minor policy shift, which the Dutch government almost certainly would have carried out fairly soon anyway; the foreign minister had committed himself to this shift before he came to office. The Japanese did not really go any further, other than a vague threat to "reconsider policies toward Israel," which remained a dead letter. The Canadian voting shift in the United Nations does not seem to have been noticed even in Canada. Moreover, all of these issues merely involve rhetoric. No country seems to have even considered ending trade with Israel or doing anything else capable of any material impact on the Middle East situation as a result of the oil weapon. Nor did the other four governments apparently pressure the United States in any unusual manner to change its policies.

There is one possible exception to this pattern: the refusal of most European countries to allow the United States to use their bases and airspace to resupply Israel with arms. In Britain this was a major issue within the government, and the decision was made early and firmly. It is not clear that it marked a major policy change, however; during the 1967 war Britain had embargoed arms to both sides, although only briefly. The Dutch case is more interesting; the Dutch government adopted a much more neutral position in 1973 than it had in 1967. The issue was not relevant for Canada or Japan. In any case, the United States found this a fairly easy decision to live with and does not seem to have felt pressured by it.

U.S. Middle East policy, however, did change during the crisis: The issue became much more important as compared to other questions, and Henry Kissinger committed himself to negotiating a series of disengagement agreements, a process eventually resulting in the Camp David accords under Jimmy Carter.

Policy Change After the Use of the Oil Weapon

In the longer run, after the embargo had ended, all five countries altered their policies toward the Arab-Israeli conflict. The Dutch government abandoned its semi-independent public policy on the Middle East. European Community policy statements and U.N. votes were increasingly sympathetic to Arab positions, although during the 1980s the Europeans seemed to abandon their efforts to establish a Middle East policy separate from that of the United States. Canada became more likely to adopt EC positions rather than those of the United States and rejected strong efforts to pass national antiboycott legislation. Japan allowed the PLO to establish a liaison office in Tokyo. Even the United States increasingly spoke favorably of the rights of the Palestinians and the necessity to include them in peace negotiations and was increasingly willing to pressure Israel to move toward some sort of settlement. The net effect was a significant rhetorical shift toward the Arab position by all countries in the study. Interestingly, there was rather little change in the relative positions of the target countries; both before and after the embargo, the United States was the most pro-Israeli, followed roughly by the Netherlands, Canada, and Britain; Japan remained the most pro-Arab country in the study. As a group, they clearly shifted toward the Arab positions, however, particularly on Palestinian issues.

Why Did the Target States Change (or Not Change) Their Policies?

As noted earlier, this question really asks for the weight of the power relationship established by the oil weapon. To answer it, we must also sketch out the other dimensions of this relationship and discuss how, if at all, they were affected by the oil weapon.

The Power Relationship

Domain. The domain can be usefully divided into actors and targets. The *actors* were the governments of the Arab oil-exporting nations that participated in the oil weapon. Formal decisions were made in a series of ad hoc meetings of oil ministers and heads of state of these countries. Even the weak organizational structure of the OAPEC Council of Ministers was not used, in order to avoid possible splits within it. In fact, these group decisions meant nothing until they were implemented by individual governments, an irregular process at best. As might have been expected in such an unstructured environment, some individuals and governments clearly had more influence than others. Among the governments, Saudi Arabia seems to have played a particularly important role both because of its oil reserves and because some members of its government were extremely interested in the use of the oil weapon. Algeria also seems to have been important because of the quality of its bureaucracy—planning and applying varied levels of sanctions to countries around the world requires an enormous amount of information processing. But the target governments often found it difficult to determine which states were in fact levying sanctions. Moreover, it was not always clear who was making decisions *within* the various actor governments; this was particularly true of Saudi Arabia.

Like the actors, the *targets* also were governments. Their identities were more obscure, however. Presumably the *primary target* was Israel, the only country that could give the Arab governments what they said they wanted. But the Arab oil producers were unable or unwilling to go to war directly with Israel, and they certainly had no economic ties to break. One alternative was to attempt to coerce *secondary targets,* countries that might be influenced to exert pressure on Israel to alter its policies. In practice, there was only one realistic secondary target; only the United States, Israel's primary ally and supplier, might be able to significantly alter important Israeli policies. But the Arab oil producers controlled only about 2 percent of America's total energy supplies in 1972, as opposed to about 47 percent of Western Europe's total energy and 57 percent of Japan's (Darmstadter and Landsberg 1976, 22). Therefore, much of the oil weapon was in fact directed at *tertiary targets,* countries that might influence the United States to influence Israel to change its policies. These countries seem to have been chosen more because they were sensitive to changes in Arab oil exports than because the producers expected them to change Israeli policy.

Scope. The scope of the power relationship was extremely unclear. The actors believed that the scope was severely limited; the Arabs had asked for no domestic changes in the target countries, only for a shift in one area of foreign policy, the Arab-Israeli dispute—an issue that was not particularly important to most of the target states and that did not affect their national security. The fact that a number of more extreme demands from the Libyan leaders were totally ignored, as discussed earlier, illustrates this limitation.

Although the Arabs limited the scope to the Arab-Israeli dispute, they were not always clear about what particular behavior they desired. The October 17 statement specified *results,* not the *means* to those results.

Also, because there was no single national actor, target states had great difficulty getting clarification about what they were supposed to do. Over time, a set of relatively modest demands emerged; no government was asked formally for troops to fight the Israelis, for example. Instead, they were asked for rhetoric, public and private. The demands varied from country to country, with more demands made on Japan than on any other state. The demands were also vague. Different spokesmen would ask for different things at different times, and, in general, the actors were willing to negotiate. This approach had its strengths and weaknesses. On the one hand, it forced the target states to try to anticipate what the Arabs wanted; for example, the decision by the Western European countries to refuse to cooperate in the American rearmament of Israel during the 1973 war did not reflect specific, formal demands by the Arab governments but instead a general assumption that such cooperation would provoke sanctions. On the other hand, the targets were encouraged to try small concessions before moving on to large ones and, in the process, they learned that large concessions were not really necessary.

A more important problem was that *each target saw the scope differently*. Henry Kissinger, and thus the U.S. government, saw the issue as an opportunity to resolve the Arab-Israeli dispute so as to reduce the likelihood of a regional confrontation between the United States and the Soviet Union. For the other four targets, relations with the United States were an important aspect of any shift in Arab-Israeli policy, although the intensity of concern varied a good deal among them. For the Dutch, the issue was not being pressured to do something they felt would be wrong (abandon Israel). The British, because of their long involvement in the Middle East, saw the oil weapon as simply more evidence of the correctness of their policy and an opportunity to persuade the Americans to adopt it. They felt under no particular pressure to change their policies. For Canada, the issues were to prevent oil shortages in eastern Canada, including Quebec, in order to avoid another strain on federation, while shifting its image from pro-Israeli to neutral in the eyes of the Arabs and maintaining good relations with the United States and Israel. The Japanese were caught in a classic dilemma, as the OECD country most dependent on the Middle East for oil and on the United States for military support and markets.

During the crisis the targets saw the scope of the oil-weapon power relationship very differently from the actors and from one another. In general, they all felt more was involved than simply a rhetorical position on the Arab-Israeli dispute; this in turn probably made it more difficult for the Arabs to persuade them to alter their policies in the short run.

After the oil crisis a new dimension was added to the scope of the relationship when the Arab oil exporters suddenly became immensely wealthy. They immediately became much more important customers on the world market. (The magnitude of this change is demonstrated by the fact that, before the embargo, Israel was the single most important Dutch customer in the Middle East.) This was a consideration for all five target states. The United States was probably influenced least because of its political involvement in the region and the unique impact of its pro-Israeli interest groups. But the availability of new markets seems to have been a relatively new factor that encouraged all five governments to at least try to avoid open political conflict with the Arab governments on issues such as the Palestinians. Interestingly enough, this was effective despite a consensus that the Arabs almost never altered important economic decisions because of political considerations.

Range of Sanctions.   The range of sanctions was the dimension most directly affected by the oil weapon. After all, the level of dependence on Arab oil had not altered too much over the preceding few years, but the Arabs had been unable to translate this dependence into a satisfactory power relationship. One reason was that the range of sanctions available to the Arabs was seen as quite limited; efforts to reduce the supply of oil to industrialized countries had been resounding failures in 1956 and 1967, and, as a result, threats to do so had little impact in 1973. Presumably, then, the whole point of the oil weapon was to demonstrate that the range of sanctions available to the Arab states was much greater than had been perceived; the oil weapon, like any coercion, is therefore more important for its ability to make *future threats* more persuasive than for its immediate impact.

What, then, was the range of sanctions in the new power relationship that emerged after 1974? In principle, the sanction was the partial or total cut-off of Arab oil exports to particular countries, and Saudi Arabia put together a complex classification system to implement the sanctions accordingly. But the actual impact was quite different from the one anticipated by this elaborate scheme. The major international oil companies responded to the oil weapon by diverting non-Arab oil to countries with shortages, roughly equalizing the loss among the OECD states. As a result, the Arabs were unable to aim sanctions with much precision. They could reduce the supply of oil to the whole world, but they were unable to get much more oil to their friends than to their enemies. Eventually it became evident that there was no clear connection between pro-Arab politics and oil imports, although this was not clear for about two months, roughly from mid-October 1973, when the oil weapon was announced, to mid-December. During this period the industrial countries seem to have assumed that the Arabs could in fact apply their threats and reacted accordingly.

Weight.   Given all these conditions, to what extent did the oil weapon alter the behavior of the target governments toward the Arab-Israeli dispute? Some of the short-term policy shifts seem to have been caused directly by the oil weapon; but, as noted earlier, they are not particularly significant. These include (1) the Dutch decision to agree to the November 6 EC statement, (2) the Japanese threat to "reconsider policy" toward Israel, and (3) the general refusal of bases by NATO allies for the U.S. rearmament of Israel. I cannot reject the idea that Canada's shift from opposition to abstention on the November 1973 U.N. Palestinian resolution was also caused by the oil weapon, but its speed and low profile make it look more like a decision that had already been made. The bases issue is the only policy shift that was potentially more than rhetoric, although as we have seen in the British case one can argue that it was not really a policy change. (The British refusal to assist the Netherlands in obtaining more oil, while not strictly related to the Arab-Israeli dispute, was another result of the oil weapon.)

The Netherlands' participation in the November 6 EC statement did not represent much of a policy shift for the Dutch. In fact, the change almost certainly would have been made in any case, given the foreign minister's previous commitment. Similarly, the Japanese statement of November 22 was essentially a reiteration of previous policies, except for the threat to "reconsider" relations, which was never carried out. The British did not alter their Middle East policy at all, and the Canadian shift was very quiet indeed.

Some observers argue that U.S. Middle East policy changed because of the oil weapon, but on balance this seems incorrect for several reasons. First, the only real change was to make the Arab-Israeli dispute a top-priority foreign policy issue. The substance of the policy itself, with its goal of a negotiated settlement based on some kind of equality among all parties, excluding the Soviets and radical Arab governments, had been established long before 1973. Second, there is very little evidence that Kissinger and those around him worried much about the oil weapon, except as an irritant. Their major concern was Soviet involvement in the Middle East and a possible superpower confrontation; they focused on the war rather than the oil weapon. As usual with Kissinger, politics took priority over economics. And, finally, interest groups and government agencies that might have been more occupied with oil supplies found themselves frozen out of Arab-Israeli policymaking at this time because of Kissinger's total control of the process, an unusual event in the normally porous United States foreign policy–making system. United States policy would probably not have been notably different if the oil embargo had not occurred.

On balance, then, the short-term impact of the oil weapon on the foreign policies of the target countries was small or negligible. The one important exception was the decision by the Western European countries to refuse to allow the United States to use their bases and airspace to resupply Israel; had the United States not been able to pressure Portugal to allow the use of the Azores bases, this might have significantly influenced the course of the war or caused a major conflict within the Western alliance.

But what of the longer-term shifts in Middle East policy evinced by the target countries? There is a clear prima facie case that the oil weapon created a new power relationship, which in turn produced these policy changes. Nonetheless, the case studies, particularly the interviews, suggest that this is too simple an explanation.

When asked to explain the policy shifts, practically no one in the United States, the Netherlands, or Canada mentioned the threat of future oil cut-offs as being important. The Netherlands and the United States are presumably the key cases here because they were actually embargoed in 1973–1974. The embargo seems to have taught the Dutch policy elite the lesson that, in practice, it is almost impossible to cut off oil to one country alone; oil supply is simply not an issue there. The change in Arab-Israeli policy seems to have come about for different reasons.

One result of the oil crisis was that the Arab states became enormously wealthy; in a very short period they became major customers for every major exporting country. The Dutch, traditionally major international traders, have an unspoken rule not to cause unnecessary trouble to major customers. Moreover, Arab wealth greatly increased personal contacts between Europeans and Arabs; as a result, more Europeans learned about the Arab point of view on Middle East issues. The increasing concern for the plight of the Palestinians coincided with a shift in Israel's image from a weak but brave little country to a regional power prepared to indulge in morally dubious acts such as the invasion of Lebanon.

The same process apparently was at work in Canada. The Arab oil-producers were seen as an important market for manufactured and high-technology exports; this concern seems to have been important, along with Trudeau's personality, in moving Canadian Middle East policy to a somewhat more pro-Arab position. The U.S. government's primary consideration, instead, seems to have been Soviet involvement in the area; American corporations certainly made a great deal of money as a result of oil riches, but this does not seem to have directly influenced U.S. foreign policy toward Israel.

In Great Britain and Japan, oil supply was an issue. The British stake in the Middle East has shifted from oil supply to export markets and support for the pound sterling, but the Arab willingness to reduce oil exports presumably would also apply to financial assets. In addition, the British expect to resume their dependence on Arab oil when the North Sea deposits are exhausted. Japan continues to worry about oil supply; it is simultaneously more dependent on Middle Eastern oil and less dependent on Middle Eastern markets than the other states. It is not really clear, however, that the Japanese government worries more because of the oil weapon. The Japanese were worried about their vulnerability to oil-supply cutoffs before the embargo. In the long run not much changed; Japan continued to believe its interests would be served by taking a position on the Arab-Israeli dispute that was simultaneously more pro-Arab than any other Western state while not breaking with the United States.

Thus, the oil weapon's effect on Middle East policy change was at best indirect. The OPEC governments were able to greatly increase the price of oil, partially because of the oil weapon; although the price rises had started earlier, the fear about possible shortages made it much easier to quadruple oil prices in a short time. Indeed, this is a unique case of the actor governments becoming rich as a result of economic sanctions. But the cause of the changing power relationship, which in turn produced the policy shifts, seems to have been the increase in wealth rather than, as the supply theory of economic sanctions would predict, the fear of future oil shortages. Thus, the apparent success of the oil weapon in changing the targets' foreign policies seems to result from the unusual fact that the sanctions were enriching, suggesting this is a precedent unlikely to be repeated.

But, as Baldwin points out, even if the target states did not *change* their policies, sanctions can be judged successful if they significantly increase the *costs* of carrying out current policy (1985, 132–133). Did the cost of maintaining the target states' various policies change as a result of the oil weapon?

The concept of political costs is rather abstract, at least when it has no obvious indicator. It is not clear, first, who would bear these costs. Baldwin implies that he is thinking of *real* costs to the *state*. In fact, given that the major international oil companies shipped non-Arab oil to make up for shortages (thus spreading the shortfall among the target countries), alternate behaviors might not have made much real difference. The critical question in predicting behavior, however, is whether the governments *believed* that they would suffer costs. On balance, it seems that they *did* expect to pay such costs, for it was not clear to them for some weeks that the oil companies would equalize the pain. Also, some governments were worried about possible *future costs*. Aside from costs to the state, it probably became more politically costly for groups within the states to resist policy change. Thus, in terms of predicting short-range behavior, the target states were generally prepared to bear significant costs to maintain their policies.

All of this raises other important questions: What constitutes significant policy change in the Arab-Israeli conflict? How do we recognize one when we see it? Clearly this involves subjective judgments, and different actors will differ on the relative importance of various acts. The key concept is *salience*, how important the act appears *to the actor*. Because the salience of different targets differs for each actor, the same act directed at different targets will possess varying degrees of importance. But that is not a problem here, because we are looking at different sorts of acts directed at a single target, Israel. We can therefore suggest a rough typology of ascending order of importance of acts by one state directed at another. Such a typology can help us to evaluate the "degree of difficulty" of the Arab governments' demands and the degree of compliance they elicited.

1. At the lowest degree of difficulty is *disagreement,* whereby an actor government states its disagreement with a specific act or policy of the target, with no mention of concrete sanctions or other issues.

2. The next degree is *improved relations with the enemy,* when the actor improves its relationship with the target's enemy. This is more substantive than disagreement, because it involves action on different sorts of issues. But by itself it involves no change of behavior toward the target.

3. *Symbolic alienation,* the next level, involves altering behavior toward the target in symbolic ways in areas other than those of the specific dispute. At this point the dispute has escalated so that the entire actor-target relationship is being altered, suggesting the risk of further division if the dispute is not resolved.

4. *Economic coercion* includes nonsymbolic actions that alter the target's capabilities, significantly raising the target's cost of continuing the disputed policy. This may take the form of reducing economic flows between the actor and target, increasing economic flows (possibly including weapons) to the target's enemy, or some combination thereof.

5. *Military coercion* involves the use of military force to persuade the target government to alter its policy.

According to this scale, the Arabs' formal demands were in classes 4 and 5, because they demanded that the international community force Israel to alter important policies toward the occupied territories, the Palestinians, and Jerusalem; realistically, to accomplish this, the industrial states would have had to undertake class 4 and class 5 actions. The Europeans and Japanese certainly felt under no pressure to send troops to oppose Israel. There was active discussion of reducing economic flows to Israel, but nothing much seems to have come of it. The lack of such demands suggests the major limitations on the Arabs' power, which were tacitly recognized by both sides; this is a very odd sort of Finlandization indeed.

Much of the discussion between the Arabs and the target governments, however, seems to have focused on the third class of action, symbolic acts that would show displeasure with Israeli policy. These demands included denial of recognition to Israel. The closest thing to compliance to these demands is the Japanese November 22 statement. But the threat to "reconsider relations" was empty and never seems to have been invoked again. There were a few second-level moves (Canada's U.N. vote shift on the Palestinian issue, Japan's aid commitment to Arab states, and the American willingness to establish economic ties with the Saudis). The level of verbal disagreement of the targets with Israel did not change much either.

This general position is reinforced by at least one authority (table 10). The Conflict and Peace Data Bank (COPDAB) classified more than 340,000 separate events from 1948 to 1978 on a 15-point conflict scale, where 1 represents voluntary unification into one nation, 15 represents total war, and 8 represents neutrality (Azar 1982, 14–16). Table 10 shows the COPDAB coding of actions taken by the Netherlands, the United Kingdom, and Japan toward Israel. The November 6 EC statement was classified as 7, actually mild *support* for Israel. The November 6 Japanese statement was coded 9, as mild verbal hostility. The November 22 Japanese statement was either given the same ranking or missed altogether (the date in the COPDAB material is November 20). The Canadian U.N. vote is not coded, presumably because it was not reported in the world press. It is difficult to determine which events refer to the Europeans' refusal to allow the United States to use their bases to send arms to Israel; however, Britain and the Netherlands each have only one or two acts during October and November that were both related to military issues and coded 10 or 11. These coded actions are somewhat offset by political statements classified as 6 or even 5. Some of the subtleties may have been missed in this coding (the Dutch November 9 statement is omitted, for example), but it does suggest that, as compared to other acts in international affairs, the behavior of these countries did not seem particularly hostile toward Israel.

So how much change was there? Observers will disagree whether the glass is half-full or half-empty; my own judgment is that there was some movement but that on balance it was less than would have been expected, given that the governments involved felt themselves under a major economic threat.

Sources of Change

Were the states that were more vulnerable to the economic effects of the oil weapon more likely to agree to the demands of the Arabs? The evidence

Table 10.   All Events from the Netherlands, the United Kingdom, and
            Japan Toward Israel Reported by the Conflict and Peace Data
            Bank, October–December 1973

	Issue Type	Conflict Score
Netherlands		
November 6	Political	7
November 11	Political	6
November 26	Military	11
December 1	Political	6
December 4	Political	10
United Kingdom		
October 8	Political-Military	6
October 10	Military	10
October 16	Military	7
November 5	Political-Military	9
November 6	Political	7
November 12	Political	5
Japan		
October 9	Economic	9
November [6]	Political	9
November 8	Political	6
November 20	Political	9
December 13	Political	10

Conflict Scale:

1	Voluntary unification into one nation	7	Mild verbal support; exchanges of minor officials
2	Major strategic alliance	8	Neutral or nonsignificant acts
3	Military, economic, or strategic support	9	Mild verbal hostility/discord
4	Nonmilitary economic, technological, or industrial exchange	10	Strong verbal hostility
5	Cultural and scientific agreement or exchange	11	Hostile diplomatic-economic actions
6	Official verbal support of goals, values, and regime	12	Political-military hostile actions
		13	Small-scale military acts
		14	Limited war acts
		15	Full-scale war

SOURCE: Azar 1982, 14–15. This source is the codebook for the Conflict and Peace Data Bank (COPDAB) data set; it also explains the conflict scale. Data in table were extracted from the COPDAB data set by the author.

NOTE: No Canadian acts toward Israel were coded during this period.

here is mixed. Japan, the most vulnerable country, did indeed go further than any other government in meeting the Arab demands: In the short run, it threatened to reconsider relations with Israel, and in the longer term it moved closer than any other state toward recognition of the PLO. Nevertheless, Japan was also the most pro-Arab state *before* the oil weapon was used. Japan's willingness to adopt pro-Arab positions, particularly on U.N. votes, was probably based on its perceived vulnerability to oil-supply cutoffs. It is not clear, however, that the actual *use* of the oil weapon made much of a difference. To put it differently, the oil weapon does not seem to have materially altered the power relationship as perceived by the Japanese government, which reflected not only Japanese dependence on imported oil but also offsetting concerns about the role of the United States.

In contrast, Canada, the least vulnerable country, was one of the quickest to change its policy, at least in U.N. voting. This shift may have been prompted by the oil weapon, but its longer-term shift toward the European position on Arab-Israeli issues was largely caused by domestic factors, particularly Trudeau's leadership. Inasmuch as Arab pressure played a role, the Canadian government became much more concerned about opening up new export markets in the Middle East; oil supply was not a serious concern. The same was true of the Dutch government. Again, we see the impact of Arab wealth, which was only indirectly the result of the oil weapon.

In general, the pattern is not one whereby the more vulnerable states move further and faster than their less vulnerable counterparts. Instead, the whole group moved in a fairly unified manner that was consistent with previous policies—contradicting the supply theory of sanctions.

Was the use of the oil weapon really an attempt at coercion? Indeed, the lack of results is so striking that some scholars argue that perhaps the Arabs did not really *want* to influence the policies of the target states. In the past actors have often imposed sanctions with other goals in mind: to influence domestic public opinion, to disarm domestic critics, to send signals or threats about subsequent foreign policy acts, or to avoid the necessity of applying further leverage. Klaus Knorr has even developed the idea of a pseudo-sanction, which is not really designed for external coercion at all (1984, 202–204), and David Deese points out that actor governments may use coercion to conceal other motives (1984, 150–155).

There are a variety of plausible explanations for using the oil weapon

other than the assumed purpose of altering the foreign policies of the industrial states. These other explanations include raising the price of oil in
order to gain more economic wealth (Rustow 1982, 156–157); saving
Egypt in the Yom Kippur war by forcing the United States to pressure
Israel to agree to a cease-fire (Rustow 1977); isolating the United States
diplomatically; reducing domestic political dissent and external political
pressure by helping the Arab cause without actually getting involved in the
1973 Arab-Israeli war; demonstrating Arab oil power to increase future
bargaining leverage (Knorr 1976, 230); asserting Saudi dominance in the
inter-Arab struggle between conservative and radical regimes (Ali 1986,
47–51); and restoring Arab self-confidence (Daoudi and Dajani 1983,
108). Indeed, some contend that no Arab public statement claimed the oil
weapon would force Israel to withdraw from the occupied territories
(Daoudi and Dajani 1983, 107; Daoudi and Dajani 1985, 149). One
might conclude from this that if the Arabs had *really* wanted to influence
the targets, they could have done so.

From the *target's* point of view, however, the question is whether or not
the Arab governments *seemed* to be trying to influence them. All of the
target governments seem to have believed that the Arabs were trying to
alter their policies by threatening to cut off exports of Arab oil and that, if
they did alter their policies, the Arabs would respond by allowing them
to purchase more oil. In fact, the initial production reductions and the
monthly cuts were actually implemented, so the target governments had
no choice but to believe for at least two months that they were the targets
of an attempt at influence and had to decide how to respond to it. In general, the officials of the target states were uncertain about the particular
demands being made, but they believed the demands were backed up by
very costly sanctions. Their behavior, therefore, is a useful guide. It is not
clear they would have behaved any differently if the Arabs had demonstrated they were "really" more determined to apply the sanctions.

Why didn't the target governments change their policies more? If the
target governments believed they were under threat for six to eight weeks
in late 1973, why didn't they change their rhetorical policy toward the
Arab-Israeli conflict? As discussed in chapter 1, we can distinguish three
possible domestic sources of policy change: (1) The leaders in office may
decide to change their policies; (2) If the leaders decide not to change,
political elites in the country may either replace them with new leaders

committed to a policy change or use this threat to force the old leaders to change their positions; and (3) If the leaders and the elites remain firm, the politically relevant public may either replace the current elites and leaders with new ones committed to a policy change or use this threat to force the old leaders and elites to change their position.

It is also worth noting that there are three, rather than two, possible responses to sanctions: The government may keep its initial position, may change in accordance with the demands of the actor, or may change in opposition to the demands of the actor. The supply theory of economic sanctions predicts that, as the future costs of economic sanctions are seen to rise, leaders, elites, and/or publics will react to the increasing cost of their policy by changing it as the actor desires. An opposing argument, often cited in discussions of coercive bargaining and economic sanctions, is that people react negatively to external pressure of this sort; the issue is transformed from the particular question under discussion (in this case, policy toward the Arab-Israeli dispute) to whether or not the target government can be forced to obey foreigners, and sanctions thus force the target to be more intransigent than it might otherwise have been. Thus, we must determine, first, why effective pressure did not stem from any of the three mentioned sources to change policy either in a pro-Arab or pro-Israeli direction during the crisis, and, second, where the pressure came from that, over time, brought about a slow pro-Arab shift.

Political Leaders.   The Arab-Israeli dispute was not a new policy issue for the political leaders. Each government had seriously considered its relationship with Israel since 1948, and the exigencies of Middle East politics had reopened the issue a number of times. Moreover, this had been a political issue, decided by politicians rather than exclusively by bureaucrats; even the Arabists of the British Foreign and Commonwealth Office had to persuade politicians to accept their arguments and did not always succeed. Although some leaders were interested in changing policy somewhat (van der Stoel and Trudeau come to mind), they had all been involved in establishing it in the first place and probably felt some commitment to it as a result. This may explain why, during the crisis itself (and indeed afterward), *no prominent individual in any of the five countries seems to have changed his or her position on the subject.*

By creating an atmosphere of crisis, the oil weapon greatly strengthened the top political leadership in each country by giving them the opportunity and incentive to focus on the issue and make decisions. In most countries the head of government chose not to be personally involved. (The exception was Britain, where major decisions were made by the cabinet system, with Heath an important actor.) The general pattern was delegation of authority to the foreign minister, despite the economic aspects of the issue.

It is not clear just how much autonomy political leaders had in this situation, but it is striking that no Western leader felt it appropriate to drastically alter policy toward the Arab-Israeli dispute as a result of the oil weapon. Political leaders certainly have more freedom of action when conflicting pressures offset one another, but offsetting pressures tend to move policy toward the center. The easiest decision in such situations is probably to keep policy constant. Thus, delegating the decision to the foreign minister usually made sense politically and personally.

The importance of the issue meant that the foreign minister himself usually took direct responsibility rather than simply delegating it to part of the bureaucracy. This was a particularly important factor because foreign ministries tend to favor more pro-Arab positions, whereas their ministers, for a variety of reasons, often do not.

Personal idiosyncrasies were not particularly important in the short run. A Dutch Socialist government more sympathetic to the Arabs than its predecessors wound up refusing to make concessions. In Britain, Prime Minister Heath, who had supported a pro-European policy, refused to aid the Netherlands. Canada's shift during the crisis was hardly noticeable, although in retrospect it was part of a general pro-Arab shift that can be traced in part to Trudeau's influence.

Japan and the United States pose less certain cases. Nakasone, head of MITI at the time, reportedly overruled his own bureaucrats and pressed for a more pro-Arab policy. But he was also supported by many other important forces inside and outside of the Japanese government, and in fact, the actual policy shift was not very large. Any U.S. secretary of state would probably have made the Middle East the primary issue of foreign policy after the 1973 war, although Kissinger may have been unusual in his lack of concern for the oil weapon as a distinct issue.

In the longer run, a policy shift has emerged and survived leadership changes in all five countries; thus, there is a prima facie case that individual factors have not mattered. The United States is an exception, however. The *direction* of American policy has not been altered since 1973, but the *intensity* with which it has been pursued has certainly varied. The willingness of Henry Kissinger and Jimmy Carter to elevate the Arab-Israeli conflict to the uppermost level of foreign policy issues, to commit large amounts of their personal time and energy, and ultimately to take considerable political risks by engaging in personal diplomacy have been critical to whatever success the United States has had in the region. The Reagan administration demonstrates what happens when such personal commitment is absent.

Why did these two men see this particular issue as so important? Kissinger's primary concern seems to have been the possibility of Soviet intervention in the Middle East rather than the oil weapon. Carter felt that the entire energy issue was crucial; it is quite possible that, without the use of the oil weapon, he would not have considered the Middle East quite so critical. Ironically, the oil weapon may thus have been partially responsible for the Camp David accords and the Egyptian-Israeli peace treaty, two events the initiators of the oil weapon strongly opposed.

Political Elites.   Elites are a possible source of a policy shift. In a democratic system, elites are found primarily in the legislature and in influential interest groups. Changes in foreign policy are more likely if the top political leadership is replaced, but the oil weapon did not make this happen. Unlike military invasion, as in Burgess's analysis of postwar Norway, economic sanctions cannot directly replace the target's leadership. Their political impact is largely felt in their ability to bring about such changes *indirectly* (as in May's discussion of the impact of bombing [1973, 125–142]), or at least threaten to do so. The oil crisis was a central issue for political discussion in all five target countries, but in *none* of them did any significant political party or political leader attack the government for failure to cope with this difficult problem. Perhaps the Netherlands is the most interesting example. At any given time, it has between seven and fourteen parties represented in its legislature. But at no time did any party argue either that the government had been too pro-Israel and had thus brought the Arab oil embargo on the Netherlands or that the Netherlands could not afford to allow itself to be coerced by oil pressure into signing the November 6 statement.

At the same time, in all five countries influential interest groups, especially those connected with business, did push for a more pro-Arab stance, a phenomenon corresponding precisely to the predictions of the supply theory of sanctions. Traditionally, these groups are particularly strong in Japan, a country with no domestic Zionist interest group and the most dependent on Middle Eastern oil. Moreover, because of weaknesses in the Japanese Foreign Office, individual business leaders served as important negotiators between Japan and the Arab governments. Business leaders also met with the prime minister and other senior politicians to argue for a pro-Arab position and were largely responsible for the prime minister's support of the November 22 resolution. Nonetheless, as noted earlier, this influence did not really change overall Japanese policy much.

One reason for this lack of impact may have been the high level of uncertainty about the range of sanctions involved. Using Wiberg's typology, we can distinguish two different, somewhat independent possible effects of sanctions: (a) reducing the supply of the target commodity, and (b) damaging the target state's economy (1969, 14). From October to December uncertainty surrounded both of these. Could the Arabs cut off oil to selected target states? What impact would this have on national economies? The uncertainty made the issue politically important, but it gave leaders an excuse to avoid changing personal positions on appropriate policy toward the Arab-Israeli dispute. Uncertainty on these points diminished by December, when it became clear that the major international oil companies were going to distribute oil supplies so as to spread out the shortages and that no national economy was severely threatened in the short run. This effectively ended the oil weapon except as a symbolic gesture.

Change was less likely because elites were already divided on the issue before the oil weapon and, like individual leaders, had staked out positions in advance. The oil weapon could change the policies only if it (a) brought significant new actors into the policy arena or (b) gave existing actors so much new ammunition that the power balance would shift. This did not happen in the five countries in the short run.

In the longer run, two groups seem to have been strengthened by the oil weapon: domestic pressure groups that had been sympathetic to Arab concerns, and the top political leadership. Business groups and the foreign ministries alike were more concerned with the Arab countries than with

Israel. (Japan was an exception, because its foreign ministry was more concerned with the United States and therefore tended to resist pro-Arab measures.) The ability of the Arab governments to damage the international economy strengthened the arguments in favor of policy changes.

In some countries pressure groups and political leaders worked in opposition to one another. Thus, increased business-group pressure in the Netherlands was countered by a refusal by political leaders to adopt positions they did not like. In Japan business groups and public pressure offset the normal pro-American bureaucracy, leaving individual politicians a good deal of room for maneuver. In the United States the Israeli lobby's normal influence was diluted by the increased importance of the issue and by Kissinger's dominance of the foreign policy machinery and his decision to focus on the Middle East. The major effect was to increase the power of the political leadership to act as it chose in the short run; interest groups (including the bureaucracies) had much less impact than usual.

Interestingly, political forces that favored Israel were significantly weakened only in the United States. The Israeli lobby's basic theory was disproved—a strong Israel had not produced peace in the Middle East. While the lobby could continue to do useful things on its own, such as sending money to Israel, the U.S. government would make the crucial decisions: Initially, what arms would be sent to Israel and who would pay for them; later, what kind of armistice terms would be supported and what sorts of pressures would be brought to bear on Israel to accept them. At precisely the time that the U.S. government assumed this crucial role, the Israeli lobby found its influence at a historically low ebb, owing first to Kissinger's dominance of the foreign policy process and later to Jimmy Carter's concern for the issue and lack of responsiveness to the lobby.

In contrast, similar groups in Canada were galvanized by the issue of the Arab economic boycott and actually developed new political power. Although they narrowly failed to get a federal antiboycott statute passed, they did get a number of provinces to pass similar laws. In Holland, Britain, and Japan, where specifically pro-Israeli pressure groups had almost no influence, there seems to have been no particular change. On balance, then, the impact of the oil weapon on pro-Israeli political forces varied a good deal for reasons that are not immediately obvious.

The long-term impact of the oil weapon on these political constellations is more difficult to measure. Over time, the political leaderships became somewhat more sympathetic to the Arab position, particularly on Palestinian issues, in the Netherlands, Canada, and the United States. (Japan and Britain, perhaps because they were more pro-Arab initially, seem to have undergone less change.) These shifts appear to have resulted from increased Arab wealth, which led both to greater economic stakes in the Arab world and to increased contact with Arabs as individuals.

I had expected new political groups to try to mobilize the high level of public concern against the government, either arguing for more concessions to the Arabs to ensure oil supplies or for fewer concessions on the grounds of nationalist resistance to coercion. This did not occur, however. During the oil crisis, no new groups entered the debate, although, as noted above, the balance among groups already participating was somewhat altered.

In the longer term, however, the increased Arab wealth did bring in at least two groups, if not for the first time then with significantly increased concerns: (a) corporations that wanted to serve this new market, and (b) government bureaucracies dedicated to increasing exports. Both groups argued strongly for political concessions on the Arab-Israeli issue and the Arab economic boycott. These groups believed the industrial nations were competing for access to the new Arab markets, and political concessions were justified as a way of gaining preferred access for their goods and services. It was not at all clear that such concessions actually increased trade much; the Arabs seemed to make economic decisions on economic criteria. Nonetheless, this concern was an important factor in the gradual shift in policy during the decade after the oil crisis.

This conclusion suggests that political elites were not important in changing policy; another interpretation is that they were important precisely because they did not pressure decision makers to make radical policy changes. There seem to have been unwritten but real limits on the kinds of concessions that were "politically possible." It was not surprising to find strong opinion in the United States and the Netherlands that Israel's continued survival was not a negotiable issue; I was somewhat more surprised to hear former government officials in Britain and Canada, as well as Japanese academics, say flatly that denying diplomatic recognition or applying sanctions to Israel under coercion was unthinkable. This may reflect a lack of imagination about how a government might react under extreme pressure, but it also suggests an unspoken moral consensus among elites that offsets the heavy rhetorical stress on national interests heard particularly often in Britain. This elite consensus, often unspoken and extremely difficult to measure, seems more important in shaping policy than public opinion in the conventional sense, even in relatively high-visibility issues such as the oil weapon.

Public Pressure. Public opinion had relatively little impact on decisions made during the oil crisis. At one level, this is not surprising, because public opinion does not usually directly influence policy during a crisis. But the oil problem went on for months, rather than for days or weeks, and it directly affected many people and was widely discussed. The public in the United States, Canada, Britain, and the Netherlands apparently disliked the idea of being coerced, as expected. Nevertheless, this attitude does not seem to have significantly influenced the behavior of national leaders.

As noted above, the five governments' policies do not differ much. It seems reasonable to divide them into countries whose leaderships accepted modest coercion (Britain, Japan, and Canada) and those that modestly resisted (the Netherlands and the United States). A government's position on the issue apparently had no impact on its political future. The obvious public concern and interest in all five countries were not effectively mobilized by anyone to oppose government policy. Instead, the public seems to have supported its leadership regardless of the decisions made. Only the British government lost power as a result of the oil crisis, but this had more to do with the government's handling of the coal miners' strike than with concern for excessive concessions to the Arabs; certainly the successor government was not much more pro-Israel.

While public opinion was worried about coercion (at least in the United States, Britain, Canada, and the Netherlands, where comparable data are available), each government seems to have responded according to its own concerns. This apparent willingness to ignore public sentiment has produced an intriguing phenomenon whereby very similar patterns of public opinion are accompanied by divergent foreign policy behavior.

On balance, then, the oil weapon confirms the conventional wisdom of theory and experience that economic sanctions by themselves are unlikely to force target states to change important policy. But why is this true? Why do economic sanctions usually fail to coerce?

Theoretical Implications: When One State Is Able to Influence Another

The single most important reason that economic sanctions usually do not coerce a target state to change its policy on an important issue is that *no foreign policy tool,* including the limited use of force, does this very well. As noted earlier, the literatures on economic sanctions and coercive diplomacy generally agree on this point. The experience of the oil weapon confirms the generalization that it is extremely difficult to force a target state to alter its policy on a topic the target feels is important. The scope of the power relationship thus becomes central to any serious analysis of coercive diplomacy.

The Scope of the Power Relationship

Basic to this discussion is deciding what makes an issue "important." We must avoid the tautology that argues that: (a) governments cannot be coerced on important issues, and (b) we know what issues are important because coercion does not work on them.

One response to this problem is that public awareness of economic sanctions *makes* the issue important by transforming the debate from the substantive question (say, policy toward the Arab-Israeli dispute) to a nationalist issue of whether or not the target state will allow itself to be coerced by an outside power. Presumably, the leadership, elites, and/or publics will become so outraged at the coercion that they will make it politically impossible for the leadership to accede to outside demands, regardless of the economic consequences. (A more sophisticated version of this position is the Coplin and O'Leary argument, cited in Mansbach and Vasquez [1981, 154], that external coercion increases negative affect, which decreases the likelihood of policy agreement between the actor and the target governments.)

This is a significant argument. It recalls an important theoretical point ignored by much of the economic sanctions literature: Political values are usually more important than economic values in making policy choices. If, for example, we could promise to industrialize India in twenty years, multiplying its per capita GNP several times and effectively eliminating poverty, on the condition that India would agree to become a British colony again, India would not accept. National independence is more important to most people than economic well-being; any government appearing to trade the former for the latter is vulnerable to attacks on its wisdom and motives by its domestic political opponents. Thus, on balance, public external coercion can make it harder, not easier, for a target government to comply.

But the logical conclusion of such an argument is that economic sanctions *never* force the target to alter its position, that they are by nature self-defeating. (Frederick Hoffman pushes the argument still further, contending that because governments know this, economic sanctions are used to signal that the actor does *not* want to alter the policy of the target government; if it wanted to do so, it would use other, effective instruments [1967].) Empirically, this does not seem to be true. Sidney Weintraub and his colleagues found examples of success as well as failure in their research on the United States' use of sanctions (1982). The best study using aggregate data found that sanctions were successful in altering the target's behavior in roughly 36 percent of the studied cases (Hufbauer and Schott 1985, 80). One can argue with particular aspects of both studies, but it seems clear that sanctions do indeed work *sometimes*. The problem is, rather, when.

It may be useful to think of issues in terms of importance not to the state as a whole, but to individual or corporate members of the decision-making unit of the government. If one or more members of this unit see the issue as central to them, and if they cannot be removed from the decision-making unit at least on this issue, it will be very difficult to alter the policy by external pressure. This means that a fairly detailed political study of the target government should be undertaken before sanctions are imposed, focusing on the political interests of major actors in the policy issue involved. "Coercion . . . must not be based on the hope that officials in another government would willingly commit political suicide" (Thies 1980, 419; for an intriguing attempt to undertake this kind of analysis on the North Vietnamese government during the early stages of the Vietnam war, see Thies 1980, 222–283).

There is a more fundamental problem here. The idea that power relationships can be accurately described by focusing on several dimensions (usually domain, scope, range, and weight), which is quite common in modern power analysis, implies that these dimensions are the same for all actors. Although this assumption may be useful for most of these dimensions, it is clearly misleading in the case of the *scope* of the issues involved. As demonstrated above, the scope of the power relationship was fundamentally different for the actors and for *each* of the target governments. The Arabs defined the scope of the power relationship in terms of the target states' positions on the Arab-Israeli conflict, but the targets themselves saw the scope as being much broader, involving issues such as their alliance with the United States or their adversarial relationships with the Soviet Union. Indeed, further disaggregation would show the scope was actually different for each of the separate individuals and organizations *within* the target governments. In fact, actor governments almost certainly possessed similar differences in scope that are ignored here.

By definition, theory involves a simplifying distortion of reality, so the existence of such differences in themselves do not tell us much; the question is whether or not such simplifications are useful. These differences in scope seem to have been crucial in predicting the actual responses of the target governments; the simplifying assumption that they are identical thus leads to misleading explanations and predictions. Instead, we will have to specify the scope involved for the actors and targets separately, probably by analyzing the importance of issues as suggested above.

This is not simply a theoretical argument; it has important policy implications. It seems logical that narrowing the scope of the power relationship will make it easier to establish; coercion should be easier when its influence is limited to a smaller area. The oil weapon experience suggests, however, that the actor cannot define the scope of the issue; instead each participant defines the scope in his or her own terms. The oil weapon is an excellent demonstration because the Arabs tried very hard to limit the scope of the power relationship; indeed it is hard to see what more they could have done to have kept the scope narrow. Perhaps this is another inherent limitation on the ability of one state to coerce another.

Can Sanctions Be Aimed at Particular Targets?

Much of the external-influence literature assumes that sanctions can be directed with some precision against selected target states, perhaps because the economic sanctions literature often assumes some sort of a blockade and the coercive bargaining material focuses on military actions. The oil weapon showed, however, that the Arabs were unable to aim their sanction with much precision. They could reduce the supply of oil to the whole world, but they were unable to get much more oil to their friends than to their enemies because the major oil companies directed non-Arab oil toward the target states. It soon became evident that there was no clear connection between pro-Arab politics and oil imports. The theoretical literature does not suggest the consequences of such undirected sanctions, either for the actor or for the target states. The present case implies that the political utility of the sanctions may be undercut; it would be interesting to compare it with other cases.

Uncertainty

Clearly, the targets experienced a very high level of uncertainty on most of the dimensions of the power relationship. The targets were often confused about which governments were applying sanctions, what the sanctions were, how much the sanctions would affect the flow of oil to their countries, and what the impact of various levels of restrictions would be on their countries' economies. Much of the theorizing about economic sanctions assumes perfect information, so the oil weapon may be unusual. For example, it hardly seems likely that the Canadian experience—in which the government spent considerable time finding out who was applying sanctions and what the sanctions were—is typical.

Much of this uncertainty, however, may be inherent in the use of economic sanctions as a foreign policy tool. Adapting Wiberg's typology somewhat, there are three levels of sanctions in such cases: the denial of a particular commodity to the target economy, the decline in the macroeconomic performance of the target, and the political costs to decision makers as a result of this decline (Wiberg 1969, 14). Each level is more difficult to implement than its predecessor. First, history shows that it is very difficult to cut off the supply of a particular commodity to a target country—the consequent rise in price makes it too profitable for individuals and groups not to supply it by hook or by crook. Moreover, elites know this, which means that threats to cut off commodities are not likely to be very powerful. Second, we know that modern economies have considerable flexibility; when the price of one commodity is raised, substitutions can often be arranged, especially over a longer period of time. Finally, even if the economy does suffer as a result of sanctions, the leadership may be able to appeal to nationalism and avoid paying any serious political costs. Although uncertainty may temporarily increase the impact of sanctions (as it did in the case of the oil weapon), these examples demonstrate ways in which elites and leaders can use this uncertainty to avoid changing their personal positions or taking politically costly stands. Thus, this uncertainty may seriously reduce the political impact of sanctions.

Replacing Decision-Makers

Probably the best way to alter the foreign policy of a government is to replace the people who hold high office. New people may hold different policy beliefs and are, at any rate, less likely to be personally committed to upholding the policies of the previous administration. The oil weapon did not cause such changes in personnel, however—a fundamental weakness of economic sanctions, albeit one shared by most other kinds of limited pressures. Even if the sanctions depress the economy sufficiently to cause elites or publics to replace the current leadership and run the risk of appearing to support a hostile government, such events are likely to take some time.

Public Opinion and Political Elites

Theorists have suggested that external influences are difficult to manipulate because their use changes the issue from the particular case in point to the more general, and volatile, question of resisting foreign domination. The implication is that public opinion might prevent a target government from making concessions, even if it desired to do so. For example, the Arab governments have been pressured by public opinion to fight Israel, even when they seemed likely to lose the resulting military conflict (Luttwak 1974, 65–67). The oil-weapon case implies that this concern may be exaggerated, at least for relatively strong governments such as the democracies studied here. Instead, the conventional interpretation regarding public opinion on foreign policy (that is, it can often be shaped by the government) seems to apply even to highly visible issues such as the use of the oil weapon.

What Is the Appropriate Target?

This in turn brings us to one of the paradoxes of limited external coercion. On the one hand, the actor wants a target government weak enough to be coerced; if the target government has a firm domestic foundation, economic sanctions will probably not weaken it much. On the other hand, a truly weak government may not be able to deliver on its concessions; it may be unable to extract the necessary resources from its population or to survive while seeming to act as the surrogate for a foreign state.

In the case of the oil weapon, there was a fundamental mismatch between most of its targets and its expressed aims. It made no sense to punish Western Europe and Japan for Israel's actions because they clearly did not have the capacity to force Israel to change its policy. The Arab states recognized this and decided to settle for rhetorical gestures (and got few enough of those). With the perversity that often characterizes international politics, the United States (a much more logical target because of its greater, although hardly unlimited, ability to influence Israel) was much less vulnerable to oil cut-offs from the Arab states. One way to square this circle, therefore, was to force Western Europe and Japan to pressure the United States to lean on Israel. But such pressure, judging from the American accounts of the time, was apparently unimportant.

On balance, then, the experience of the oil weapon confirms the generally accepted position that economic sanctions by themselves cannot usually force a change in an important policy of the target government. It also implies, however, that the results of such sanctions can be better explained by focusing less on the sanction itself and more on its impact on the government and society of the target state. The weakness of sanctions is that they cannot create the domestic pressures that are required to change policy. In particular, it is not clear how sanctions can directly lead to the replacement of top-level decision makers; indeed, they may be used by these individuals to reinforce their own political positions.

As David Baldwin eloquently argues, however, the weaknesses inherent in economic sanctions are shared by almost all other limited techniques used by one state to influence another (1985). To some extent, economic sanctions have suffered from being implicitly compared to an imaginary alternative that could effectively and cheaply bring about change in the target state. This suggests that it is a mistake to separate the study of, for example, economic sanctions and coercive diplomacy. In particular, the focus on the internal politics of the target state, which coercive diplomacy emphasizes, must become integral to the more general analysis of external influence in foreign policy.

Policy Implications

Prospects for Another Arab Oil Embargo

A major Arab-Israeli war would almost certainly result in the use of the oil weapon again. Most journalistic analyses of the 1973–1974 experience have stressed its apparent success, and domestic Arab pressures would force governments to apply it again (Daoudi and Dajani 1985, 171–172).

It is true that the oil weapon was not used during the Israeli invasion of Lebanon in 1979, despite a call from Libya to do so. But in that instance the only Arab armies committed to battle were those of Syria and the PLO. It has also been suggested that the oil glut made any such embargo impossible to apply (Meir 1986, 49). But, although this factor would partially determine the *impact* of another use of the oil weapon, the decision to *use* it would probably be made on the basis of domestic Arab politics. Sheikh R. Ali makes a persuasive argument that relations between the United States and Saudi Arabia are so close now that, unless the Saudi regime is overthrown, the oil weapon cannot be used again, but he also notes that another Arab-Israeli war would trigger another embargo (1986, 51–55, 110, 133).

This is not to say that the oil glut is irrelevant: As long as the elites of both the actor and the target states are aware of an oversupply of oil, the threat of an oil embargo is much less persuasive and is much less likely to be used in more limited conflicts where Arab domestic pressures will not be so great.

If the oil weapon is used again, it will be unlikely to force the target governments to significantly alter their policies much. It is useful to again adopt Wiberg's three criteria for a successful embargo—effectiveness, efficiency, and success—to see how the situation has changed since 1973.

Effectiveness. The Arab ability to deny significant amounts of oil to the target states (effectiveness) has undoubtedly been increased by the Iranian revolution. Despite its ongoing war with Iraq, Iran would probably support an Arab use of the oil weapon against the "Great Satan" and its allies. At the same time, however, the current world oversupply of oil fundamentally undercuts Arab leverage. Thus, any analysis has to assume that at some point this glut, like others in the past, will end; until then, no embargo is likely to have much impact, at least in the short term.

As opposed to 1973, the international oil companies now control a much smaller percentage of worldwide oil in transit; one set of estimates shows them with 100 percent of this oil in 1960, 70 percent in 1970, and 50 percent in 1980 (Kemezis and Wilson 1984, 86). The political implications of this fact are not altogether clear. On the one hand, the important role of the major oil companies made it easier for the Arabs to monitor oil shipments a decade ago; the rise of the spot market, for example, probably makes Arab enforcement much more difficult. On the other hand, state trading companies, which are moving an increasing percentage of world oil, may be more vulnerable to combined pressure from both the Arabs and their home governments to comply (although the behavior of British Petroleum, Compagnie Française des Pétroles, Elf-Erap, and Sonatrach during the embargo does not support this argument [Stobaugh 1976, 189–190]).

In a sophisticated analysis, William R. Brown maintains that the Arabs have instituted a strategy whereby they increasingly control the transportation and marketing of their own oil by using bilateral state trading arrangements, requiring petroleum shippers to provide discharge certificates showing the destinations of shipments, buying more tankers, requiring that a percentage of their oil be transported on their own ships, and increasing their refining capacity (1982, 310–313). He argues that this will produce uncertainty, which will raise fears of inadequate supply. But uncertainty about world supplies does not mean the Arabs will be able to reduce oil supplies to selected target countries. Nor is it clear that these strategies will alleviate the fundamental weakness of the Arab position—the lack of control over oil exports from other OPEC countries and the significant amount of non-OPEC oil production. It seems likely that, in a future oil crisis, the price of oil in embargoed countries will rise, eventually attracting non-Arab oil. Formal allocations by major corporations would certainly smooth this transfer, but it is not clear that they are absolutely essential for the transfer to occur.

Efficiency. In terms of the perceived impact of an effective embargo on the target economies (efficiency), the Arab position is again probably not as strong as in 1973. In the first place, this is not a new issue, as it was then: A good deal of effort has been devoted to trying to discover the precise relationship between oil supply and economic prosperity. Although the results are certainly not definitive, there is at least an illusion of knowledge that should prevent a recurrence of the almost panic-stricken reaction in the industrial world in 1973. Moreover, our experience with rising and falling oil prices suggests that variations in supply, while certainly important economically, can be dealt with if necessary. The accuracy of such assumptions is less important than whether or not elites and publics generally accept them, at least in the short run.

Success. One can argue that the Arabs are now more capable of translating damage to industrial economies into political concessions (success). The target countries have all moved in a more pro-Arab direction, which may make further shifts "thinkable." Moreover, the 1973–1974 experience and its consequences permanently altered the power balance within the target countries' governments by strengthening pro-Arab groups such as corporations and some government departments and weakening the positions of the pro-Israeli groups, most of which are outside of government. Finally, Israel's own moral standing in the West has diminished considerably as a result of its Palestinian policies and the Lebanese invasion.

Notwithstanding these pro-Arab shifts of influence, policy vis-à-vis the Arab-Israeli conflict has not changed significantly, and another use of the oil weapon might actually have *less* domestic political impact than the 1973–1974 experience. Arab wealth would probably not increase from its current base by the same percentage as it did earlier, and it is not clear that other, new pro-Arab groups could be mobilized. The Western perception of the Arab states has not improved much since 1974 either (the exception is Egypt, but this new image rests on its government's desire for peace). The Israeli lobby, particularly in the United States, remains strong. Perhaps most important, the U.S. government remains fundamentally committed to Israel, and its allies know it and accept it, albeit with considerable misgivings.

Presumably a lot depends on what kind of demands would be made in the event of another embargo. It might be possible to coerce Japan and some Western European nations to recognize the PLO, for example, in part because there is significant domestic pressure to do so now. But denying diplomatic recognition to Israel or voting to expel it from the United Nations would be much more difficult, although even these would be only rhetorical gestures. Persuading these governments to take concrete actions to alter the Middle Eastern power balance, such as applying sanctions to Israel, seems out of the question.

On balance, then, use of the oil weapon under current economic circumstances would have relatively little political impact. After the oil glut ends, the oil weapon might produce the kind of rhetorical shifts seen in 1973–1974; it seems unlikely to do more under the most favorable circumstances.

Other Third World "Resource Weapons"

The weaknesses of the oil weapon are not inherent in the fact that oil was the sanctioned commodity. Indeed, on balance oil appears to be an unusually potent weapon; it is a critical resource for industrial economies and is difficult to stockpile or substitute, especially in the short run. One could therefore argue that, if the oil weapon did not work, no other commodity is likely to be more useful politically.

A major weakness of the oil weapon was that the Arab states did not control most of the world's oil exports and were not able to persuade any other exporting countries to join in the embargo (although these other countries were quite willing to cooperate in the price increases). Another commodity exclusively controlled by a few Third World states might be more useful as a political lever, but it is hard to think of a commodity that fits this description.

In addition, the Arabs possessed an unusual degree of at least rhetorical political unity in their common opposition to Israel. Such unity is critical if actor states are to suffer economic costs for a common political gain. The Arabs, however, were unable to maintain unity (Iraq, for example, never joined the embargo, and there were persistent rumors of leakage in the embargo, particularly from Libya). Although the Third World countries have been able to maintain impressive and unexpected unity in rhetorical gestures, such as voting in international conferences in the Group of 77 (Rothstein 1976), it is not clear that such unity could be extended to enduring significant economic sacrifice. On balance, it seems unlikely that other Third World countries can use the oil-weapon model to gain significant political concessions.

Reducing Vulnerability to "Resource Weapons"

Much effort has been devoted to analyzing what measures could be taken to prevent an effective or efficient embargo; that is, to assure that the target countries either receive adequate oil supplies or can function economically without them. Less attention has been given to the question of political vulnerability.

Perhaps the most important resource in decreasing political vulnerability is information. More than anything else, the lack of knowledge weakened the political leaderships in the target countries in 1973–1974. Much of this knowledge was essentially economic: How much oil was available, what impact would an oil supply reduction have on the economy, and so forth. Clearly we know more than we did then, although important gaps in our knowledge base remain. Perhaps more important is the fact that we think we know more than we did then, which should reduce panic.

But leaders also need information about the *political* consequences of such a crisis, and this has been in much shorter supply. How will the oil weapon affect the domestic political scene, to which the leaders are accountable? How will it alter the behavior of other governments with which their countries are allied? The experiences of five governments in 1973–1974 cannot definitely answer these questions, but they are at least suggestive.

On balance, this study suggests that domestic political constraints coming from outside the government are not likely to be critical. Although there was certainly a great deal of public concern over the issue, opposition groups were not successful in mobilizing effective political opposition to the policies of the governments, regardless of what those policies were; indeed, almost no such attempts were made. Internally, the major result was that high-level political decision makers had to become concerned with the issue, but they retained a good deal of flexibility with which to make a decision.

Internationally, this study suggests that target governments are not likely to change their policies radically, even on issues that are not particularly important and when the actor governments credibly threaten to inflict considerable economic damage. There was a lot of discussion about how Western governments engaged in "beggar-thy-neighbor" or *sauve qui peut* policies as a result of the oil weapon, but in fact no government markedly changed its policies at all. This experience suggests that, at least among the industrial nations, a government can stand firm against external coercion without much fear of being undercut by its own allies.

WORKS CITED

Adams, Michael, and Christopher Mayhew. 1975. *Publish It Not . . . The Middle East Cover-Up*. London: Longman.

Adams, William. 1981. *Television Coverage of the Middle East*. Norwood, N.J.: ABLEX.

Adler-Karlsson, Gunnar. 1968. *Western Economic Warfare, 1947–1967*. Stockholm: Almqvist and Widsell.

Ahrari, Mohammed. 1979. "OAPEC and 'Authoritative' Allocation of Oil: An Analysis of the Arab Oil Embargo." *Studies in Comparative International Development* 14 (Spring): 9–21.

———. 1986. *OPEC: The Failing Giant*. Lexington: University Press of Kentucky.

Akao, Nobutoshi. 1983. "Japan's Search for Economic Security." In *Japan's Economic Security*, edited by Nobutoshi Akao, 245–272. New York: St. Martin's Press.

Akins, James. 1973. "The Oil Crisis: This Time the Wolf Is Here." *Foreign Affairs* 51 (Spring): 462–490.

al-Mani, Saleh A. 1983. *The Euro-Arab Dialogue: A Study in Associative Diplomacy*. New York: St. Martin's Press.

al-Sowayegh, Abdulaziz H. 1980. "Saudi Oil Policy During King Faisal's Era." In *King Faisal and the Modernization of Saudi Arabia*, edited by Willard A. Beling, 202–229. Boulder, Colo.: Westview Press.

———. 1984. *Arab Petropolitics*. New York: St. Martin's Press.

Ali, Sheikh Rustum. 1976. *Saudi Arabia and Oil Diplomacy*. New York: Praeger.

———. 1986. *Oil, Turmoil, and Islam in the Middle East*. New York: Praeger.

Allen, David. 1978. "The Euro-Arab Dialogue." *Journal of Common Market Studies* 16: 323–342.

Allen, David, and Alfred Pijpers. 1984. *European Foreign Policy-Making and the Arab-Israeli Conflict*. The Hague: Martinus Nijhoff.

Allen, David, and Michael Smith. 1984. "Europe, the United States and the Middle East: A Case Study in Comparative Policymaking." *The International Spectator* 19 (January–March): 42–55.

Allison, Graham T. 1971. *Essence of Decision: Explaining the Cuban Missile Crisis*. Boston: Little, Brown.

AlRoy, Gil Carl. 1975. *The Kissinger Experience: American Policy in the Middle East*. New York: Horizon Press.

Atherton, Alfred L., Jr. 1984. "Arabs, Israelis—And Americans: A Reconsideration." *Foreign Affairs* 62 (Summer): 1194–1209.

Ayubi, Shaheen, Richard E. Bissell, Nana Amu-Brafih Korsah, and Laurie A. Lerner. 1982. "Economic Sanctions in U.S. Foreign Policy." Philadelphia Policy Papers. Philadelphia: Foreign Policy Research Institute.

Azar, Edward E. 1982. "Conflict and Peace Data Bank (COPDAB), 1948–1978: Daily Event Records, International and Domestic Files (ICPSR 7767)." 2d ICPSR ed. Ann Arbor, Mich.: Inter-university Consortium for Political and Social Research.

Babayan, Kathryn. 1985. "Trends in Japanese Trade Relations with the Middle East." In *Japan and the Middle East in Alliance Politics,* edited by Ronald Morse, 97–109. Asia Program/International Security Studies Program. Washington, D.C.: Wilson Center.

Baehr, Peter R. 1974. "Parliamentary Control over Foreign Policy in The Netherlands." *Government and Opposition* 9 (Spring): 165–188.

———. 1978. "The Foreign Policy of the Netherlands." In *The Foreign Policy of the Netherlands,* edited by J. H. Leurdijk, 3–27. Alphen aan den Rijn, Netherlands: Sijthoff and Noordhoff.

———. 1980. "The Dutch Foreign Policy Elite: A Descriptive Study of Perceptions and Attitudes." *International Studies Quarterly* 24 (June): 223–261.

———. 1982. "Democracy and Foreign Policy in the Netherlands." Paper prepared for the Symposium on Democracy and Foreign Policy, Leeuwenhorst Congress Center, Noordwijkerhout, the Netherlands, March 25–26.

Baehr, Peter R., Philip P. Everts, Fred Grünfeld, Alfred van Staden, and Gijs de Vries. 1982. "The Foreign Policy Process in the Netherlands. Two Cases of Foreign Policy-Making: The Impact of the Middle-East Crisis of 1973; The Modernization of Nuclear Weapons in Western Europe." Paper presented at the International Political Science Association Twenty-second World Congress, Rio de Janeiro, August 9–14.

Baerwald, Hans H. 1977. "The Diet and Foreign Policy." In *The Foreign Policy of Modern Japan,* edited by Robert A. Scalapino, 37–54. Berkeley and Los Angeles: University of California Press.

Bahbah, Bishara A. 1982. "The United States and Israel's Energy Security." *Journal of Palestine Studies* 11 (Winter): 111–131.

Bailey, Richard. 1983. *The European Connection: Implications of EEC Membership.* New York: Pergamon Press.

Bailey, Sydney D. 1985. *The Making of Resolution 242.* Dordrecht: Martinus Nijhoff.

Baker, Michael S. 1978. "Grappling With Dependence: Japanese Oil Policy, 1970–1974." East Asian Institute Certificate Essay, Columbia University, New York.

Baldwin, David A. 1985. *Economic Statecraft.* Princeton: Princeton University Press.

Balta, Paul. 1974. "La dossier arabe sur des Pays-Bas." *Maghreb-Machrek* 61 (January–February): 14–16.

Barber, James. 1976. *Who Makes British Foreign Policy?* Milton Keynes, U.K.: Open University Press.

———. 1979. "Economic Sanctions as a Policy Instrument." *International Affairs* (London) 55 (July): 367–401.

Barberis, Mary A. 1976. "The Arab-Israeli Battle on Capitol Hill." *Virginia Quarterly Review* 52 (Spring): 203–223.

Ben-Zvi, Abraham. 1984. "Alliance Politics and the Limits of Influence: The Case of the US and Israel, 1975–1983." Paper no. 25, Jaffee Center for Strategic Studies, Tel Aviv University. Boulder, Colo.: Westview Press.

Benesh, John. 1979. "Canadians' Images of the Middle East Conflict and Canada's Middle East Foreign Policy to 1977." Research Essay, Norman Paterson School of International Affairs, Carleton University, Ottawa. (Benesh is executive director of the Canadian Law Information Council in Ottawa.)

Bessin, Shira Herzog, and David Kaufman. 1979. "Canada-Israeli Friendship: The First Thirty Years." Toronto: Canada-Israel Committee.

Binnendijk, Hans. 1974. "The U.S. and Japan: Fine Tuning a New Relationship." *Pacific Community* 6 (October): 22–37.

Blechman, Barry M., and Stephen S. Kaplan. 1978. *Force Without War: U.S. Armed Forces as a Political Instrument.* Washington, D.C.: Brookings Institution.

Bones, Alan. 1985. "Zionist Interest Groups and Canadian Foreign Policy." In *Canada and the Arab World,* edited by Tareq Ismael, 151–172. Edmonton: University of Alberta Press.

Bonham, G. Matthew, Michael J. Shapiro, and Thomas I. Trumble. 1979. "The October War: Changes in Cognitive Orientation Toward the Middle East Conflict." *International Studies Quarterly* 23 (March): 3–44.

Boon, H. N. 1976. *Afscheidsaudientie: Tien Studies Uit de Diplomatieke Prakiyk.* Rotterdam: A. D. Donker.

Bouchuiguir, Sliman Brahim. 1979. "The Use of Oil as a Political Weapon: A Case Study of the 1973 Arab Oil Embargo." Ph.D. diss., George Washington University; #7915599, University Microfilms International, Ann Arbor, Mich.

Brecher, Michael. 1980. *Decisions in Crisis: Israel, 1967 and 1973.* Berkeley and Los Angeles: University of California Press.

"Britain's Foreign Office." 1982. *The Economist* 245 (November 27): 19–25.

Brooks, William. 1985. "The Politics of Japan's Foreign Aid to the Middle East." In *Japan and the Middle East in Alliance Politics,* edited by Ronald Morse, 81–91. Asia Program/International Security Studies Program. Washington, D.C.: Wilson Center.

Brown, George. 1971. *In My Way.* London: Victor Gollancz.

Brown, Seyom. 1983. *The Faces of Power: Constancy and Change in United States Foreign Policy from Truman to Reagan.* New York: Columbia University Press.

Brown, William R. 1982. "The Oil Weapon." *Middle East Journal* 36 (Summer): 301–318.

Brzezinski, Zbigniew. 1974–1975. "Recognizing The Crisis." *Foreign Policy* 17 (Winter): 63–74.

———. 1983. *Power and Principle: Memoirs of the National Security Adviser, 1977–1981*. New York: Farrar Straus Giroux.

Bundy, William P. 1984. "The Last Dozen Years: What Might We Learn?" *Foreign Affairs* 62 (Summer): 1210–1237.

Burgess, Philip M. 1968. *Elite Images and Foreign Policy Outcomes: The Case of Norway*. Columbus: Ohio State University Press.

Burns, William J. 1985. *Economic Aid and American Policy Toward Egypt, 1955–1981*. Albany: State University of New York Press.

Buzan, Barry, and R. J. Barry Jones. 1981. *Change and the Study of International Relations: The Evaded Dimension*. New York: St. Martin's Press.

Byers, R. B. 1974. "External Affairs and Defense." In *Canadian Annual Review of Politics and Public Affairs, 1973,* edited by John Saywell, 221–292. Toronto: University of Toronto Press.

———. 1978. "Defense and Foreign Policy in the 1970s: The Demise of the Trudeau Doctrine." *International Journal* 33 (Spring): 312–338.

Byers, R. B., and David Leyton-Brown. 1977. "Canadian Elite Images of the International System." *International Journal* 32 (Summer): 608–639.

Byrd, Peter. 1973. "Trade and Commerce in External Relations." In *The Management of Britain's External Relations,* edited by Robert Boardman and A. J. R. Groom, 173–200. London: Macmillan.

Cable, James. 1971. *Gunboat Diplomacy: Political Applications of Limited Naval Force*. Studies in International Security 16, International Institute for Strategic Studies. London: Chatto & Windus.

Caldwell, Martha Ann. 1981a. "Petroleum Politics in Japan: State and Industry in a Changing Policy Context." Ph.D. diss., University of Wisconsin, Madison; #8117506, University Microfilms, Ann Arbor, Mich.

———. 1981b. "The Dilemmas of Japan's Oil Dependency." In *The Politics of Japan's Energy Strategy: Resources—Diplomacy—Security,* edited by Ronald A. Morse, 65–84. Research Papers and Policy Studies 3. Berkeley: Institute of East Asian Studies, University of California.

Campbell, Colin, Jr. 1980. "Political Leadership in Canada: Pierre Elliott Trudeau and the Ottawa Model." In *Presidents and Prime Ministers,* edited by Richard Rose and Ezra N. Suleiman, 50–93. Washington, D.C.: American Enterprise Institute for Public Policy.

Campbell, William R. 1984. "Japan and the Middle East." In *Japan's Foreign Relations: A Global Search for Economic Security,* edited by Robert S. Ozaki and Walter Arnold, 133–152. Boulder, Colo.: Westview Press.

Canada. 1985. Senate. Standing Committee on Foreign Affairs. "Report on Canada's Relations with the Countries of the Middle East and North Africa." Ottawa: Ministry of Supply and Services.

Caradon, Lord, Arthur J. Goldberg, Mohamed H. el-Azyyat, and Abba Eban. 1981. "U.N. Security Council Resolution 242: A Case Study in Diplomatic Ambiguity." Edmund A. Walsh School of Foreign Service, Georgetown University. Washington, D.C.: Institute for the Study of Diplomacy.

Carter, Jimmy. 1982. *Keeping Faith: Memoirs of a President*. New York: Bantam Books.

Carvely, Susan Carol. 1985. "Dual Dependence: Japanese Policy Toward the Middle East, 1973 to 1984." Ph.D. diss., George Washington University.

Chafets, Ze'ev. 1985. *Double Vision: How the Press Distorts America's View of the Middle East*. New York: William Morrow.

Charney, Leon H. 1984. *Special Counsel*. New York: Philosophical Library.

Clark, E. 1932. *Boycotts and Peace*. New York: Harper & Brothers.

"Close Up on the Cabinet." 1982. *The Economist* 282 (February 6): 34–35.

Cobham, Helena. 1983. "The British Press." In *The Middle East: Press Perspectives and National Policies*, edited by Luci E. Richardson, 8–11. Proceedings of a seminar cosponsored by the Center for Middle East Policy and the Royal Institute of International Affairs. Washington, D.C.: Center for Middle East Policy. (Publication available at Georgetown University library, Washington, D.C.)

Cohen, Bernard C. 1977–1978. "Political Systems, Public Opinion, and Foreign Policy: The United States and the Netherlands." *International Journal* 33 (Winter): 195–216.

Cohn, Werner. 1979. "English and French Canadian Public Opinion on Jews and Israel: Some Poll Data." *Canadian Ethnic Studies* 11 (2): 31–48.

Commission on Economic Coercion and Discrimination. 1977. *The Arab Boycott in Canada*. Report of the Commission on Economic Coercion and Discrimination, in association with the Center for Law and Public Policy, Montreal.

Congressional Quarterly. 1979. *The Middle East: U.S. Policy, Israel, Oil and the Arabs*. 4th ed. Washington, D.C.: Congressional Quarterly.

Corwin, Edward S. 1957. *The President: Office and Powers*. New York: New York University Press.

Craig, Gordon A., and Alexander L. George. 1983. *Force and Statecraft: Diplomatic Problems of Our Time*. New York: Oxford University Press.

Crossman, Richard. 1976. *The Diaries of a Cabinet Minister*. 3 vols. London: Hamish Hamilton & Jonathan Cape.

Curtis, Gerald L. 1975. "Big Business and Political Influence." In *Modern Japanese Organization and Decision-Making*, edited by Ezra F. Vogel, 33–70. Berkeley and Los Angeles: University of California Press.

Curtiss, Richard H. 1982. *A Changing Image: American Perceptions of the Arab-Israeli Dispute*. Washington, D.C.: American Educational Trust.

Dahl, Robert A. 1963. *Modern Political Analysis*. Englewood Cliffs, N.J.: Prentice-Hall.

Daoudi, Mohammed S. 1981. "Political Dynamics of Economic Sanctions: A Case Study of Arab Oil Embargoes." Ph.D. diss., University of South Carolina; #8129493, University Microfilms, Ann Arbor, Mich.

Daoudi, M. S., and M. S. Dajani. 1983. *Economic Sanctions: Ideals and Experience*. London: Routledge & Kegan Paul.

———. 1985. *Economic Diplomacy: Embargo Leverage and World Politics*. Boulder, Colo.: Westview Press.

Darmstadter, Joel, and Hans H. Landsberg. 1976. "The Economic Background." In *The Oil Crisis,* edited by Raymond A. Vernon, 15–38. New York: W. W. Norton.

de la Serre, Françoise. 1974. "L'Europe des neuf et le conflit Israelo-Arabe." *Revue Française de Science Politique* 24 (August): 801–811.

Deboutte, Jan, and Alfred van Staden. 1978. "High Politics in the Low Countries." In *Foreign Policy Making in Western Europe,* edited by William Wallace, 56–82. Farnborough: Saxon House.

Deese, David. 1984. "The Vulnerability of Modern Economies." In *Strategic Dimensions of Economic Behavior,* edited by Gordon H. McCormick and Richard E. Bissell, 148–180. New York: Praeger.

Delvoie, L. A. 1976. "Growth in Economic Relations of Canada and the Arab World." *International Perspectives* (November–December): 29–33.

Destler, I. M., Hideo Sato, Priscilla Clapp, and Haruhiro Fukui. 1976. *Managing an Alliance: The Politics of U.S.-Japanese Relations.* Washington, D.C.: Brookings Institution.

Deutsch, Karl W. 1968. *The Analysis of International Relations.* Englewood Cliffs, N.J.: Prentice-Hall.

Dewitt, David B., and John J. Kirton. 1982. "Canadian Foreign Policy Towards the Middle East, 1947–1982: International, Domestic, and Governmental Determinants." Department of Political Science, University of Toronto. Mimeo.

———. 1983. "Canadian Foreign Policy Towards the Middle East, 1947–1982: International, Domestic and Governmental Determinants." In *The Middle East at the Crossroads,* edited by Janice Gross Stein and David B. Dewitt, 176–199. Oakville, Ont.: Mosaic Press.

Djonovich, Dusan J., comp. and ed. 1976. *United Nations Resolutions.* Series 1. Resolutions Adopted by the General Assembly, vol. 13, 1970–1971. Dobbs Ferry, N.Y.: Oceana Publications.

———. 1978. *United Nations Resolutions.* Series 1. Resolutions Adopted by the General Assembly, vol. 14, 1972–1974. Dobbs Ferry, N.Y.: Oceana Publications.

Douglas-Home, Sir Alec. *See* Home, Lord.

Doxey, Margaret P. 1980. *Economic Sanctions and International Enforcement.* 2d ed. New York: Oxford University Press.

Dowty, Alan. 1984. *Middle East Crisis: U.S. Decision-Making in 1958, 1970, and 1973.* Berkeley and Los Angeles: University of California Press.

Dunsmore, Barrie. 1980. "Television Hard News and the Middle East." In *The American Media and the Arabs,* edited by Michael C. Hudson and Ronald G. Wolfe, 73–76. Washington, D.C.: Center for Contemporary Arab Studies, Georgetown University.

Eayrs, James. 1983. "Canada: The Department of External Affairs." In *The Times Survey of Foreign Ministries of the World,* edited by Zara Steiner, 95–117. London: Times Books.

Economist Intelligence Group. 1973–1974. *Quarterly Economic Review: Saudi Arabia, Jordan.* Various issue numbers.

Edwards, Geoffrey. 1984. "National Approaches to the Arab-Israeli Conflict: Britain." In *European Foreign Policy-Making and the Arab-Israeli Conflict,* edited by David Allen and Alfred Pijpers, 47–59. The Hague: Martinus Nijhoff.

Ellings, Richard J. 1985. *Embargoes and World Power: Lessons from American Foreign Policy.* Boulder, Colo.: Westview Press.

Emerson, Steven. 1985. *The American House of Saud: The Secret Petrodollar Connection.* New York: Franklin Watts.

Erskine, Hazel. 1969–1970. "The Polls: Western Partisanship in the Middle East." *Public Opinion Quarterly* 33 (Winter): 627–640.

Everts, Philip P. 1985. *Controversies at Home: Domestic Factors in the Foreign Policy of the Netherlands.* Dordrecht: Martinus Nijhoff.

Feith, Douglas J. 1980. "Love and Oil." *New Republic* 183 (November 22): 20–23.

———. 1981. "The Oil Weapon De-Mystified." *Policy Review* 15 (Winter): 19–39.

Feld, Werner J. 1978. "West European Foreign Policies: The Impact of the Oil Crisis." *Orbis* 22 (Spring): 63–88.

Feuerwerger, Marvin C. 1979. *Congress and Israel: Foreign Aid Decision-Making in the House of Representatives, 1969–1976.* Westport, Conn.: Greenwood Press.

Franck, Thomas M., and Edward Weisband. 1974. "Introduction: Executive Secrecy in Three Democracies: The Parameters of Reform." In *Secrecy and Foreign Policy,* edited by Thomas M. Franck and Edward Weisband, 3–12. New York: Oxford University Press.

———. 1979. *Foreign Policy by Congress.* New York: Oxford University Press.

Frankel, Joseph. 1975. *British Foreign Policy, 1945–1973.* London: Oxford University Press.

Freedman, Maurice. 1974. "Perspectives." In *The Yom Kippur War: Israel and the Jewish People,* edited by Moshe Davis, 162–173. New York: Arno Press.

Freedman, Robert O. 1970. *Economic Warfare in the Communist Bloc.* New York: Praeger.

Friedman, Edward, Paul Seabury, and Aaron Wildavsky. 1975. *The Great Détente Disaster: Oil and the Decline of American Foreign Policy.* New York: Basic Books.

Fukui, Haruhiro. 1975. "Okinawa Reversion: Decision Making in the Japanese Government." Paper prepared for the Conference on Okinawa Reversion, Hakome, Japan, January 10–13.

———. 1977. "Policy-Making in the Japanese Foreign Ministry." In *The Foreign Policy of Modern Japan,* edited by Robert Scalapino, 3–35. Berkeley and Los Angeles: University of California Press.

———. 1978. "The GATT Tokyo Round: The Bureaucratic Politics of Multilateral Diplomacy." In *The Politics of Trade: US and Japanese Policymaking for the GATT Negotiations,* edited by Michael Blaker, 75–169. Occasional Papers Series, Project on Japan and the United States in Multilateral Diplomacy. New York: East Asian Institute, Columbia University.

Gallup, George. 1972. *The Gallup Poll: Public Opinion, 1935–1971.* New York: Random House.

———. 1976. *The Gallup International Public Opinion Polls: Great Britain, 1937–1975.* 2 vols. New York: Random House.

———. 1978. *The Gallup Poll: Public Opinion, 1972–1977.* 2 vols. Wilmington, Del.: Scholarly Resources.

———. 1980. *The International Gallup Polls: Public Opinion, 1978.* Wilmington, Del.: Scholarly Resources.

———. 1981. *The International Gallup Polls: Public Opinion, 1979.* Wilmington, Del.: Scholarly Resources.

Gallup Opinion Index. 1978a. Issue no. 153 (April).

———. 1978b. Issue no. 156 (July).

———. 1978c. Issue no. 158 (September).

———. 1979a. Issue no. 163 (February).

———. 1979b. Issue no. 166 (May).

"Gallup Poll: More Urge Israeli Pullback." 1973. *Daily Telegraph* (London), December 22.

Gallup Report. 1982. Issue no. 203 (August).

Galtung, Johan. 1967. "On the Effects of International Economic Sanctions with Examples from the Case of Rhodesia." *World Politics* 19 (April): 378–416.

Garfinkle, Adam M. 1983. "Western Europe's Middle East Diplomacy and the United States." Philadelphia Policy Papers. Philadelphia: Foreign Policy Research Institute.

Garnham, David. 1976. "The Oil Crisis and U.S. Attitudes Toward Israel." In *Arab Oil: Impact on Arab Countries and Global Implications,* edited by Naiem A. Sherbiny and Mark A. Tessler, 295–304. New York: Praeger.

Garratt, Joan. 1982. "Euro-American Energy Diplomacy in the Middle East, 1970–80: The Pervasive Crisis." In *The Middle East and the Western Alliance,* edited by Steven L. Spiegel, 82–103. London: George Allen & Unwin.

George, Alexander, David K. Hall, and William R. Simons. 1971. *The Limits of Coercive Diplomacy: Laos, Cuba, Vietnam.* Boston: Little, Brown.

Gilboa, Eytan. 1985. "Effects of the War in Lebanon on American Attitudes Toward Israel and the Arab-Israeli Conflict." *Middle East Review* 18 (Fall): 30–44.

———. 1986. *American Public Opinion Toward Israel and the Arab-Israeli Conflict.* Lexington, Mass.: Lexington Books.

Gilpin, Robert. 1981. *War and Change in World Politics.* New York: Cambridge University Press.

Glick, Edward Bernard. 1982. *The Triangular Connection: America, Israel, and American Jews.* London: George Allen & Unwin.

Goldsmith, Arthur A. 1986. "Power Politics and Technological Change in Agriculture: The Case of India, 1951–1966." Paper prepared for the American Political Science Association, Washington, D.C. (Paper available through Department of Management, University of Massachusetts at Boston.)

Gordon, Avishag H. 1975. "The Middle East October 1973 War as Reported by the American Networks." *International Problems* 14 (Fall): 76–85.

Gordon Walker, Patrick. 1970. *The Cabinet: Political Authority in Britain*. New York: Basic Books.

———. 1974. "Secrecy and Openness in Foreign Policy Decision-Making: A British Cabinet Perspective." In *Secrecy and Foreign Policy*, edited by Thomas M. Franck and Edward Weisband, 42–52. New York: Oxford University Press.

Grossman, Laurence I. 1978. "The Shifting Canadian Vote on Mideast Questions at the UN." *International Perspectives* (May–June): 9–13.

Gruen, George E. 1975–1976. "Arab Petropower and American Public Opinion." *Middle East Review* 8 (Winter): 33–40.

Grünfeld, Fred. 1984. *Nederland en het Midden-Oosten*. Leiden: Instituut voor Internationale Studiën.

———. 1985. "The Netherlands and the Middle East—Two Foreign Policy Crises." In *Controversies at Home: Domestic Factors in the Foreign Policy of the Netherlands*, edited by Philip P. Everts, 157–178. Dordrecht: Martinus Nijhoff.

Guichard, Louis. 1930. *The Naval Blockade, 1914–1918*. New York: Appleton.

Haig, Alexander M., Jr. 1984. *Caveat: Realism, Reagan, and Foreign Policy*. New York: Macmillan.

Halliday, Fred. 1977. *Mercenaries: "Counter-Insurgency" in the Gulf*. Nottingham, U.K.: Spokesman.

Harf, James E., David G. Hoovler, and Thomas E. James, Jr. 1974. "Systemic and External Attributes in Foreign Policy Analysis." In *Comparing Foreign Policies: Theories, Findings and Methods*, edited by James N. Rosenau, 235–250. Beverly Hills, Calif.: Sage.

Hastings, Elizabeth Hann, and Philip K. Hastings. 1981. *Index to International Public Opinion, 1979–1980*. Westport, Conn.: Greenwood Press.

———. 1983. *Index to International Public Opinion, 1981–1982*. Westport, Conn.: Greenwood Press.

Heldring, J. L. 1975. "The Impact of the Energy Crisis on Dutch Policy and Politics." Paper prepared for the Conference on the Middle East and the Crisis in Relations Among Industrial States, Atlantic Institute for International Affairs, Paris, May 15–16.

———. 1978. "Between Dreams and Reality." In *The Foreign Policy of the Netherlands*, edited by J. H. Leurdijk, 307–327. Alphen aan den Rijn, Netherlands: Sijthoff & Noordhoff.

Hermann, Charles F. 1972. *International Crisis: Insights from Behavioral Research*. New York: Free Press.

Hersh, Seymour M. 1983. *The Price of Power: Kissinger in the Nixon White House*. New York: Summit Books.

Hill, Christopher, and James Mayall. 1983. "The Sanctions Problem: International and European Perspectives." EUI working paper no. 59. Florence: European University Institute.

Hirasawa, Kazushige. 1976. "Japan's Tilting Neutrality." In *Oil, the Arab-Israeli Dispute, and the Industrial World*, edited by J. C. Hurewitz, 138–146. Boulder, Colo.: Westview Press.

Hoffmann, Frederick. 1967. "The Functions of Economic Sanctions: A Comparative Analysis." *Journal of Peace Research* 4 : 140–160.

Hollis, Rosemary. 1987. "Great Britain." In *The Powers in the Middle East: The Ultimate Strategic Arena*, edited by Bernard Reich, 179–225. New York: Praeger.

Holsti, K. J. 1982. *Why Nations Realign: Foreign Policy Restructuring in the Postwar World*. London: George Allen & Unwin.

Holsti, Ole, Randolph Siverson, and Alexander George. 1980. *Change in the International System*. Boulder, Colo.: Westview Press.

Home, Lord. 1976. *The Way the Wind Blows*. London: William Collins Sons.

Horvat, Andrew. 1973. "Lessons of War." *Mainichi Daily News*, November 8.

Howe, Russell Warren, and Sarah Hays Trott. 1977. *The Power Peddlers: How Lobbyists Mold America's Foreign Policy*. Garden City, N.Y.: Doubleday.

Hudson, Michael C. 1980. "The Media and the Arabs: Room For Improvement." In *The American Media and the Arabs*, edited by Michael C. Hudson and Ronald G. Wolfe, 91–103. Washington, D.C.: Center for Contemporary Arab Studies, Georgetown University.

Hudson, Valerie M. 1983. "The External Predisposition Component of a Model of Foreign Policy Behavior." Ph.D. diss., Department of Political Science, Ohio State University; #8403532, University Microfilms, Ann Arbor, Mich.

Hufbauer, Gary Clyde, and Jeffrey J. Schott. 1983. "Economic Sanctions in Support of Foreign Policy Goals." Policy Analyses in International Economics 6. Washington, D.C.: Institute for International Economics.

———. 1985. *Economic Sanctions Reconsidered: History and Current Policy*. Washington, D.C.: Institute for International Economics.

Ikeda, Akifumi. 1986. "Japan's Perception of Jews and Israel." *Forum on the Jewish People, Zionism and Israel* 59 (Spring): 73–83.

Insight Team of the Sunday Times. 1974. *Insight on the Middle East War*. London: André Deutsch.

International Institute for Strategic Studies. 1974. *Strategic Survey, 1973*. London: International Institute for Strategic Studies.

Ismael, Tareq Y. 1976. "Canada and the Middle East." In *Canada and the Third World*, edited by Peyton V. Lyon and Tareq Y. Ismael, 240–276. Toronto: Macmillan.

———. 1986. *International Relations of the Contemporary Middle East: A Study in World Politics*. Syracuse: Syracuse University Press.

Itayim, Fuad. 1974. "Strengths and Weaknesses of the Oil Weapon." In *The Middle East and the International System*, vol. 2: *Security and the Energy Crisis*, 1–7. Adelphi Paper no. 115. London: International Institute for Strategic Studies.

Jenkins, Peter. 1970. "Alec in Earnest." *Guardian*, November 3.

Johnson, Chalmers. 1978. "Japan's Public Policy Companies." AEI-Hoover Policy Studies 24. Washington, D.C.: American Enterprise Institute for Public Policy.

(Copublished as Hoover Institution Studies 60. Stanford: Hoover Institution on War, Revolution, and Peace.)

———. 1982. *MITI and the Japanese Miracle: The Growth of Industrial Policy, 1925– 1975*. Stanford: Stanford University Press.

Joyner, Christopher C. 1984. "The Transnational Boycott as Economic Coercion in International Law: Policy, Place, and Practice." *Vanderbilt Journal of Transnational Law* 17 (Spring): 206–286.

Joyner, Nelson T. 1976. "Arab Boycott/Anti-Boycott: The Effect on U.S. Business." 2d ed. McLean, Va.: Rockville Consulting Group. (Report available at Johns Hopkins University library.)

Juster, Kenneth. 1976. "Japanese Foreign Policy Making During the Oil Crisis." B.A. thesis, Department of Government, Harvard University. (Thesis available at Harvard University library, but it cannot be borrowed. Juster can be contacted at Arnold and Porter, Washington, D.C.)

———. 1977. "Foreign Policy-Making During the Oil Crisis." *Japan Interpreter* 11 (Winter): 293–312.

Kaplan, Stephen S. 1981. *Diplomacy of Power: Soviet Armed Forces as a Political Instrument*. Washington, D.C.: Brookings Institution.

Karsten, Peter, Peter D. Howell, and Artis Frances Allen. 1984. *Military Threats: A Systematic Historical Analysis of the Determinants of Success*. Westport, Conn.: Greenwood Press.

Kassis, Jihad Georges. 1981. "U.S. Foreign Policy-Making in Middle East Crises: The 1973–74 Oil Embargo." Ph.D. diss., University of Notre Dame; #8118574, University Microfilms International, Ann Arbor, Mich.

Kay, Zachariah. 1978. *Canada and Palestine: The Politics of Non-Commitment*. Jerusalem: Israel Universities Press.

Kelly, J. B. 1976. "Saudi Arabia and the Gulf States." In *The Middle East: Oil, Conflict & Hope*, edited by A. L. Udovitch, 427–462. Critical Choices for Americans, vol. 10. Lexington, Mass.: Lexington Books.

———. 1980a. *Arabia, the Gulf and the West*. New York: Basic Books.

———. 1980b. "Saudi Censors." *New Republic* 182 (May 17): 14–16.

Kemezis, Paul, and Ernest J. Wilson III. 1984. *The Decade of Energy Policy: Policy Analysis in Oil-Importing Countries*. New York: Praeger.

Keohane, Robert O., and Joseph S. Nye. 1977. *Power and Interdependence: World Politics in Transition*. Boston: Little, Brown.

Kirton, John J. 1978. "Foreign Policy Decision-Making in the Trudeau Government: Promise and Performance." *International Journal* 33 (Spring): 287–311.

Kissinger, Henry A. 1982. *Years of Upheaval*. Boston: Little, Brown.

Korany, Baghat. 1986. *How Foreign Policy Decisions Are Made in the Third World: A Comparative Analysis*. Boulder, Colo.: Westview Press.

Knorr, Klaus. 1975. *The Power of Nations: The Political Economy of International Relations*. New York: Basic Books.

———. 1976. "The Limits of Economic and Military Power." In *The Oil Crisis*, edited by Raymond A. Vernon, 229–244. New York: W. W. Norton.

———. 1984. "Economic Relations as an Instrument of National Power." In *Strategic Dimensions of Economic Behavior,* edited by Gordon H. McCormick and Richard E. Bissell, 183–207. New York: Praeger.

Kohler, Beate. 1982. "Euro-American Relations and European Political Cooperation." In *European Political Cooperation: Towards a Foreign Policy for Western Europe,* edited by David Allen, Reinhardt Rummel, and Wolfgang Wessels, 83–94. London: Butterworth Scientific.

Kubursi, Atif. 1985. "Canada's Economic Relations with the Arab World: Patterns and Prospects." In *Canada and the Arab World,* edited by Tareq Ismael, 45–68. Edmonton: University of Alberta Press.

Kuroda, Yasumasa. 1983. "Japan's Middle East Policy: Changing Relationships Between Japan and West Asia." Paper presented to the International Studies Association—West, Berkeley, California, March 25–27.

———. 1984. "Japan and the Arabs: The Economic Dimension." *Journal of Arab Affairs* 3 (1): 1–17.

———. 1985. "Japanese Perceptions of the Arab World: Their Nature and Scope." In *Japan and the Middle East in Alliance Politics,* edited by Ronald Morse, 25–40. Asia Program/International Security Studies Program. Washington, D.C.: Wilson Center.

Laqueur, Walter. 1974. *Confrontation: The Middle East and World Politics.* New York: Quadrangle/New York Times Book Company.

Lasswell, Harold D., and Abraham Kaplan. 1950. *Power and Society: A Framework for Political Inquiry.* New Haven: Yale University Press.

Lauren, Paul Gordon. 1979. "Theories of Bargaining with Threats of Force." In *Diplomacy: New Approaches in Theory, History and Policy,* edited by Paul Lauren, 183–211. New York: Free Press.

Lebow, Richard Ned. 1981. *Between Peace and War: The Nature of International Crisis.* Baltimore: Johns Hopkins University Press.

Lenczowski, George. 1976a. "Middle East Oil in a Revolutionary Age." National Energy Study no. 10. Washington, D.C.: American Enterprise Institute for Public Policy Research.

———. 1976b. "The Oil-Producing Countries." In *The Oil Crisis,* edited by Raymond A. Vernon, 59–72. New York: W. W. Norton.

Leyton-Brown, David. 1981. "External Affairs and Defense." In *Canadian Annual Review of Politics and Public Affairs, 1979,* edited by R. B. Byers, 183–265. Toronto: University of Toronto Press.

Licklider, Roy. 1982a. "The Arab Oil Embargo as Economic Coercive Bargaining." Paper prepared for the International Studies Association, Cincinnati, Ohio, March.

———. 1982b. "The Failure of the Arab Oil Weapon in 1973–1974." *Comparative Strategy* 3 (4): 365–380.

———. 1985. "Arab Oil and Japanese Foreign Policy." *Middle East Review* 18 (Fall): 23–29.

————. 1987. "The Arab Oil Weapon of 1973–74." In *The Utility of International Economic Sanctions,* edited by David Leyton-Brown, 165–181. New York: St. Martin's Press.

————. 1988. "The Power of Oil: The Netherlands, the United Kingdom, Canada, Japan and the United States." *International Studies Quarterly* 32 (June).

Lieber, Robert J. 1976. "Oil and the Middle East War: Europe in the Energy Crisis." Harvard Studies in International Affairs no. 35. Cambridge: Center for International Affairs, Harvard University.

————. 1982. "Cohesion and Disruption in the Western Alliance." In *Global Insecurity: A Strategy for Energy and Economic Renewal,* edited by Daniel Yergin and Martin Hillenbrand, 320–348. Boston: Houghton Mifflin.

————. 1986. *The Oil Decade: Conflict and Cooperation in the West.* Lanham, Md.: University Press of America.

Lijphart, Arend. 1975. *The Politics of Accommodation: Pluralism and Democracy in the Netherlands.* Berkeley and Los Angeles: University of California Press.

Lindemann, Beate. 1982. "European Political Cooperation at the UN: A Challenge for the Nine." In *European Political Cooperation: Towards a Foreign Policy for Western Europe,* edited by David Allen, Reinhardt Rummel, and Wolfgang Wessels, 110–133. London: Butterworth Scientific.

Lindsay, James M. 1986. "Trade Sanctions as Policy Instruments: A Re-Examination." *International Studies Quarterly* 30 (June): 153–174.

Lipset, Seymour Martin, and William Schneider. 1977. "Carter vs. Israel: What the Polls Reveal." *Commentary* 64 (November): 21–29.

Losman, Donald K. 1979. *International Economic Sanctions.* Albuquerque: University of New Mexico Press.

Lucas, Noah. 1981. "Congress and Foreign Policy: Ritual and Substance—The Case of Arms Sales to the Middle East, May 1978." In *Congress and American Foreign Policy,* edited by Goran Rystad, 79–92. Lund Studies in International Studies 13, Scandinavian University Books. Lund, Sweden: Esselte Studium.

Luttwak, Edward N. 1974. *The Political Uses of Sea Power.* Baltimore: Johns Hopkins University Press.

————. 1976. *The Grand Strategy of the Roman Empire.* Baltimore: Johns Hopkins University Press.

————. 1979. "Sea Power in the Mediterranean: Political Utility and Military Constraints." *Washington Papers* 61. Beverly Hills, Calif.: Sage.

Lyon, Peyton V. 1982. "Canada's Middle East Tilt." *International Perspectives* (September–October): 3–5.

————. 1985. "Canada's National Interest and the Middle East." In *Canada and the Arab World,* edited by Tareq Ismael, 26–36. Edmonton: University of Alberta Press.

Maachou, Abdelkader. 1983. *OAPEC: An International Organization for Economic Cooperation and an Instrument for Regional Integration.* Translated by Antony Melville. New York: St. Martin's Press.

Mack, Andrew J. R. 1975. "Why Big Nations Lose Small Wars: The Politics of Asymmetric Conflict." *World Politics* 27 (January): 175–200.

Maclennan, R. A. R. 1974. "Secrecy and the Right of Parliamentarians to Know and Participate in Foreign Affairs." In *Secrecy and Foreign Policy,* edited by Thomas M. Franck and Edward Weisband, 132–143. New York: Oxford University Press.

Macrae, Norman. 1983. "British Media and Energy Policy." In *Energy Coverage—Media Panic,* edited by Nelson Smith and Leonard J. Theberge, 123–137. New York: Longman.

Maghroori, Ray, and Stephen M. Gorman. 1981. *The Yom Kippur War: A Case Study in Crisis Decision-Making in American Foreign Policy.* Washington, D.C.: University Press of America.

Mandel, Robert. 1985. "The Effectiveness of Gunboat Diplomacy." Paper presented at the International Studies Association, Washington, D.C. (Paper available through Department of Political Science, Lewis & Clark College, Portland, Oregon.)

Mangold, Peter. 1978. *Superpower Intervention in the Middle East.* London: Croom Helm.

Manoharan, S. 1974. *The Oil Crisis: End of an Era.* New Delhi: S. Chand.

Mansbach, Richard W., and John A. Vasquez. 1981. *In Search of Theory: A New Paradigm for Global Politics.* New York: Columbia University Press.

Maull, Hanns. 1975. "Oil and Influence: The Oil Weapon Examined." Adelphi Paper no. 117. London: International Institute for Strategic Studies.

———. 1976. "The Strategy of Avoidance: Europe's Middle East Policies After the October War." In *Oil, the Arab-Israeli Dispute, and the Industrial World,* edited by J. C. Hurewitz, 110–137. Boulder, Colo.: Westview Press.

May, Ernest R. 1973. *"Lessons" of the Past: The Use and Misuse of History in American Foreign Policy.* New York: Oxford University Press.

McCormick, Gordon H., and Richard E. Bissell, eds. 1984. *Strategic Dimensions of Economic Behavior.* New York: Praeger.

Mearsheimer, John J. 1983. *Conventional Deterrence.* Ithaca: Cornell University Press.

Meir, Shemuel. 1986. "Strategic Implications of the New Oil Reality." Paper no. 4, Jaffee Center for Strategic Studies, Tel Aviv University. Boulder, Colo.: Westview Press.

Miller, Linda B. 1974. "The Limits of Alliance: America, Europe, and The Middle East." Jerusalem Papers on Peace Problems 10, Hebrew University of Jerusalem. Jerusalem: Leonard Davis Institute for International Relations.

Milward, Allan S. 1984. "Restriction of Supply as a Strategic Choice." In *Strategic Dimensions of Economic Behavior,* edited by Gordon H. McCormick and Richard E. Bissell, 44–58. New York: Praeger.

Minerbi, Sergio I. 1974. "Israel and the Enlarged European Economic Community." *International Problems* 13 (January): 383–399.

Miyoshi, Shuich. 1974. "Oil Shock." *Japan Quarterly* 21 (April–June): 143–151.

Moïsi, Dominique. 1984. "Tensions Within the West: The Middle East." In *NATO—The Next Generation,* edited by Robert E. Hunter, 217–226. Boulder, Colo.: Westview Press.

Momoi, Makoto. 1974. "The Energy Problem and Alliance Systems: Japan." In *The Middle East and the International System,* vol. 2, *Security and the Energy Crisis,* 25–31. Adelphi Paper no. 115. London: International Institute for Strategic Studies.

Monroe, Elizabeth. 1981. *Britain's Moment in the Middle East, 1914–1971.* Baltimore: Johns Hopkins University Press.

Moore, Raymond A. 1984. "The Carter Presidency and Foreign Policy." In *The Carter Years: The President and Policy Making,* edited by M. Glenn Abernathy, Dilys M. Hill, and Phil Williams, 54–83. New York: St. Martin's Press.

Morehouse, Geoffrey. 1977. *The Diplomats: The Foreign Office Today.* London: Jonathan Cape.

Morris, Roger. 1977. *Uncertain Greatness: Henry Kissinger and American Foreign Policy.* New York: Harper & Row.

Morse, Ronald A. 1981. "The Politics of Japan's Energy Strategy: Resources—Diplomacy—Security." Research Papers and Policy Studies 3. Berkeley: Institute of East Asian Studies, University of California.

———. 1982. "Japanese Energy Policy." In *After the Second Oil Crisis: Energy Policies in Europe, America and Japan,* edited by Wilfrid L. Kohl, 255–270. Lexington, Mass.: Lexington Books.

Mueller, Peter G., and Douglas A. Ross. 1975. *China and Japan: Emerging Global Powers.* New York: Praeger.

Murakami, Teruyasu. 1982. "The Remarkable Adaptation of Japan's Economy." In *Global Insecurity: A Strategy for Energy and Economic Renewal,* edited by Daniel Yergin and Martin Hillenbrand, 138–167. Boston: Houghton Mifflin.

Naff, Thomas. 1981. *The Middle East Challenge, 1980–1985.* Middle East Research Institute, University of Pennsylvania. Carbondale and Edwardsville: Southern Illinois University Press.

Nau, Henry R. 1974–1975. "U.S. Foreign Policy in the Energy Crisis." *Atlantic Community Quarterly* 12 (Winter): 426–439.

———. 1977. "The International Political Economy of Food and Energy." *Journal of International Affairs* 31 (Fall–Winter): 327–339.

———. 1980a. "The Evolution of U.S. Foreign Policy in Energy: From Alliance Politics to Politics-as-Usual." In *International Energy Policy,* edited by Robert M. Lawrence and Martin O. Heisler, 37–64. Lexington, Mass.: Lexington Books.

———. 1980b. "Japanese-American Relations During the 1973–74 Oil Crisis." In *Oil and the Atom: Issues in US-Japan Energy Relations,* edited by Michael Blaker, 1–36. Occasional Papers Series, Project on Japan and the United States in Multilateral Diplomacy. New York: East Asian Institute, Columbia University.

Nelson, Walter Henry. 1978. "The 'Byers Bill' Is Needed." *Britain and Israel* (126–134 Baker Street, London) 78 (January).

Nelson, Walter Henry, and Terence C. F. Prittie. 1977. *The Economic War Against the Jews.* New York: Random House.

Neustadt, Richard E. 1966. "White House and Whitehall." *Public Interest* 2 (Winter): 56–69.

Nincic, Miroslav, and Peter Wallensteen. 1983. *Dilemmas of Economic Coercion: Sanctions in World Politics.* New York: Praeger.

Nish, Ian. 1983. "Japan: The Foreign Ministry." In *The Times Survey of Foreign Ministries of the World,* edited by Zara Steiner, 327–344. London: Times Books.

Nixon, Richard. 1978. *RN: The Memoirs of Richard Nixon.* New York: Grosset & Dunlap.

Noble, Paul C. 1985a. "From Refugees to a People? Canada and the Palestinians, 1967–1973." In *Canada and the Arab World,* edited by Tareq Ismael, 85–106. Edmonton: University of Alberta Press.

———. 1985b. "Where Angels Fear to Tread: Canada and the Status of the Palestinian People, 1967–1982." In *Canada and the Arab World,* edited by Tareq Ismael, 107–149. Edmonton: University of Alberta Press.

Nossal, Kim Richard. 1984. "Bureaucratic Politics and the Westminister Model." In *International Conflict and Conflict Management: Readings in World Politics,* edited by Robert O. Matthews, Arthur G. Rubinoff, and Janice Gross Stein, 120–127. Scarborough: Prentice-Hall of Canada.

Novik, Nimrod. 1986. *The United States and Israel: Domestic Determinants of a Changing U.S. Commitment.* Boulder, Colo.: Westview Press.

Ohira, Masayoshi. 1979. *Brush Strokes: Moments from My Life.* Tokyo: Foreign Press Center.

Ojimi, Yoshihisa. 1975. "A Government Ministry: The Case of the Ministry of International Trade and Industry." In *Modern Japanese Organization and Decision-Making,* edited by Ezra F. Vogel, 101–112. Berkeley and Los Angeles: University of California Press.

Okita, Saburo. 1974. "Natural Resource Diplomacy and Japanese Foreign Policy." *Foreign Affairs* 52 (Summer): 714–724.

Olson, Richard Stuart. 1979. "Economic Coercion in World Politics: With a Focus on North-South Relations." *World Politics* 31 (July): 471–494.

Oppenheim, A. N., and Ian Smart. 1973. "The British Diplomat." In *The Management of Britain's External Relations,* edited by Robert Boardman and A. J. R. Groom, 75–116. London: Macmillan.

Osbaldeston, Gordon. 1982. "Reorganizing Canada's Department of External Affairs." *International Journal* 37 (Summer): 453–466.

Paarlberg, Robert L. 1985. *Food Trade and Foreign Policy: India, the Soviet Union, and the United States.* Ithaca: Cornell University Press.

Pajak, Roger F. 1979. "West European and Soviet Arms Transfer Policies in the Middle East." In *Great Power Intervention in the Middle East,* edited by Milton Leitenberg and Gabriel Sheffer, 134–164. New York: Pergamon Press.

Pearson, Frederic S. 1979. "Netherlands Foreign Policy and the 1973–74 Oil Embargo—The Effects of Transnationalism." In *Transnationalism in World Politics and Business,* edited by Forest L. Grieves, 114–140. New York: Pergamon Press.

Peck, Malcolm C. 1980. "The Saudi-American Relationship and King Faisal." In *King Faisal and the Modernization of Saudi Arabia,* edited by Willard A. Beling, 230–247. Boulder, Colo.: Westview Press.

Pfaltzgraff, Robert L., Jr. 1978. "Resource Constraints and Arms Transfers: Implications for NATO and European Security." In *Arms Transfers to the Third World: The Military Buildup in Less Industrial Countries,* edited by Uri Ra'anan, Robert L. Pfaltzgraff, Jr., and Geoffrey Kemp, 159–212. Boulder, Colo.: Westview Press.

———. 1980. "Energy Issues and Alliance Relationships: The United States, Western Europe and Japan." Special Report. Cambridge, Mass.: Institute for Foreign Policy Analysis.

Phillips, Charlotte A. 1979. "The Arab Boycott of Israel: Possibilities for European Cooperation with U.S. Antiboycott Legislation." Document 79–215F. Washington, D.C.: Congressional Research Service. (1980. *Major Studies and Issue Briefs of the Congressional Research Service: 1979–80 Supplement.* Reel #6. Washington, D.C.: University Publications of America. Microfilm.)

Pijpers, Alfred. 1983. "The Netherlands: How to Keep the Spirit of Fouchet in the Bottle." In *National Foreign Policies and European Political Cooperation,* edited by Christopher Hill, 166–182. London: George Allen & Unwin.

Pollock, David. 1982. *The Politics of Pressure: American Arms and Israeli Policy Since the Six-Day War.* Westport, Conn.: Greenwood Press.

Prittie, Terence. 1980. "The British Media and the Arab-Israeli Dispute." *Middle East Review* 12 (4): 67–71.

———. 1982a. "Britain and the Arab Boycott-1." *Britain and Israel* (126–134 Baker Street, London) 129 (June/July).

———. 1982b. "Britain and the Arab Boycott-2." *Britain and Israel* (126–134 Baker Street, London) 130 (June/July).

———. 1983. "Britain and the Boycott." *Jewish Chronicle,* July 8.

Prittie, Terence, and R. I. Jones. 1970. "Britain and Israel" and "Future Frontiers." *Britain and Israel* (126–134 Baker Street, London) 1 (November).

Prittie, Terence, and Walter Henry Nelson. 1978. "The Call For Government Action." *Britain and Israel* (126–134 Baker Street, London) 85 (September).

Quandt, William B. 1973. "Domestic Influences on U.S. Foreign Policy in the Middle East: The View from Washington." In *The Middle East: Quest for an American Policy,* edited by Willard A. Beling, 263–285. Albany: State University of New York Press.

———. 1975. "Kissinger and the Arab-Israeli Disengagement Negotiations." *Journal of International Affairs* 29 (Spring): 33–48.

———. 1976a. "Soviet Policy in the October 1973 War." R-1864-ISA. Santa Monica, Calif.: Rand Corp.

———. 1976b. "U.S. Energy Policy and the Arab-Israeli Conflict." In *Arab Oil: Impact on Arab Countries and Global Implications,* edited by Naiem A. Sherbiny and Mark A. Tessler, 279–294. New York: Praeger.

———. 1977. *Decade of Decisions: American Policy Toward the Arab-Israeli Conflict, 1967–1976.* Berkeley and Los Angeles: University of California Press.

———. 1978. "Influence Through Arms Supply: The U.S. Experience in the Middle East." In *Arms Transfers to the Third World: The Military Buildup in Less Industrial Countries,* edited by Uri Ra'anan, Robert L. Pfaltzgraff, Jr., and Geoffrey Kemp, 121–130. Boulder, Colo.: Westview Press.

———. 1980. "The Middle East Crises." *Foreign Affairs* 58 (3): 540–562.

———. 1982a. "Saudi Arabia's Oil Policy." Staff Paper. Washington, D.C.: Brookings Institution.

———. 1982b. "The Western Alliance in the Middle East: Problems for US Foreign Policy." In *The Middle East and the Western Alliance,* edited by Steven L. Spiegel, 9–17. London: George Allen & Unwin.

———. 1986. *Camp David: Peacemaking and Politics.* Washington, D.C.: Brookings Institution.

Raab, Earl. 1974. "Is Israel Losing Popular Support? The Evidence of the Polls." *Commentary* 57 (January): 26–29.

Ramazani, R. K. 1977. "Beyond the Arab-Israeli Settlement: New Directions for U.S. Policy in the Middle East." Cambridge, Mass.: Institute for Foreign Policy Analysis.

———. 1978. "The European Community and the Middle East." In *Middle East Contemporary Survey, Volume I, 1976–77,* edited by Colin Legum, 49–57. London: Holmes and Meier.

Rand, Christopher. 1975. *Making Democracy Safe for Oil.* Boston: Little, Brown.

Reich, Bernard. 1977. *Quest for Peace: United States–Israel Relations and the Arab-Israeli Conflict.* New Brunswick, N.J.: Transaction Books.

Reich, Bernard, and Patrick Coquillon. 1987. "Europe." In *The Powers in the Middle East: The Ultimate Strategic Arena,* edited by Bernard Reich, 151–178. New York: Praeger.

Renwick, Robin. 1981. "Economic Sanctions." Harvard Studies in International Affairs 45. Cambridge: Center for International Affairs, Harvard University.

Richards, Peter G. 1973. "Parliament and the Parties." In *The Management of Britain's External Relations,* edited by Robert Boardman and A. J. R. Groom, 245–262. London: Macmillan.

Roehm, John Francis, Jr. 1980. "Congressional Participation in Foreign Policy: A Study of Congress' Role in U.S. Middle East Policy Vis-à-vis the Confrontation States in the Arab-Israeli Conflict from the Yom Kippur War to the End of the Ford Administration." Ph.D. diss., University of Pittsburgh; #8028051, University Microfilms, Ann Arbor, Mich.

———. 1981. "Congressional Participation in U.S.-Middle East Policy, October 1973–1976: Congressional Activism vs. Policy Coherence." In *Congress, the Presidency and American Foreign Policy,* edited by John Spanier and Joseph Nogee, 22–43. New York: Pergamon Press.

Rosenau, James N. 1980. *The Scientific Study of Foreign Policy.* Rev. and enl. ed. New York: Nichols.

Rothstein, Robert L. 1976. "Foreign Policy and Development Policy: From Non-alignment to International Class War." *International Affairs* 52 : 598–616.

Royal Institute of International Affairs. 1938. *International Sanctions.* London: Oxford University Press.

———. 1956. *The War and the Neutrals.* Survey of International Affairs, vol. 9. London: Oxford University Press.

Rubin, Barry. 1983. "The Reagan Administration and the Middle East." In *Eagle Defiant: United States Foreign Policy in the 1980s,* edited by Kenneth A. Oye, Robert J. Lieber, and Donald Rothchild, 367–389. Boston: Little, Brown.

———. 1984. "U.S. Policy on the Middle East in the Period Since Camp David." In *The Middle East Since Camp David,* edited by Robert O. Freedman, 59–76. Boulder, Colo.: Westview Press.

Rustow, Dankwart A. 1975. "Petroleum Politics, 1951–1974." *Dissent* 21 (Spring): 144–153.

———. 1977. "U.S.-Saudi Relations and the Oil Crises of the 1980s." *Foreign Affairs* 55 (Spring): 494–516.

———. 1982. *Oil and Turmoil: America Faces OPEC and the Middle East.* New York: W. W. Norton.

Saeki, Kiichi. 1976. "Japan's Energy-Security Dilemma." In *Oil, the Arab-Israeli Dispute, and the Industrial World,* edited by J. C. Hurewitz, 258–265. Boulder, Colo.: Westview Press.

Safran, Nadav. 1978. *Israel: The Embattled Ally.* Cambridge: Belknap Press of Harvard University Press.

Sahr, Naomi. 1974. "Arab Embargo Checks Expansion in Dutch Trade." *Middle East Economic Digest* 18 (November 6): 1334–1336.

Said, Edward S. 1981. *Covering Islam: How the Media and the Experts Determine How We See the Rest of the World.* New York: Pantheon.

Sampson, Anthony. 1974a. "The Oil War." *Observer,* November 17.

———. 1974b. "Oil War: The Arabs' Grand Slam." *Observer,* November 24.

———. 1979. *The Seven Sisters: The Great Oil Companies and the World They Shaped.* New York: Bantam Books.

Sankari, Farouk A. 1976. "The Character and Impact of Arab Oil Embargoes." In *Arab Oil: Impact on Arab Countries and Global Implications,* edited by Naiem A. Sherbiny and Mark A. Tessler, 265–278. New York: Praeger.

Sasagawa, Masahiro. 1982. "Japan and the Middle East." In *The Middle East and the Western Alliance,* edited by Steven L. Spiegel, 33–46. London: George Allen & Unwin.

Saywell, John. 1976. *Canadian Annual Review of Politics and Public Affairs, 1975.* Toronto: University of Toronto Press.

Scheinman, Lawrence. 1976. "US International Leadership." In *Oil, the Arab-Israeli Dispute, and the Industrial World,* edited by J. C. Hurewitz, 9–20. Boulder, Colo.: Westview Press.

Schelling, Thomas C. 1966. *Arms and Influence.* New Haven: Yale University Press.

Schmidt, Dana Adams. 1974. *Armageddon in the Middle East.* New York: John Day.

Schneider, Steven A. 1983. *The Oil Price Revolution.* Baltimore: Johns Hopkins University Press.

Scolnick, Joseph M., Jr. 1985. "The Value of Foreign Threats for the Governments of Threatened States." Paper prepared for the International Studies Association, Washington, D.C. (Paper available through Department of Political Science, Clinch Valley College, University of Virginia.)

Seabury, Paul. 1974. "Thinking About an Oil War." *New Leader* 57 (November 11): 5–8.

Shadid, Mohammed K. 1981. *The United States and the Palestinians.* London: Croom Helm.

Shamma, Amal Mustafa. 1980. "The 1973 Oil Embargo: Arab Oil Diplomacy." M.A. thesis, Department of Political Science, Western Michigan University.

Sheehan, Edward F. 1976. *The Arabs, Israelis, and Kissinger: A Secret History of American Diplomacy in the Middle East.* New York: Thomas Y. Crowell.

Shillony, Ben-Ami. 1985. "Japan and Israel: A Special Relationship." In *Japan and the Middle East in Alliance Politics,* edited by Ronald Morse, 69–79. Asia Program/International Security Studies Program. Washington, D.C.: Wilson Center.

———. 1985–1986. "Japan and Israel: The Relationship that Withstood Pressures." *Middle East Review* 18 (Winter): 17–24.

Shlaim, Avi, Peter Jones, and Keith Sainsbury. 1977. *British Foreign Secretaries Since 1945.* London and North Pomfret, Vt.: David and Charles.

Shlaim, Avi, and Raymond Tanter. 1978. "Decision Process, Choice, and Consequences: Israel's Deep-Penetration Bombing in Egypt, 1970." *World Politics* 30 (July): 483–516.

Shonfeld, Andrew. 1973. "Dutch Oil Ban a Test of Faith for EEC." *The Times,* November 11, 18.

Shubik, Martin, and Paul Bracken. 1984. "Strategic Purpose and the International Economy." In *Strategic Dimensions of Economic Behavior,* edited by Gordon H. McCormick and Richard E. Bissell, 208–228. New York: Praeger.

Shwadran, Benjamin. 1977. *Middle East Oil: Issues and Problems.* Cambridge, Mass.: Schenkman.

Sicherman, Harvey. 1976. "The Yom Kippur War: End of Illusion?" *The Foreign Policy Papers* 1 (4): 1–91.

———. 1978. "Broker Or Advocate? The U.S. Role in the Arab-Israeli Dispute, 1973–1978." Monograph no. 25. Philadelphia: Foreign Policy Research Institute.

———. 1980. "Politics of Dependence: Western Europe and the Arab-Israeli Conflict." *Orbis* 23 (Winter): 845–857.

Singer, S. Fred. 1978. "Limits to Arab Oil Power." *Foreign Policy* 30 (Spring): 53–67.

Smart, Ian. 1976. "Uniqueness and Generality." In *The Oil Crisis,* edited by Raymond A. Vernon, 259–281. New York: W. W. Norton.

———. 1977. "Oil, the Super-Powers, and the Middle East." *International Affairs* 53 (January): 17–35.

Smith, Geoffrey, and Nelson W. Polsby. 1981. *British Government and Its Discontents.* New York: Basic Books.

Smith, Terence. 1983. "The U.S. Press (I)." In *The Middle East: Press Perspectives and National Policies,* edited by Luci E. Richardson, 18–23. Proceedings of a seminar cosponsored by the Center for Middle East Policy and the Royal Institute of International Affairs. Washington, D.C.: Center for Middle East Policy. (Publication available at Georgetown University Library, Washington, D.C.)

Snyder, Glen H., and Paul Diesing. 1977. *Conflict Among Nations: Bargaining, Decision-Making, and System Structure in International Crises.* Princeton: Princeton University Press.

Soetendorp, R. B. 1982. *Het Nederlandse beleid ten aanzien van het Arabisch-Israelisch conflict, 1947–1977.* Groningen: Krips Repro Meppel.

———. 1983. *Pragmatisch of Principieel. Het Nederlandse beleid ten aanzien van het Arabisch–Israelisch conflict.* Leiden: Martinus Nijhoff.

———. 1984. "National Approaches to the Arab-Israeli Conflict: The Netherlands." In *European Foreign Policy-Making and the Arab-Israeli Conflict,* edited by David Allen and Alfred Pijpers, 37–46. The Hague: Martinus Nijhoff.

Solomon, Harold. 1978. "Japan—Pro-Arab?" *Midstream* (January): 14–26.

Sorley, Lewis. 1983. *Arms Transfers Under Nixon: A Policy Analysis.* Lexington: University Press of Kentucky.

Spiegel, Fredelle Z. 1981–1982. "The Arab Lobby." *Middle East Review* 14 (Fall–Winter): 69–73.

Spiegel, Steven L. 1985. *The Other Arab-Israeli Conflict: Making America's Middle East Policy, from Truman to Reagan.* Chicago: University of Chicago Press.

Stanislawski, Howard Jerry. 1981. "Elites, Domestic Interest Groups, and International Interests in the Canadian Foreign Policy Decision-Making Process: The Arab Economic Boycott of Canadians and Canadian Companies Doing Business With Israel." Ph.D. diss., Department of Politics, Brandeis University; #812689, University Microfilms International, Ann Arbor, Mich.

———. 1983. "Ethnic Interest Group Activity in the Canadian Foreign Policy Making Process: A Case Study of the Arab Boycott." In *The Middle East at the Crossroads,* edited by Janice Gross Stein and David B. Dewitt, 200–218. Oakville, Ont.: Mosaic Press.

———. 1984. "Domestic Interest Groups and Canadian and American Policy: The Case of the Arab Boycott." In *International Conflict and Conflict Manage-*

ment: Readings in World Politics, edited by Robert O. Matthews, Arthur G. Rubinoff, and Janice Gross Stein, 137–147. Scarborough: Prentice-Hall of Canada.

———. 1987. "Impact of the Arab Boycott of Israel on the United States and Canada." In *The Utility of International Economic Sanctions,* edited by David Leyton-Brown, 223–254. New York: St. Martin's Press.

Starr, Harvey. 1984. *Henry Kissinger: Perceptions of International Politics.* Lexington: University Press of Kentucky.

Stein, Janice Gross. 1976a. "Canada: Evenhanded Ambiguity." In *Oil, the Arab-Israeli Dispute, and the Industrial World,* edited by J. C. Hurewitz, 96–109. Boulder, Colo.: Westview Press.

———. 1976b. "Canadian Foreign Policy in the Middle East After the October War." *Social Praxis* 4 (3–4): 271–297.

———. 1982. "Alice in Wonderland: The North Atlantic Alliance and the Arab-Israeli Dispute." In *The Middle East and the Western Alliance,* edited by Steven L. Spiegel, 49–81. London: George Allen & Unwin.

———. 1983. "The Politics of Alliance Policy: Europe, Canada, Japan, and the United States Face the Arab-Israeli Conflict." In *The Middle East at the Crossroads,* edited by Janice Gross Stein and David B. Dewitt, 145–175. Oakville, Ont.: Mosaic Press.

Steinbach, Udo. 1979. "Western European and EEC Policies Towards Mediterranean and Middle Eastern Countries." In *Middle East Contemporary Survey, Volume II, 1977–78,* edited by Colin Legum, 40–48. London: Holmes and Meier.

Steinberg, Blema. 1983. "American Foreign Policy in the Middle East: A Study in Changing Priorities." In *The Middle East at the Crossroads,* edited by Janice Gross Stein and David B. Dewitt, 110–145. Oakville, Ont.: Mosaic Press.

Stobaugh, Robert B. 1976. "The Oil Companies in the Crisis." In *The Oil Crisis,* edited by Raymond A. Vernon, 179–202. New York: W. W. Norton.

Stockholm International Peace Research Institute. 1974. *Oil and Security.* New York: Humanities Press.

Stookey, Robert W. 1975. *America and the Arab States: An Uneasy Encounter.* New York: John Wiley & Sons.

Suleiman, Michael W. 1980. "American Public Support of Middle Eastern Countries: 1939–1979." In *The American Media and the Arabs,* edited by Michael C. Hudson and Ronald G. Wolfe, 13–36. Washington, D.C.: Center for Contemporary Arab Studies, Georgetown University.

Sus, Ibrahim. 1974. "Western Europe and the October War." *Journal of Palestine Studies* 3 (Winter): 65–83.

Sylvan, David J. 1983. "Ideology and the Concept of Economic Security." In *Dilemmas of Economic Coercion: Sanctions in World Politics,* edited by Miroslav Nincic and Peter Wallensteen, 211–241. New York: Praeger.

Szyliowicz, Joseph S. 1975. "The Embargo and U.S. Foreign Policy." In *The Energy Crisis and U.S. Foreign Policy,* edited by Joseph S. Szyliowicz and Bard E. O'Neill, 183–232. New York: Praeger.

Szyliowicz, Joseph S., and Bard E. O'Neill. 1977. "The Oil Weapon and American Foreign Policy." *Air University Review* 28 (March–April): 41–52.

Takach, George. 1980. "Clark and the Jerusalem Embassy Affair: Initiative and Constraint in Canadian Foreign Policy." M.A. thesis, Norman Paterson School of International Affairs, Carleton University, Ottawa.

Tapsell, Peter. 1975. "United Kingdom." Paper prepared for conference on "The Middle East and the Crisis in Relations Among Industrial States," Atlantic Institute for International Affairs, Paris, May 15–16.

Taras, David. 1983. "Parliament and the Arab-Israeli Conflict: Politics and Government During the Yom Kippur War." Department of Political Science, Scarborough College, University of Toronto. Mimeo.

Taylor, Alan R. 1978. "The Euro-Arab Dialogue: Quest for an Interregional Partnership." *Middle East Journal* 32 (Autumn): 429–456.

Taylor, Trevor. 1973. "Britain's Middle Eastern Role: Diminished But Still Vital." *New Middle East* 57 (June): 12–19.

Theberge, Leonard J. 1982. *TV Coverage of the Oil Crises: How Well Was the Public Served?* 3 vols. Washington, D.C.: Media Institute.

Thies, Wallace J. 1980. *When Governments Collide: Coercion and Diplomacy in the Vietnam Conflict, 1964–1968.* Berkeley and Los Angeles: University of California Press.

Thomas, Hugh. 1970. *The Suez Affair.* rev. ed. Harmondsworth: Penguin Books.

Tillman, Seth P. 1982. *The United States in the Middle East: Interests and Obstacles.* Bloomington: Indiana University Press.

Trice, Robert H. 1976a. "American Interest Groups After October 1973." In *Oil, the Arab-Israeli Dispute, and the Industrial World,* edited by J. C. Hurewitz, 79–95. Boulder, Colo.: Westview Press.

———. 1976b. "Interest Groups and the Foreign Policy Process: U.S. Policy in the Middle East." Sage Professional Papers in International Studies, vol. 4, series no. 02–047. Beverly Hills, Calif.: Sage.

———. 1977. "Congress and the Arab-Israeli Conflict: Support for Israel in the U.S. Senate, 1970–1973." *Political Science Quarterly* 92 (Fall): 443–463.

———. 1978. "Foreign Policy Interest Groups, Mass Public Opinion and the Arab-Israeli Dispute." *Western Political Quarterly* 31 (June): 238–252.

———. 1979. "The American Elite Press and the Arab-Israeli Conflict." *Middle East Journal* 33 (Summer): 304–325.

———. 1981. "Domestic Interest Groups and the Arab-Israeli Conflict." In *Ethnicity and U.S. Foreign Policy,* edited by Abdul Aziz Said, 121–142. rev. ed. New York: Praeger.

Tsurumi, Yoshi. 1976. "Japan." In *The Oil Crisis,* edited by Raymond A. Vernon, 113–128. New York: W. W. Norton.

Tucker, Michael. 1980. *Canadian Foreign Policy: Contemporary Issues and Themes.* Toronto: McGraw-Hill Ryerson.

Turner, Louis. 1974. "The Politics of the Energy Crisis." *International Affairs* 50 (July): 404–415.

———. 1978a. "European and Japanese Energy Policies." *Current History* (March): 104–108, 129, 136.

———. 1978b. *Oil Companies in the International System*. Royal Institute of International Affairs. London: George Allen & Unwin.

Twitchett, Kenneth J. 1976. *Europe and the World: The External Relations of the Common Market*. Published for the David Davies Memorial Institute of International Studies. London: Europa.

United Kingdom. 1972–1973. Commons. *Parliamentary Debates*. 5th ser., vols. 861–864. London: Her Majesty's Stationery Office.

———. 1974. Commons. *Parliamentary Debates*. 5th ser., vol. 867. London: Her Majesty's Stationery Office.

———. 1983. Foreign and Commonwealth Office. "Background Brief, Britain and the Arab-Israeli Dispute Since 1967." (January).

U.S. Congress. 1973–1974. House. Committee on Foreign Affairs. Subcommittee on Asian and Pacific Affairs. *Hearings on Oil and Asian Rivals: Sino-Soviet Conflict; Japan and the Oil Crisis*. 93d Cong., 1st and 2d sessions.

———. 1975a. Senate. Committee on Foreign Relations. Subcommittee on Multinational Corporations. *Hearings on Multinational Corporations and United States Foreign Policy*. Pt. 9. 93d Cong., 2d sess.

———. 1975b. Senate. Committee on Foreign Relations. Subcommittee on Multinational Corporations. *U.S. Oil Companies and the Arab Oil Embargo: The International Allocation of Constricted Supplies*. Prepared by the Office of International Energy Affairs, Federal Energy Administration. 94th Cong., 1st sess.

"U.S. Steering Toward Special Relationship with Saudi Arabia." 1973. *Platt's Oilgram News Service* 51 (188): 1, 3.

"Using the Oil Crisis." 1973. *Oriental Economist* 41 (December): 8–13.

Vance, Cyrus. 1983. *Hard Choices: Critical Years in America's Foreign Policy*. New York: Simon and Schuster.

Venn, Fiona. 1986. *Oil Diplomacy in the Twentieth Century*. New York: St. Martin's Press.

Vernon, Raymond A. 1976. *The Oil Crisis*. New York: W. W. Norton.

Vielvoye, Roger. 1973. "Government 'Satisfied' With Oil Distribution." *Times*. November 24, 1.

Vogel, Ezra F. 1975. "Introduction: Toward More Accurate Concepts." In *Modern Japanese Organization and Decision-Making*, edited by Ezra F. Vogel, xiii–xxv. Berkeley and Los Angeles: University of California Press.

Volker, Edmond. 1976. *Euro-Arab Cooperation*. Leyden: A. W. Sijthoff.

von Riekhoff, Harald. 1978. "The Impact of Prime Minister Trudeau on Foreign Policy." *International Journal* 33 (Spring): 267–286.

Voorhoeve, J. J. C. 1979. *Peace, Profits and Principles: A Study of Dutch Foreign Policy*. The Hague: Martinus Nijhoff.

Wagner, Charles H. 1973. "Elite American Newspaper Opinion and the Middle East: Commitment vs. Isolation." In *The Middle-East: Quest for an American*

Policy, edited by Willard A. Beling, 306–334. Albany: State University of New York Press.

Wagner, R. Harrison. 1986. "Economic Dependence, Bargaining Theory, and Economic Sanctions." Paper prepared for the American Political Science Association, Washington, D.C. (Paper available through Department of Government, University of Texas at Austin.)

Wallace, William. 1973. "The Role of Interest Groups." In *The Management of Britain's External Relations,* edited by Robert Boardman and A. J. R. Groom, 263–288. London: Macmillan.

———. 1975. *The Foreign Policy Process in Britain.* London: Royal Institute of International Affairs.

———. 1978. *Foreign Policy Making in Western Europe.* Farnborough: Saxon House.

———. 1982. "National Inputs into European Political Cooperation." In *European Political Cooperation: Towards a Foreign Policy for Western Europe,* edited by David Allen, Reinhardt Rummel, and Wolfgang Wessels, 46–59. London: Butterworth Scientific.

———. 1983. "Introduction: Cooperation and Convergence in European Foreign Policy." In *National Foreign Policies and European Political Cooperation,* edited by Christopher Hill, 1–17. London: George Allen & Unwin.

Wallace, William, and David Allen. 1977. "Political Cooperation: Procedure as Substitute for Policy." In *Policy-Making in the European Communities,* edited by Helen Wallace, William Wallace, and Carole Webb, 227–248. London: John Wiley & Sons.

Wallensteen, Peter. 1968. "Characteristics of Economic Sanctions." *Journal of Peace Research* 5: 248–267.

———, ed. 1969. "International Sanctions: Theory and Practice." Report no. 1. Uppsala: Department of Peace and Conflict Research, Uppsala University.

Watt, David. 1973a. "A Clash of Interests and Sympathies." *Financial Times,* October 19, 23.

———. 1973b. "A Pressure Too Strong To Be Ignored." *Financial Times,* November 9, 23.

Watt, D. Cameron. 1982. "British Diplomacy in the 1980s: From Power to Influence." Speech delivered at Chatham House, Royal Institute of International Affairs, London, March 18. (Transcript in Chatham House library.)

———. 1984. *Succeeding John Bull: America in Britain's Place, 1900–1975.* Cambridge: Cambridge University Press.

Weintraub, Sidney, ed. 1982. *Economic Coercion and U.S. Foreign Policy: Implications of Case Studies from the Johnson Administration.* Boulder, Colo.: Westview Press.

Wels, C. B. 1983. "The Netherlands: The Foreign Policy Institutions in the Dutch Republic and the Kingdom of the Netherlands, 1579 to 1980." In *The Times Survey of Foreign Ministries of the World,* edited by Zara Steiner, 363–389. London: Times Books.

Wessels, Wolfgang. 1982. "European Political Cooperation: A New Approach to

Foreign Policy." In *European Political Cooperation: Towards a Foreign Policy for Western Europe,* edited by David Allen, Reinhardt Rummel, and Wolfgang Wessels, 1–20. London: Butterworth Scientific.

Wheelock, Thomas R. 1978. "Arms for Israel: The Limit of Leverage." *International Security* 3 (Fall): 123–137.

Whetten, Lawrence L. 1976–1977. "The Arab-Israeli Dispute: Great Power Behaviour." Adelphi Paper no. 128. London: International Institute for Strategic Studies.

Wiberg, Hakan. 1969. "Notes on the Efficiency of Economic Sanctions." In *International Sanctions: Theory and Practice,* Report no. 1, edited by Peter Wallensteen, 13–50. Uppsala: Department of Peace and Conflict Research, Uppsala University.

Wildavsky, Aaron. 1975. *Perspectives on the Presidency.* Boston: Little, Brown.

Wilkins, Mira. 1976. "The Oil Companies in Perspective." In *The Oil Crisis,* edited by Raymond A. Vernon, 159–178. New York: W. W. Norton.

Williams, Phil. 1976. *Crisis Management: Confrontation and Diplomacy in the Nuclear Age.* New York: John Wiley & Sons.

Wilson, Harold. 1981. *The Chariot of Israel: Britain, America and the State of Israel.* New York: W. W. Norton.

"World Energy Demands and the Middle East." 1972. 2 parts. Proceedings of the twenty-sixth annual conference of the Middle East Institute, Washington, D.C., September 29–30.

Wu, Yuan-li. 1977. *Japan's Search For Oil: A Case Study on Economic Nationalism and International Security.* Hoover Institution Publication 165. Stanford, Calif.: Hoover Institution Press.

Yanaga, Chitoshi. 1968. *Big Business in Japanese Politics.* New Haven: Yale University Press.

Yasunobe, Hisao. 1983. "The Japanese Press." In *The Middle East: Press Perspectives and National Policies,* edited by Luci E. Richardson, 12–17. Proceedings of a seminar cosponsored by the Center for Middle East Policy and the Royal Institute of International Affairs. Washington, D.C.: Center for Middle East Policy. (Publication available at Georgetown University library, Washington, D.C.)

Yorke, Valerie. 1983. "Oil, the Middle East and Japan's Search for Security." In *Japan's Economic Security,* edited by Nobutoshi Akao, 45–70. New York: St. Martin's Press.

Yoshitsu, Michael M. 1984. *Caught in the Middle East: Japan's Diplomacy in Transition.* Lexington, Mass.: D. C. Heath.

Young, Oran. 1968. *The Politics of Force.* Princeton: Princeton University Press.

Zinnes, Dina A. 1984. "Conflict Processes and the Breakdown of International Systems: Merriam Seminar Series on Research Frontiers," vol. 20, bk. 2. Monograph Series in World Affairs. Denver: Graduate School of International Studies, University of Denver.

INDEX

Aaron, David, 262
Abu Dhabi, embargoes the Netherlands, 34
Action Committee on American Arab Relations (ACAAR), 235–236
Actors, 274–275
Afghanistan, 209
Agnew, Spiro, 231, 256, 257
Akins, James, 194, 238
Albania, invaded by Yugoslavia, 21
Aldington, Lord, 73
Algeria, 275; and Canada, 111, 137; and the Netherlands, 34, 38; and oil production halt, 196; and the United States, 201–202
Ambassadors, Arab: in Japan, 150–151, 173; in the Netherlands, 33
American Arab Affairs Council, 235
American Arab Anti-Discrimination Committee, 235
American Council for Judaism, 240
American Educational Trust, 235
American-Israel Public Affairs Committee (AIPAC), 226–235
American Jewish Alternatives to Zionism, 240
American Jewish Committee, 228
American Jewish Congress, 228
American Petroleum Institute, 236–237
Anglo-Persian Oil Company. See British Petroleum
Anti-Defamation League of B'nai B'rith, 206, 228
Antimonopoly law (Japan), 178
Arab ambassadors. See Ambassadors, Arab
Arab American community, 235–236
Arab-American Oil Company (ARAMCO), 196, 198, 237, 238–239
Arab economic boycott of Israel, 21; and Canada, 88, 117–121, 126, 130,

131–132, 136, 141–142, 143–144, 274, 291; definition of, 14–15; and France, 77, 80; and Great Britain, 75–77, 80, 87–88, 104, 105; and Japan, 170; and the Netherlands, 43–44; and the United States, 88, 206, 227–228, 246–247
Arabian Oil Company (Japan), 150, 167–168, 172–173. See also Mizuno, Sohei
Arab Information Center, 130
Arab-Israeli war (1967): and Canada, 108; and Great Britain, 66; and the Netherlands, 30; and the United States, 191, 224
Arab-Israeli war (1973–1974): and Canada, 141; and Great Britain, 69; and the Netherlands, 32; and the United States, 194–195, 196–200, 213, 224, 268, 286
Arabists: in Great Britain, 95, 97; in Japan, 175; in the United States, 250, 252. See also Expertise on Middle East
Arab League: and Canada, 130; and the Netherlands, 34; and the United States, 235
Arab oil embargo (1967), and Britain, 66–67
Arab oil-importing companies, 13
Arab oil weapon: declaration of, 1; definition of, 11–12; future use of, 300–305; history of, 11–14; as unique case, 25–27
Arafat, Yasser: and Japan, 156, 171; and the United States, 244
Armed suasion. See Coercive bargaining
Arms sales, transfers, and embargoes: British, 67, 70, 74, 75, 83, 104; Dutch, 30; French, 80; Japanese, 152–153, 157;

United Nations General Assembly Resolution 2628 (1970): France and, 148; Greece and, 148; Japan and, 148, 152; the Netherlands and, 31–32
United Nations General Assembly Resolution 2792D (1971), Japan and, 148
United Nations General Assembly Resolution 2949 (1972), the Netherlands and, 31–32, 36
United Nations General Assembly Resolution 3089D (1973), Canada and, 113
United Nations peacekeeping, Canada and, 107, 112, 113, 137, 141
United Nations Relief and Works Agency for Palestine Refugees in the Near East. *See* Foreign aid: Japanese
United Nations Security Council, and location of embassies in Israel, 43
United Nations Security Council Resolutions 242 and 338: Canada and, 108, 113, 121, 133, 283; different interpretations of, 30; and EPC, 31, 36; Great Britain and, 66, 67–68, 73, 100; Japan and, 147–148, 150, 152, 153, 154, 155–156, 172, 186–187; the Netherlands and, 30, 32, 34; proposal to alter, 209; the United States and, 191–193, 196, 204, 207
United States: and aerial bombing, 9; and Canada, 107, 122–123; denied use of European bases for arms airlift to Israel, 71, 211–212, 273, 278, 279; and Euro-Arab dialogue, 40; and Great Britain, 69, 80, 103; and Japan, 146–147, 152, 158–160, 162–163, 180, 182, 184, 186–187; and Lebanon, 209–210; and location of embassy in Israel, 120; and the Netherlands, 29, 46; and oil embargo, 11–13; reasons for studying, 17; use of economic sanctions by, 21; vulnerability of, 26–27; warned of embargo, 35; and Western Europe, 42
Unit of analysis, 13

Values, 22–23
Vance, Cyrus, 263–264; and Canada, 123; and Andrew Young, 209
van der Stoel, Max, 54, 58–60, 287; and Arab ambassadors, 33, 60; parliamentary support for, 55; and policy toward Palestinians, 36, 56; refuses to see oil ministers, 38; and Gerritt Wagner (Royal Dutch Shell), 50–51
van Lynden, D. W., 58–59; and oil ministers, 38–39
Variable clusters, 15–16
Veliotes, Nicholas, 253
Venice Declaration, 42; and Canada, 122; and Great Britain, 74, 102; and Japan, 156
Vietnam war, 4, 9, 24, 194, 258, 295
Vulnerability, 17–18, 25, 275, 283–285, 299, 301, 304–305; and Canada, 109–110, 112, 113, 122, 142–143; and Great Britain, 64, 66–67, 68–69, 104, 105, 146; and Japan, 25–27, 146–147, 186, 187, 276, 280; and the Netherlands, 35, 146; and the United States, 26–27; and Western Europe, 25–27

Wagner, Gerritt, 50–51
Walker, Peter, 98
War of attrition, 67, 192–193
Washington Energy Conference, 184
Watergate, 244–245, 247, 256–258
Weight of power relationship, 4, 278–283. *See also* Impact of oil weapon
Weinberger, Casper, 265, 266, 270
Western Europe, vulnerability of, 25–27. *See also* European Community
Wheeler, Earl, 252
Wilson, Harold, 66, 68, 75, 98–101; and Arab economic boycott, 76; and creation of British National Oil Company, 90
World Events Interaction Survey (WEIS), 17
World War I, 21; Palestine and Great Britain during, 65

Yamagata, Eiji, 177
Yamani, Sheik Ahmed Zaki, 189, 194, 196, 201
Yamashita, Hideaki, 177, 178, 179
Yom Kippur war. *See* Arab-Israeli war (1973–1974)
Yost, Charles, 195
Young, Andrew, 209, 233
Yugoslavia, invades Albania, 21

Zaikai, 165, 168, 169
Zaire, 68–69

Compositor: G & S Typesetters, Inc.
Printer: Braun-Brumfield, Inc.
Binder: Braun-Brumfield, Inc.
Text: 10/13 Galliard
Display: Friz Quadrata